Internet of Things
Technologies and Applications for a New Age of Intelligence

Internet of Things
Technologies and Applications for a New Age of Intelligence
Second Edition

Vlasios Tsiatsis
Ericsson, Stockholm, Sweden

Stamatis Karnouskos
SAP, Walldorf, Germany

Jan Höller
Ericsson, Stockholm, Sweden

David Boyle
Dyson School of Design Engineering,
Imperial College London,
London, United Kingdom

Catherine Mulligan
Imperial College London,
London, United Kingdom

ACADEMIC PRESS
An imprint of Elsevier

Library of Congress Cataloging-in-Publication Data
A catalog record for this book is available from the Library of Congress

British Library Cataloguing-in-Publication Data
A catalogue record for this book is available from the British Library

ISBN: 978-0-12-814435-0

For information on all Academic Press publications
visit our website at https://www.elsevier.com/books-and-journals

Working together
to grow libraries in
developing countries

www.elsevier.com • www.bookaid.org

Publisher: Mara Conner
Acquisition Editor: Tim Pitts
Editorial Project Manager: Andrea Gallego Ortiz
Production Project Manager: Bharatwaj Varatharajan
Designer: Greg Harris

Typeset by VTeX

Contents

PART 1 THE EVOLVING IOT LANDSCAPE

PART 2 IOT TECHNOLOGIES AND ARCHITECTURES

About the Authors

Vlasios Tsiatsis is a Senior Researcher at Ericsson Research, Ericsson AB and has been working on the Internet of Things (IoT) for 20 years, on subjects ranging from energy-efficient communication algorithms on 8-bit microcontrollers to streaming data analytics in the cloud and recently to IoT Security. He has contributed to several research projects on Wireless Sensor Networks by DARPA, United States, European Union research projects such as RUNES, SENSEI, IoT-i, and CityPulse as well as internal Ericsson corporate research projects around machine/man/mobile-to-machine and IoT services. Vlasios has extensive theoretical and practical experience on IoT technologies and deployments and his research interests include security, system architecture, IoT system management, machine intelligence, and analytics. He holds a PhD in the area of Networked Embedded Systems from the University of California, Los Angeles.

Stamatis Karnouskos is an expert on the IoT at SAP, Germany. He investigates the added value and impact of emerging technologies in enterprise systems. For over 20 years, he has led efforts in several European Commission and industry-funded projects related to IoT, Cyber-Physical Systems, Industrie 4.0, manufacturing, smart grids, smart cities, security, and mobility. Stamatis has extensive experience in research and technology management within the industry as well as the European Commission and several national research funding bodies (e.g., in Germany, France, Switzerland, Denmark, Czech Republic, and Greece). He has served on the technical advisory board of the Internet Protocol for Smart Objects (IPSO) Alliance and the Permanent Stakeholder Group of the European Network and Information Security Agency (ENISA).

Jan Höller is a Research Fellow at Ericsson Research, where he has a responsibility to define and drive technology and research strategies and to contribute to the corporate strategies for the IoT. He established Ericsson's research activities in IoT over a decade ago, and he has been contributing to several European Union research projects including SENSEI, IoT-i, and Citypulse. Jan has held various positions in Strategic Product Management and Technology Management and has, since he joined Ericsson Research in 1999, led different research activities and research groups. He has served on the Board of Directors at the IPSO Alliance, the first IoT alliance formed back in 2008. He currently serves on the Board of Directors of OMA SpecWorks and is a cochair of the Networking Task Group in the Industrial Internet Consortium.

David Boyle is a Lecturer in the Dyson School of Design Engineering at Imperial College London. He has more than 14 years experience developing IoT technologies across academia and industry. His research interests lie at the intersection of complex sensing, actuation, and control systems (Cyber-Physical Systems), IoT and sensor network applications, data analytics, and digital economy. David was awarded his PhD in Electronic and Computer Engineering from the University of Limerick, Ireland, in 2009, following his B.Eng. (Hons) in Computer Engineering in 2005. His work has been recognized and awarded internationally and published in leading technical journals, including the IEEE Transactions on Industrial Electronics (TIE) and Informatics (TII). He actively participates in a number of Technical Programs and Organizing Committees for the premier conferences in the field. Before joining the Dyson School of Design Engineering in 2018, David was a Research Fellow in the Department of Electrical and Electronic Engineering at Imperial College London since 2012. Previously, he worked with the Wireless Sensor Network and Microelectronics Applications Integration Groups in the Microsystems Centre at Tyndall National Institute, and the Embedded Systems Research Group, University College Cork, Ireland. Prior to this, he was with France Telecom R&D Orange Labs, France, and a Visiting Postdoctoral Scholar at the Higher Technical School of Telecommunications Engineering, Technical University of Madrid (ETSIT UPM), Spain.

Dr Cathy Mulligan is a Visiting Researcher at Imperial College and was a founding Co-Director of the ICL Centre for Cryptocurrency Research and Engineering. She is also a Senior Research Associate at University College where she is Chief Technology Officer of the GovTech Lab and DataNet, which focuses on the potential and application of blockchain, AI and advanced communications technologies as a foundational part of the world's economy. Cathy is an expert and fellow of the World Economic Forum's Blockchain council and has recently become a member of the United Nations Secretary General's High Level Panel on Digital Co-Operation. She holds a PhD and MPhil from the University of Cambridge and is the author of several books on telecommunications including EPC and IoT.

Foreword to the First Edition by Zach Shelby

I grew up in a time when the Internet was used by computer science students using Gopher to browse their course syllabus. We ran private bulletin-board systems using ANSI text over 2400 baud modems over fixed phone lines, and we transferred news and mailing lists overnight through USENET. Think of this as an analogy to where we have been with automation systems and M2M over the past decade. The same incredible growth of people using the Internet in the 1990s is now being repeated by things using the Internet in the 2010s.

It is wonderful to see this book published during the peak of the IoT hype cycle, where most writing is in Tweets and blog entries. The deployment of traditional IP networks, security technology, and Web infrastructure requires a lot of knowledge and skill, and understanding the Internet of Things requires a similar breadth of knowledge. Today we take that knowledge for granted because we have trained the world through books and teaching over several decades. Luckily most of the knowledge we have gained from building today's Internet and Web services can be applied to IoT. There are, however, many aspects of IoT technologies that are new, including IPv6 over low-power networks, new applications of TLS security, efficient web transfer protocols, and techniques for managing and using devices through commonly understood data objects.

System and network architects, administrators, and software developers will find this book useful as an overview of IoT architecture and technology. At the same time, business and product managers will find the book useful as an introduction to the market segments, applications, and requirements as input for a successful IoT product or service. Finally, the technology overview is a great starting place to find the information needed to dive deeper into a particular area, and the architecture overview covers a wide range of design paradigms. One important point made is that without trust and security built into IoT technology and systems in a holistic way, we will not see an Internet of Things, but continue to see silos of things.

The technology is available today to build an Internet of Things where devices and services can be developed and deployed for the benefit of society and industry as a whole. The challenge now is for us to educate people.

Zach Shelby
Vice President for IoT, ARM Inc.

Foreword to the First Edition
by Geoff Mulligan

Taking notice of the IoT – Since the age of 9, when I started programming, I thought *computers* were cool. At 15, I was hired to hack networks, catching the attention of some newspapers, and thought *networks* were cool. As a Lieutenant in the Air Force at the Pentagon, I was helping build the Arpanet and still thought *networks* were cool. In 1996, while helping design IPv6, I wrote the first implementation of v6 for the PC and in 2001 rewrote it for an 8-bit microcontroller and realized that *embedded networks* were cool. Most recently helping found and grow the IP for Smart Objects Alliance and now serving as White House Presidential Innovation Fellow working on Cyber-Physical Systems and the Internet of Things, I'm seeing everyone take notice of how CPS, the IoT, and M2M will reshape our world and that is *really cool*.

In 1999 Scott McNealy quipped, "You have zero privacy anyway... Get over it." We should not get over it, but instead deal with it. It is critical that we think about it – privacy – experiment with it, and work to get out in front of the issues rather than play catch up. Books like this one are important in bringing the concepts and ideas related to this new emerging smarter world into focus for discussion and debate. According to a recent survey, the United States now has more Internet-connected gadgets, sensors, controllers, phones, and light bulbs than the 311 million people that live in the US. Understanding architectural design trade-offs with the application to specific implementation scenarios is important if we are to get this right.

It is fundamentally important that the Internet of Things and these Machine-to-Machine networks are built using open standard protocols – especially IP. Jari Arkko, the current IETF Chairman, started describing "permissionless innovation" whereby new businesses, new systems, and new business models can be created without having to ask permission from others. Open protocols and open standards set the stage for these opportunities. When Vint Cerf and others created the Internet, they didn't plan for YouTube or Facebook, but their layered network design and freely available protocols allowed for these types of innovation.

There are quite a few books about the Internet of Things, but few of them provide an accessible description of a vision for the connected world and the basic building blocks necessary to bring this vision to reality. But this book goes beyond those fundamentals to sharing specific examples in Asset Management, Industrial Automation, the Smart Grid, Commercial Building Automation, Smart Cities (a particular favorite since it aligns with my Presidential Innovation Fellow project – The SmartAmerica Challenge), and Participatory Sensing. Teasing apart the important nuances that differentiate each of the application spaces is critical in understanding how to apply sound design to each. Within each segment are differing requirements for latency, security, privacy, determinism, throughput, and speed. Understanding these differences is critical for a proper system design and successful installation and deployment. This book can provide just such necessary information.

The next generation of devices that will become part of the Internet of Things will not just sense and report, but will control. Whether is it the connected vehicle, a Building Automation System, an agile manufacturing robot, a thermostat, or a door lock, these new connected machines will have a greater impact on our lives. The protection of the control data and operating instructions will be critical as we allow greater control and autonomy so as to ensure our safety and security. Privacy and security "by design" is imperative and must not be an afterthought.

As we rush toward 2020 and the 50 billion Internet of Things as predicted by Ericsson, we need to be thoughtful and clue-full and have a plan so as not to be crushed by the onslaught of Device Management, privacy concerns, and the avalanche of data. It's been over a decade since Kevin Ashton first used the term the Internet of Things. Progress has been slowed by the deployment of islands of proprietary protocols and the gateway required to interconnect them; the proliferation of pseudo-open standard (but yet proprietary) protocols and yet more gateways are required for interconnection and the continued quest for new and "better" protocols. We have the necessary tools at hand. It is the application of sound design and open standards that will allow us to march into this new era of the connected everything with the confidence that it holds the promise of a safer and more efficient world and society – it will be *awesomely cool*.

Geoff Mulligan
Presidential Innovation Fellow, Founder
IPSO Alliance, LoRa Alliance

Foreword to the Second Edition by Geoff Mulligan

Here we are in 2018 and still much of the promise of the IoT has not yet been delivered – though progress has been made. "Smart" systems are being deployed, the IoT is being built, but much of the profound changes predicted with the coming wave of IoT haven't materialized. Was it just overhyped 5+ years ago or some other systemic problem that has delayed the future promised by the IoT? This update to the book brings attention to new technologies and their impact as well as some of the continuing issues revolving around the wide-scale deployment of the IoT. This is a welcome refresh to the concepts and ideas presented in the first edition.

This update brings new information around market forecasts and the understanding of the current siloed approaches to IoT systems and necessity and issues of bringing them together. Additionally, as with any technology and especially with the IoT, the underlying connectivity tools and protocols continue to evolve, and the authors have researched and explained the changing landscape with a technology agnostic viewpoint. This is critically important for engineers to understand the pros and cons of each of the alternatives so as to best apply the proper technology and avoid the old adage – "When all you have is a hammer, everything looks like nail."

And finally, one section that was missing from the first edition was coverage of security. As we have seen all too often, recently, product and system designers fail to adequately embrace the issues surrounding good security practices – "security by design". It is good to see that this topic is now being addressed in this updated version.

Let us hope that in 5 years' time, we can all be leading better lives using more eco-friendly systems because the IoT has lived up to its potential and that this book can help readers understand some of the next steps and understand some of the necessary trade-offs to start building these world changing systems.

Geoff Mulligan
Former White House Presidential Innovation Fellow on IoT
Founder Skylight Digital Consultancy

Preface

INTRODUCTION

The Internet of Things (IoT) is rapidly becoming part of our everyday lives, from consumer solutions to industrial-scale ones. Interest in IoT is therefore increasing – in particular how to create robust, real-world solutions based on the broad spectrum of standards and technologies available. In addition, companies and governments are seeking solutions that are both technically and economically viable as well as appropriate frameworks to design and implement them.

The number of "connected devices" (i.e., devices connected to the Internet) is growing and is expected to continue to grow exponentially as people increase the numbers of devices they purchase. Worldwide, mobile phone subscriptions have increased almost 20% since this book went into press for the 1st edition (from 6.7 billion in 2013 to around 8 billion in 2018) according to the Nov 2017 Ericsson Mobility report [2]. According to the same report there are currently 7 billion IoT devices (short- and wide-area; IoT devices are connected devices with different communication technologies apart from PCs, laptops, tablets, mobile, and fixed phones) and in 2023 there will be about 20 billion IoT devices with a 12% (wide-area)/88% (short-range) split with respect to communication technology. End-users have also started using multiple devices (e.g., tablets, e-book readers, mobile handsets, digital TVs) over the years at an increasing pace. This takes the number of subscriptions to a few tens of billions when considering connected devices in residential and corporate buildings; across cities, regions, and nations in public and private infrastructure. For example, millions of such connected devices will be used within public transport to improve services and information delivery to citizens. This increased efficiency is expected to help reduce carbon emissions and generate innovation around the data created by IoT platforms. This explosive growth is unprecedented within not just the communications industries, but also the wider global economy.

In addition to all this, IoT solutions and services have a wider role to play in the future of our world. In 2015 about 54% of the world's population was living in cities rather than rural areas according the World Cities Report[1] by the UN HABITAT and by 2050 this percentage is expected to rise to 66%. The infrastructure of cities and nations must therefore adapt accordingly, from roads and lighting to metro/commuter trains and pipelines, to name just a few. Much of this infrastructure will be instrumented with sensors and actuators for more efficient management, and all these devices associated with infrastructure will be connected to large-scale data analysis and management systems, the data of which needs effective capture, analysis, and visualization in order to be applied effectively in the development of smart, sustainable societies and cities.

The unprecedented numbers of devices foreseen, in combination with the vertical nature of many M2M applications, create an interesting set of barriers to success for anyone wishing to implement a solution based on these technologies. The deployment and operational costs of traditional telecom platforms adapted to handle the traffic load from tens of billions of additional connected devices would be

---

[1] http://wcr.unhabitat.org

prohibitively high. Moreover, due to the specialized nature of the cases where Machine-to-Machine (M2M) technologies will be applied, a fragmented ecosystem is emerging in each of the solution "silos". Such industrial dynamics create barriers to entry for individuals and companies wanting to develop M2M applications or services, from supporting a mix of diverse devices and billing to handling settlement and commission across the value chain. Understanding how corporations and governments should respond to these changes is therefore a critical need for corporations, cities, and governments.

This book provides a thorough and high-level analysis for anyone wishing to learn about the state of IoT today. When the first edition of the book was being written around 2012–2013, the concepts of M2M and IoT were competing for fame in the world and a piece of the market. M2M was an established but small market segment focusing on simple integrated communication solutions for reaching remote machines and IoT was a fusion of the academic research on Wireless Sensor Networks and academic/industrial research and development on RFID and related identification technologies.

With the first edition we aimed at distinguishing these terms and showing the way of the future, a converged world with the term IoT encompassing both these terms. Five years later we produced the second edition of the book which now includes only a few traces of the term M2M and its technologies. The market has grown almost exponentially and every major IT player in the market is handicapped if they do not have a convincing story and supporting products based on IoT.

IoT has also grown from almost being a hobbyist or enthusiast-driven "weekend project" community to being a major industrial application area for traditional industries such as manufacturing, automotive, and utilities. This is mainly an *upgrade* of M2M with new communication technologies and cloud services and its spread in every imaginable enterprise. This can be evidenced from the creation of the Industrial Internet Consortium and RAMI 4.0, as well as the multitude of industrial efforts.

The first edition of the book has been well received, cited, and used in a variety of international universities as a textbook. Since its publication in 2014, however, a variety of new technical advances mean that it is now an appropriate time to update the original content.

STRUCTURE OF THE BOOK

Part 1: The Evolving IoT Landscape

Part 1 outlines the global context of IoT, including technology and business drivers.

Chapter 1: Why the Internet of Things?

Chapter 1 provides an overview of the market and technical drivers for the Internet of Things as a motivation for the book.

Chapter 2: Origins and IoT Landscape

Chapter 2 provides an overview of the origins and landscape of IoT, its main characteristics and features as a set of technologies, and some of the types of problems IoT addresses to solve, including drivers based on selected megatrends.

Chapter 3: IoT – A Business Perspective

Chapter 3 provides an overview of the market drivers, industrial structures, value chains, and example business models for IoT.

Chapter 4: An Architecture Perspective

Chapter 4 provides an introduction to architecture and system design and some of the main functional elements of an IoT architecture, as well as a basic understanding of standardization considerations for IoT.

Part 2: IoT Technologies and Architectures

In Part 2, the technology building blocks of IoT solutions are presented, including security, privacy, and trust, as well as the state-of-the-art in architecture and reference models.

Chapter 5: Technology Fundamentals

Chapter 5 presents an overview of technology fundamentals – the building blocks upon which the IoT rests, including: Devices and Device Gateways, Local and Wide Area Networking, Data Management, Business Processes, Cloud Technologies, Machine Intelligence, and Distributed Ledgers.

Chapter 6: Security

Chapter 6 describes the basic mechanisms required to provide security against malicious actors and outlines a number of potential threats against IoT systems. It suggests some mitigation schemes following a layered approach. An overview of the security mechanisms specified by the main standards bodies is presented, in addition to discussing safety and privacy aspects critical to building trustworthy IoT applications and systems, finishing with a view to future developments in IoT security.

Chapter 7: Architecture and State-of-the-Art

Chapter 7 provides parts and fragments of an architecture maintained by standards development organizations, alliances, and technologies communities. The chapter does not claim full coverage of the possible outlets which develop parts and whole architectures but it attempts to cover the major organizations and groups focusing on different aspects of IoT.

Chapter 8: Architecture Reference Model

Chapter 8 provides the most relevant Architecture Reference Models (ARMs) in IoT today, namely IoT-A (IoT ARM) and the Industrial Internet Consortium (IIC) Reference Architecture (IIRA). While the IoT-A ARM presents an Information Technology (IT) reference architecture, the IIRA presents the Operational Technology (OT) counterpart.

Chapter 9: Designing the Internet of Things for the Real-World

Chapter 9 outlines design constraints that need to be taken into account when developing real-world technical solutions.

Part 3: IoT Use Cases

Part 3 covers real-world implementation examples of IoT solutions.

Chapter 10: Asset Management

Chapter 10 discusses Asset Monitoring, which enables the remote tracking and management of inventory in the field. Typically such functionality involves the collection of the exact location and state of assets at regular intervals for the purposes of improving the business (e.g., preventing stock-outs) or reducing risks (e.g., of getting lost).

Chapter 11: Industrial Automation

Chapter 11 covers the emerging approach in industrial environments, which is to create system intelligence by a large population of intelligent, small, networked, embedded devices at a high level of granularity, as opposed to the traditional approach of focusing intelligence on a few large and monolithic applications within industrial solutions.

Chapter 12: Smart Grid

Chapter 12 covers the Smart Grid, a revolution currently transforming the electricity system. Rapid advances in IT are increasingly being integrated in several infrastructure layers of the electricity grid and its associated operations. IoT interactions create new capabilities in the monitoring and management of the electricity grid and the interaction between its stakeholders.

Chapter 13: Commercial Building Automation

Chapter 13 covers commercial buildings and the use of IoT. The purpose of a Building Automation System is typically to reduce energy and maintenance costs, as well as to increase control, comfort, reliability, and ease of use for maintenance staff and tenants. IoT plays an increasingly important role within Commercial Building Automation.

Chapter 14: Smart Cities

Chapter 14 covers Smart Cities, an emerging and increasingly important field of application for IoT. This includes how sensors and associated IoT systems are being applied and linked to other paradigms (e.g., open data initiatives).

Chapter 15: Participatory Sensing

Chapter 15 covers Participatory Sensing (PS), or Urban, Citizen, People-Centric Sensing or Social Sensing. This is a form of citizen engagement for the purpose of capturing the city surrounding environment and daily life. This chapter covers a few examples of such scenarios.

Chapter 16: Autonomous Vehicles and Systems of Cyber-Physical Systems

Chapter 16 describes the state-of-the-art in autonomous vehicles, broadly defined, and discusses how their interactions via the so-called IoT are contributing to emerging systems of Cyber-Physical Systems.

Chapter 17: Logistics

Chapter 17 outlines the main roles and actors in Logistics Management, briefly refers to the main involved technologies, and outlines an example scenario on food transport in which traditional Logistics technologies (RFID, barcodes, EPCIS, etc.) benefit from the introduction of IoT technologies such as sensing, local processing, and potential local actuation.

Chapter 18: Conclusions and Looking Ahead

Chapter 18 provides a brief outlook on the future for IoT.

Appendix A: ETSI M2M

Appendix A contains a summary of the ETSI Machine-to-Machine architecture and interfaces. Since the architecture and interface specifications are merged to oneM2M specifications and evolved since the conclusion of ETSI work in 2012, the material of this chapter is only of historical importance.

V. Tsiatsis
S. Karnouskos
J. Höller
D. Boyle
C. Mulligan
November 2018

Acknowledgments

A work of this nature is not possible without others' support and input. The authors would like to gratefully acknowledge the contribution of many of our colleagues at Ericsson, SAP, and Imperial College London, as well as our colleagues across industry and academia.

This is an enhanced revision of the "From Machine-to-Machine to the Internet of Things: Introduction to a New Age of Intelligence" [1] published in 2014. In addition to the acknowledgments in that edition, we would also like to thank Jennifer Zhu-Scott for contributing Section 5.7. We would like to thank Stefan Avesand, who coauthored the 1st edition of this book, and although due to other engagements he could not participate, his touch is still evident in this revised version.

We would also like to thank our 1st edition readers for their support and comments that made this edition possible.

Dr. Mulligan would like to acknowledge Olavi Luotonen (EU Commission), Omar Elloumi (Bell-Labs), and the Open Agile Smart Cities (OASC).

We would also like to acknowledge our colleagues at Ericsson for many good discussions and support, in particular: Sara Mazur, Eva Fogelström, Hans Eriksson, Göran Selander, John Mattsson, Francesca Palombini, Peter von Wrycza, Ramamurthy Badrinath, Nanjangud Narendra, P. Karthikeyan, Carlos Azevedo, Klaus Raizer, Ricardo Souza, Sandeep Akhouri, Ari Keränen, Jaime Jiménez, and András Veres.

We would also like to thank our families as writing this book would not have been possible without their generosity and support throughout this process.

V. Tsiatsis
S. Karnouskos
J. Höller
D. Boyle
C. Mulligan
November 2018

THE EVOLVING IOT LANDSCAPE

PART

1

Part 1 of this book provides an overview of the vision and market conditions for the Internet of Things (IoT). Here we discuss the global context within which IoT exists and the business and technical drivers at work in both technology and industry. This part also provides the basics of the IoT-Architecture and the principles behind them, preparing the reader for Part 2, which outlines in detail an architecture reference model for IoT.

PART

THE EVOLVING IoT LANDSCAPE

WHY THE INTERNET OF THINGS?

1

This book provides a thorough overview for anyone wishing to learn about the technology aspects of the Internet of Things (IoT), and how IoT solutions are being implemented and deployed in various industries and in society at large. This chapter provides a brief introduction to the necessary bigger picture of IoT and the topics covered.

Since the inception of the Internet and its inflection point back in the 1980s, and followed by the introduction of the World Wide Web in the early 1990s, the Internet and the Web have redefined a number of businesses such as media, travel, retail, and finance. For instance, the music industry moved from analog to digital encoding of audio, and once digital, the Internet became a natural distribution channel for music. This resulted in a fundamental transformation of an entire industry that moved from selling tangible products, i.e., vinyl records and compact discs, to selling intangible products, like mp3-encoded music files, and then later to a subscription model based on streaming of music from actors like Spotify and Apple Music. The implication was a complete change in how music was distributed, sold, and enjoyed, which effectively led to the collapse and simplification of the music industry value chain as well as the underlying business model. Today, the Internet provides the complete means for producing, distributing, marketing, and consuming music. From a consumer as well as a business-to-business perspective, the travel industry is similarly transformed and integrated with how booking services, e.g., combined travel and accommodation, are provided. The same can also be seen in retail with online shopping as a global phenomenon with Amazon and Alibaba as prime examples. The IoT is another such wave of fundamental transformation that is redefining business processes and practices across a number of different industry and society sectors, like energy, manufacturing, transportation, and healthcare. What is different with IoT is that it adds the dimension of the real world of machines, things, and spaces as first class citizens to the existing Internet by embedding sensors to capture physical properties, and actuators to control their states. IoT is in essence about enabling intelligent operations involving real-world assets and machines, whether they are in the consumer, enterprise or industry domains. Intelligent operations are about using software to gather insights about the real world and to automate processes for transformational outcomes of different kinds.

The World Economic Forum (WEF) has studied the industrial aspects and implications of IoT and outlined [3] how the Industrial IoT is transformative. It will have an impact on competition and how industry borders will change and it will create new business opportunities including the emergence of new disruptive companies just as the Internet to date has done so. WEF has identified that the key business opportunities are to be found in four areas. Firstly, it is about significant improvements in operational efficiencies, such as resource utilization and improved equipment uptime via remote management, and predictive maintenance of assets, i.e., to be able to predict and schedule when machine servicing is needed. Secondly, it is the emergence of an outcome economy, which implies that businesses will increasingly shift from selling *products* to selling the *value* their customers expect from the products. Thirdly, ecosystems will be connected using software platforms that enable online collaboration based on the exchange of data and information, which then become tradeable assets. This

Internet of Things. https://doi.org/10.1016/B978-0-12-814435-0.00012-2

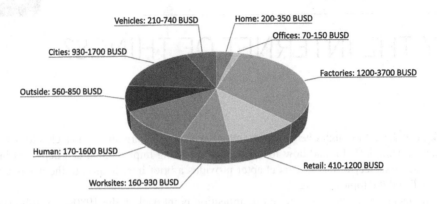

Vehicles: 210-740 BUSD
Home: 200-350 BUSD
Offices: 70-150 BUSD
Cities: 930-1700 BUSD
Factories: 1200-3700 BUSD
Outside: 560-850 BUSD
Human: 170-1600 BUSD
Retail: 410-1200 BUSD
Worksites: 160-930 BUSD

FIGURE 1.1

The economic potential of IoT across different settings.

will further increase customer value and efficiency and scale in its delivery. Lastly, it will also enable new means for collaboration between people and machines, to augment workers, increase safety and efficiency, and also hopefully make work more engaging and inspiring.

To understand the potential and impact IoT can have, McKinsey Global Institute studied the economic reach of IoT solutions across a number of different settings [4]. The study estimated the total potential economic impact of IoT to be in the range of 3.9–11.1 trillion US dollars per year in 2025. This can be compared to the World Bank projected global Gross Domestic Product (GDP) of 99.5 trillion US dollars in 2025, i.e., IoT could have a potential about as high as 11 percent of the total world economy. Note that this value is the estimated economic transformative impact IoT can have and does not represent the value of revenue from sales of IoT products, solutions or services.

The settings in the McKinsey study include IoT use in different environments essentially representing physical spaces, such as worksites and homes, rather than in various vertical markets, for example, consumer electronics or automotive. Nine different settings were defined, each with its own estimated range of economic impact. The potential value across the different settings is illustrated in Figure 1.1. The chart is based on the median value per setting as the range varies across the settings.

In order to provide a high-level understanding what type of opportunities these settings contain, the following is a summary of some key objectives and application examples per setting. The interested reader is recommended to consult [4] for more details.

- **Human.** This represents devices attached to or inside the human body, e.g., wearables and ingestibles. Applications include human health and fitness, monitoring and treatment of illness, increasing wellness, and proactive lifestyle management. This setting also includes increased human productivity using, e.g., augmented reality to assist in tasks, as well as the use of sensors and cameras for skills training. Human health and safety when working in hazardous environments is yet another application example of this setting.
- **Home.** This setting is about buildings where people live. Home-based IoT applications include automation of domestic chores and energy management, as well as security and safety. These are applications with a direct benefit to consumers, but also with benefits to other stakeholders, such as utility companies.

- **Retail.** This setting includes the spaces where consumers engage in commerce and is not only related to products but also to services. The spaces included are stores and showrooms with a focus on products, as well as spaces where services are purchased, like banks, restaurants, and various arenas. It includes applications like self and automated checkouts, in-store offers, and inventory optimization.
- **Offices.** Offices are defined as spaces where knowledge workers work. Similar to the home setting, energy and work environment management, as well as security, are typical applications. Another area is the increase of human productivity and performance, including for mobile workers.
- **Factories.** Factories are here defined as standardized production environments. Factories include discrete manufacturing and process industry plants. It is broadly defined also to include other sites where repetitive work routines apply, for instance, farms in agriculture or hospitals. Examples of applications in the factories setting include condition-based maintenance of equipment and automated quality monitoring. Other applications include the autonomous operation of parts of a process, e.g., robot manufacturing of components or irrigation in agriculture, and also optimization of a supply chain of materials.
- **Worksites.** This setting covers custom production environments where each site is unique and no two projects are the same in terms of streamlining operations. An example domain is natural resource extraction, such as mining, oil, and gas. Another is a construction site. Common characteristics include a constantly changing and many times unpredictable environment. Usually, operations involve costly and complex machinery, such as drill rigs and giant haulers. Again, applications target Predictive Maintenance of expensive machines to ensure high utilization, operations optimization, and worker safety. Increasing in importance is also sustainability and minimizing environmental impacts.
- **Vehicles.** The Vehicles setting includes vehicles on the road, rail, and sea and in the air and focuses on the value of using IoT in, to, and between the vehicles themselves. Example applications include autonomous vehicles, remote diagnostics for planned servicing, and also the monitoring of the behavior and usage of a vehicle in order to aid in the vehicle development and design process.
- **Cities.** A city is an urban environment that is a combination of public spaces and different infrastructures, e.g., for energy, water, and transportation. Large densely populated areas require smooth operations of transportation of people and goods, efficient use of resources, and ensuring a healthy and safe environment. The "smart city" is hence opportunity-rich in a variety of IoT applications that require sensing, actuation, and intelligent operations.
- **Outside.** This final setting is about IoT usage outside urban environments and the other settings. A prime example is logistics of produced goods in both supply chain and online retail where track and trace is a key IoT application. The second major application in this setting is autonomous passenger vehicles outside the urban setting, whether on rail, on road, on sea or in the air.

It must be noted that the mentioned value capture and creation in part require that a set of barriers are overcome [4,3]. The barriers can be of different types, such as technology, organizational, regulatory, and even emotional. Examples include a lack of technical interoperability and a lack of security and trust, sometimes even just a perceived lack of trust. Also, access to data, understanding of data, and ownership of data are other barriers. Many times, data tend to remain in "silos" that cannot be accessed easily across system or organizational boundaries.

By now we have started to see the full potential of IoT and the broad spectrum of different use cases. We have also seen the economic potential across the global economy including the industrial, enterprise, consumer, and public sectors. There are different ways to structure and size the market and also to structure the different applications and use cases. As can be seen, there are indeed recurring applications from across sectors, like resource optimization, predictive maintenance of equipment, and autonomous operations. Further drivers and enablers for IoT are covered in Chapter 2, and use cases are the subject of Part 3 of this book.

Another way to look at IoT is the popular view of IoT as a plethora of different devices and gadgets that will be connected to the Internet. Looking at the evolution of telecommunication and data communication networks, it took about 100 years to connect around 1 billion places with fixed phones. It took another 25 years to connect 5 billion people with mobile devices, more than half of them being smartphones capable of running Internet applications. Mobile network generations of GSM, 2G, 3G, and 4G/LTE have primarily focused on human users with different mobile devices. However, with the advent of 5G, the target is to fully support a wide range of different IoT applications, ranging from massive numbers of low-power sensors to ultra-reliable low latency connectivity for mission-critical industrial applications. The next step in this evolution of connecting things is to connect the rest of the real world – the machines, objects, and spaces – which is the Internet of Things. Projected numbers of connected IoT devices vary depending on the source, but they all show the general trend of exponential growth with a total number of tens of billions of devices in the coming years even in the more modest forecasts. The Ericsson Mobility Report [2], which is issued on a yearly basis, gives one example of the rapid growth rate of IoT devices, as illustrated in Figure 1.2.

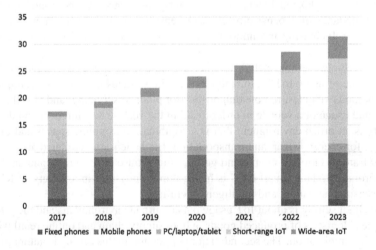

FIGURE 1.2

Forecast of different types of connected devices (billions) (adapted from Ericsson).

As can be seen, around 20 billion connected IoT devices are forecast by 2023, which outnumbers the rest of connected devices by a factor of two. Between 2017 and 2023, connected IoT devices are expected to increase at a compound annual growth rate of 19 percent, mainly driven by new use cases and affordability. This growth shall be compared to the modest growth rate of mobile phones and the saturated growth of PCs, laptops, and tablets. What can also be seen is that IoT devices will not be connected by a single networking technology, but a combination of different technologies. The wide-area segment consists of devices using cellular connections, as well as other low-power long-reach technologies, for instance, LoRa (see details in Section 5.2). The short-range segment largely consists of devices connected by technologies with a typical range of up to 100 meters, such as Wi-Fi and Bluetooth. This segment also includes devices connected over wired Local Area Networks and powerline technologies.

To summarize, IoT is a game changer. It is a perfect example of innovation coming from the interplay of technology and business. From a technical perspective, IoT is not a single technology but a systemic approach of combining devices, networks, compute infrastructures, and software for extracting insights and automation. Real-world objects and places are instrumented with devices containing sensors and actuators to capture and control physical properties. Different types of networks are needed to collect data from devices and provide remote control, all depending on the type of object of interest. Specialized software is required to process the data to extract insights, to do reasoning, and to automate processes involving the various physical objects, all depending on the needs and objectives of the involved stakeholders. Building IoT solutions requires proper system architecture design guidelines and best practices. Standards are also needed to ensure that systems can be efficiently built and to ensure interoperability inside a specific deployment as well as on the global level. IoT solutions will be applied to numerous varieties of use cases in a range of different industrial, enterprise, consumer, and public sectors and will have a profound impact on how markets evolve. This book covers all of these necessary aspects.

ORIGINS AND IOT LANDSCAPE

2

CHAPTER OUTLINE

2.1 INTRODUCTION

Our world is on the verge of an amazing transformation; one that will affect every person, city, company, and thing that forms the basis of our society and economy. In the same way that the Internet redefined how we communicate, work, and play, a new revolution is unfolding that will again challenge us to meet new business demands and embrace the opportunities of technical evolution. Old and new industries, cities, communities, and individuals alike will need to adapt, evolve, and help create the new patterns of engagement that our world desperately needs. In response to these issues, we are moving towards a new era of intelligence – one driven by rapidly growing technical capabilities.

Anything in the physical realm that is of interest to observe and control for people, businesses, or organizations will be connected and will offer services via the Internet. The physical entities can be of any nature, such as buildings, farmland, natural resources like air, and even such personal real-world concepts as my favorite hiking route through the forest or my route to work. The Internet of Things (IoT) is not a single new technology or phenomenon. It is a set of technologies that combined deliver the promise of IoT. The origin of IoT is the Internet itself that connects computers and mobile devices, but connecting things is actually not new. Machine-to-Machine (M2M) communication has been around for some 20 years already and has now carried over to the Internet. In the same way, technologies to build sensor networks are not new, neither are capabilities to process data that originate from things. The IoT can be viewed as a coming together of a set of different technology disciplines and usages.

Internet of Things. https://doi.org/10.1016/B978-0-12-814435-0.00013-4

2.2 EVOLVING TO AN INTERNET OF THINGS

IoT as the concept of sharing data and information about things and machines using devices and systems can be unfolded as an evolving family of technologies and practices; see Figure 2.1.

Wireless Sensor Networks (WSNs) are interconnected sensor nodes that collect data about the physical condition of an environment and are built by the networking of embedded systems. The use of control systems to operate industrial processes has been around for a long time and is a core capability in Cyber-Physical Systems (CPS) that intertwine software and physical components of various machines. The introduction of the Internet in industrial applications, and especially in manufacturing, has given rise to Industrie 4.0 as a new trend of automation. Industrie 4.0 can further be viewed as a part of a larger concept dubbed the Industrial Internet, which can be seen as the combination of Information Technology (IT) and Operational Technology (OT). OT is a denomination of industrial machinery containing software.

Common to all these approaches is a centricity on data and information about the things and the associated processes. In parallel, technologies to process data of different types and characteristics have been evolving, and these include means to analyze data, handling of large amounts of data, popularly referred to as Big Data, and lately Artificial Intelligence (AI) to provide automation and smartness to the things.

Finally comes the perspective of actually connecting all the devices with the software systems that contain the necessary application logic, algorithms, and data processing. This is where the use of the Internet or Internet-based technologies come in, and also where M2M comes in, two of the IoT origins. The focus of M2M has been and still is to provide the necessary connectivity to various devices across application domains, primarily using the mobile network infrastructures.

As can be understood by now, this is a family of both close and distant relatives, coming from different generations, but all having the same meaning in life: the meaning of interacting with the

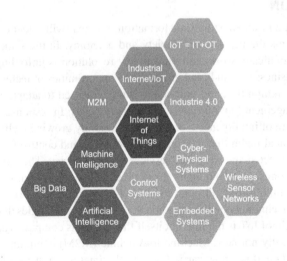

FIGURE 2.1

The family and relationships of IoT.

physical reality in various ways. The terms introduced above are covered further in this book, including the shifting characters each one has. Necessary references for further deeper studies are also provided.

2.2.1 A BRIEF BACKGROUND

The IoT is a result of the technological progress over the last decades, including not just the spectacular uptake of the Internet Protocol (IP) and the broad adoption of the Internet, but also the decreasing costs and sizes of semiconductor components and sensor technologies. The application opportunities for such solutions are limited only by our imaginations; however, the role that IoT will have in industry and broader society is just starting to emerge for a series of interacting and interlinked reasons.

As already mentioned in Chapter 1, the Internet has undoubtedly had a profound impact across society and industries over the past two decades. Starting off as ARPANET connecting remote computers together, the introduction of the TCP/IP protocol suite, and later the introduction of services like email and the World Wide Web (WWW) created a tremendous growth of usage and traffic. In conjunction with innovations that dramatically reduced the cost of semiconductor technologies and the subsequent extension of the Internet at a reasonable cost via mobile networks, billions of people and virtually all enterprises are now interacting and conducting business over the Internet. Quite simply, no industry and no part of society have remained untouched by this technical revolution.

The other technology revolution that has been unfolding is the use of sensors, electronic tags, and actuators to digitally observe, identify, and control objects in the physical world. Rapidly decreasing costs of discrete sensor and actuator components have meant that where such components previously cost several euros each, they are now at the cents level. In addition, embedded computing technologies have also evolved to smaller and cheaper chips costing less than a euro, but still with enough processing and memory capacity to host a tiny web server and with a power consumption that can make a device operate for years on ordinary AA batteries. And they can also be networked using different radio technologies like Bluetooth, WiFi or cellular at low or very low additional costs. As a result, IoT devices can be embedded in ordinary commodity things like lightbulbs or power tools at low relative costs.

So, while we have seen for instance M2M solutions for quite some time, we are now entering a period of time where the uptake of IoT solutions will increase dramatically. The reasons for this are three-fold:

- An increased need for understanding the physical environment in its various forms, from industrial installations through to public spaces and consumer demands. These requirements are often driven by efficiency improvements, sustainability objectives, or improved health and safety [4].
- The availability of technologies and services to more cheaply collect and analyze data via improved networking and analytics tools.
- Reduced costs of components for IoT devices that are instrumenting everyday objects with sensing and computational capabilities.

What has made the IoT market attractive and take off is the intersection in time of (i) technology and cost maturity and (ii) enterprise and societal needs.

2.2.2 A SIMPLE ENTERPRISE IOT SOLUTION OVERVIEW

To provide an overview of a typical simple IoT solution of today, we describe it from an enterprise perspective. Here, the IoT solution is used to remotely monitor and control enterprise assets of various kinds and to integrate those assets into the business processes of the enterprise in question. The asset can be of a wide range of types (e.g., vehicle, freight container, building, or electricity meter), all depending on the type of business. A typical simple enterprise IoT system solution consists of IoT devices, communication networks that provide remote connectivity for the devices, an application platform, the IoT application itself, and the integration of the IoT application into the enterprise IT system of business applications; see Figure 2.2.

In our example, the IoT system components are as follows:

- IoT Device. The asset of interest is instrumented with the IoT device and provides the sensing and actuation capabilities. The IoT device is here generalized, as there are a number of different realizations of these devices, ranging from low-end simple sensor nodes to high-end complex devices with multimodal sensing capabilities.
- Network. The purpose of the network is to provide remote connectivity between the IoT device and the application backend and enterprise IT system. Many different network types can be used, including both Wide Area Networks (WANs) and Local Area Networks (LANs), sometimes also referred to as capillary networks. Examples of WANs are public cellular mobile networks, fixed private networks, or even satellite links.
- IoT application platform. Within the generalized system solution outlined above, the concept of a separate application platform is also introduced. This platform provides generic functionality that is common across a number of different applications. Its primary purpose is to reduce cost for implementation and improve ease of application development.
- IoT application. The IoT application is a realization of the highly specific monitor and control process of the asset. The application is further integrated into the overall enterprise IT system. The IoT application can be of many different types, for instance, remote car diagnostics or electricity meter data management. The corresponding enterprise business application could then be a car service scheduling application or an invoicing application for consumer electricity consumption.

The different technologies and system solution aspects outlined in this example are further elaborated in Chapter 4 and covered in detail in Part 2.

Existing IoT applications cover numerous industry sectors and here a few examples are mentioned. Connected cars represent a fast growing sector and typical applications include navigation, remote vehicle diagnostics, pay-as-you-drive insurance schemes, road charging, and stolen vehicle recovery. Another sector is the utility industry where massive rollouts of different metering applications are

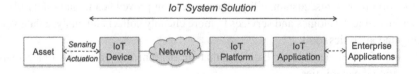

FIGURE 2.2

A simple IoT system solution.

ongoing, primarily for remote meter management and data collection for electricity, gas, and water consumption. Healthcare is another rapidly growing sector, where remote patient monitoring is an example. The logistics sector has a number of applications like goods track and trace, vehicle positioning, and monitoring of perishable goods like food. Automated Teller Machines (ATMs) and Point of Sales (POS) terminals are examples from the finance and retail sectors. In the consumer domain, home automation and wearables for lifestyle and fitness are prime examples.

2.2.3 THE INTERNET OF THINGS AHEAD

Given the previous example, IoT might look like solutions that have been around for many years in various sectors, for instance Telemetry or M2M, so what is really new? Simply answered, it is the use of open and standardized Internet and Web technologies instead of industry-specific technologies, it is the use of the Internet itself as an infrastructure instead of separately deployed and isolated network infrastructures, it is a common fabric for mixing devices, data, and applications, it is the richness and complexity in applications, it is the interconnectedness of IoT systems, and it is the collaboration and joint innovation of all engaged stakeholders.

What is underway is an IoT ecosystem not dissimilar to the current Internet, allowing things and real-world objects to connect, communicate, interact among themselves and with people, and engage in diverse applications in the same way humans do via the web today. Increased understanding of the complexity of the systems in question, economies of scale, and methods for ensuring interoperability, in conjunction with key business drivers and governance structures across value chains, will create wide-scale adoption and deployment of IoT solutions. We cover this in more detail in Chapter 3.

So, no longer will the Internet be only about people, media, and content, but it will also include all real-world assets as intelligent creatures exchanging information, interacting with people, supporting business processes of enterprises, enabling automation, and creating knowledge. The IoT is not a new Internet, it is an extension to the existing Internet.

The IoT is also about how data from the many different devices can be shared and integrated in numerous applications, i.e., instead of a closed-off "one device – one application" to solve a point problem, the IoT is about an open-ended multiplexity approach "many devices – many applications" as indicated in Figure 2.3. This multiplexity approach enables true open innovation and also drives how technology is defined by enabling interoperability and specific points in the overall IoT system design that allow this horizontal integration.

IoT is about the choice of technologies, but also about in what contexts they are applied. IoT can have a focus on the open innovative promises of the technologies at play and also on advanced and complex processing inside very targeted and closed environments such as industrial automation. When employing IoT technologies in more closed environments, an alternative interpretation of IoT could then be "Intranet of Things".

Early visions put forward (e.g., [5]) have included notions like a global open fabric of sensor and actuator services that integrate numerous WSN deployments and provide different levels of aggregated sensor and actuator services in an open manner for application innovation and for use in not only pure monitor and control types of applications, but also to augment or enrich other types of services with contextual information. IoT applications will not only rely on data and services from sensor and actuators alone. Equally important is the blend-in of other information sources that have relevance from the viewpoint of the physical world. These can be data from Geographic Information Systems (GISs)

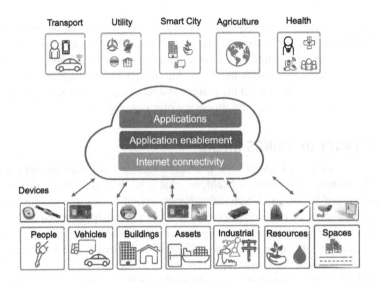

FIGURE 2.3

The Internet of Things.

like road databases and weather forecasting systems, and these can be of both a static nature and a real-time nature. Even information extracted from social media like Twitter feeds or Facebook status updates that relate to real-world observations can be fed into the same IoT system. Several use cases have emerged and examples are covered in Part 2 of this book.

Looking towards the applications and services in the IoT, we see that the application opportunities are open-ended, and only imagination will set the limit of what is achievable. One easily foresees emerging application domains that are driven from very diverse needs from across industry, society, and people, and that can be of both local interest and global interest. Applications can focus on safety, convenience, or cost reduction, optimizing business processes, or fulfilling various requirements on sustainability and assisted living. Listing all possible application segments is futile, as is providing a ranking of the most important ones. Prominent IoT settings were already covered in Chapter 1. We can point to examples of emerging application domains that are driven by different trends and interests (Figure 2.4). As can be seen, they are very diverse and can include applications like urban agriculture, robots, and food safety tracing, and we will give brief explanations of what these three examples might look like.

Urban Agriculture. Our world is now a "city planet" with more than 50% of the world's population living in urban areas and cities. The increased attention on sustainable living includes reducing the environmental impact from transportation of people and goods and, in the case of food production, reducing the needs for pesticides and leakage of fertilizers to rivers and lakes. The prospect of producing food at the place where it is consumed (i.e., in urban areas) is a promising example. By using IoT technologies, urban agriculture could be highly optimized. Sensors and actuators can monitor and control the plant environment and tailor the conditions according to the needs of the specific specimen. Water supply through a combination of rain collection and remote feeds can be combined on demand. City or urban districts can have separate infrastructures for the provisioning of different fertilizers. Drainage

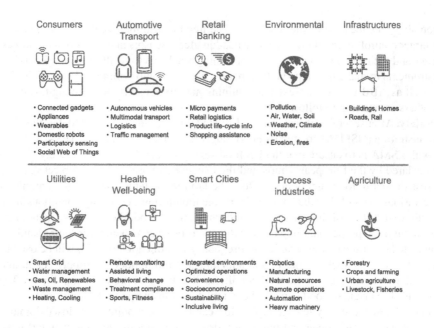

FIGURE 2.4

Emerging IoT application domains.

can be provided so as not to spoil crops growing on facades and rooftops of buildings, as well as to take care of any recyclable nutrients. Weather and light can be monitored, and necessary blinds that can shield and protect, as well as create greenhouse microclimates, can be automatically controlled. Fresh air generated by plants can be collected and fed into buildings, and tanks of algae that consume waste can generate fertilizers. Urban agriculture can be a mix of highly industrialized deployments with vertical greenhouses [6] and collective efforts by individuals in apartments by the use of more do-it-yourself style equipment. The latter can also foster a new business model of micro-markets based on communities, all in the spirit of democratized marketplaces[1]. A vision of urban agriculture is to be a self-sustaining system and can be an example of the previously mentioned circular economy.

Mining. The mining industry is undergoing a change for the future. Production rates must be increased, the cost per produced unit must decrease, and the lifetime of mines and sites must be prolonged. In addition, human workforce safety must be higher, with fewer or no accidents, and environmental impact must be decreased by reducing energy consumption and carbon emissions. The mining industry answer to this is to turn each mine into a fully automated and remotely controlled operation. The process chain of the mine involving blasting, crushing, grinding, and ore processing will be highly automated and interconnected. The heavy machinery used will be remotely controlled and monitored, mine sites will be connected, and shafts will be monitored in terms of air and gases. Since up to 50% of energy consumption in a mine can come from ventilation, energy savings can be done by very precise ventilation where the diesel vehicles are operating, and sensors in the mine can provide

---

[1] https://en.wikipedia.org/wiki/Sharing_economy

information about the location of the machines. The trend is also that local control rooms will be replaced by larger control rooms at the corporate headquarters. Sensors and actuators to remotely control both the sites and the massive robots in terms of mining machines for drilling, haulage, and processing are the instruments to make this happen. Companies like Rio Tinto with their Mine of the Future program[2], as well as ABB with their Next Level mining activity are currently driving this development, but many others are starting to follow suit.

Food Safety. After several outbreaks of food-related illnesses in the United States, the US Food and Drug Administration (USFDA) created its Food Safety and Modernization Act (FSMA) [7]. The main objective with FSMA is to ensure that the US food supply is safe. Similar food safety objectives have also been declared by the European Union and the Chinese authorities. These objectives will have an impact across the entire food supply chain, from the farm to the table, and require a number of actors to integrate various parts of their businesses. From the monitoring of farming conditions for plant and animal health, registration of the use of pesticides and animal food, the logistics chain all the way to retailers to monitor the proper handling and conditions as produce is being transported and stored – all will be connected end-to-end. Tags like Radio Frequency Identification (RFID) will be used to identify the items so they can be tracked, traced, and authenticated throughout the supply chain, and sensors will monitor that necessary environmental conditions like temperature and humidity are kept within specified levels. The origin of food can also be completely transparent to the consumers.

As can be seen by these very few examples, IoT can target very point- and closed domain-oriented applications, as well as very open and innovation-driven applications enabling new business models. Applications can stretch across an entire value chain and support entire lifecycle perspectives. Applications can be for business-to-business (B2B) as well as for business-to-consumer (B2C), and they can be complex and involve numerous actors, as well as large sets of heterogeneous data sources.

We will progress to see how IoT is driven by a set of diverse needs and how, based on those needs, one can arrive at a set of different needed, recurring capabilities. We will also see how different technologies emerge that will enable building the IoT, as well as a generalized model, or architecture, for how to build different target IoT solutions.

2.3 IOT IN A GLOBAL CONTEXT

IoT solutions have become quite common in many different scenarios. While the need to remotely monitor and control assets – personal, enterprise or other – is not new, a number of concurrent things are now converging to create drivers for change not just within the technology industry, but within the wider global economy and society. Our planet is facing massive challenges – environmental, social, and economic. The changes that humanity needs to deal with in the coming decades are unprecedented. Many of them are happening at the same time, ranging from constraints on natural resources to a reconfiguration of the world's economy, and people are increasingly looking to technology to assist with these issues.

Essentially, a set of megatrends are in combination creating both expectations of capabilities to solve the challenges and technologies that can enable one to address the expected capabilities. From this one can further derive a set of IoT Technology and Business Drivers. This is illustrated in Figure 2.5.

---

[2]http://www.riotinto.com/australia/pilbara/mine-of-the-future-9603.aspx

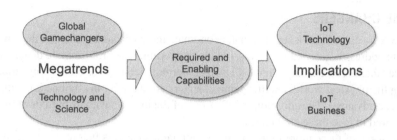

FIGURE 2.5

Megatrends, capabilities, and implications.

A megatrend is a pattern or trend that will have a fundamental and global impact on society at a macrolevel over several generations. It is something that will have a significant impact on the world in the foreseeable future. We here include both game changers and challenges, as well as technology and scientific advancements that can be used to meet these challenges. A full description of megatrends is beyond the scope of this book, and interested readers are directed to the many excellent books and reports available on this topic, including publications from the National Intelligence Council [8], European Internet Foundation [9], Frost & Sullivan [10], and McKinsey Global Institute [11]. In the following section, we focus on the megatrends that have implications for IoT. For the sake of simplicity, we also provide Table 2.1 as a summary of the main game changers, technology and science trends, capabilities, and implications for IoT.

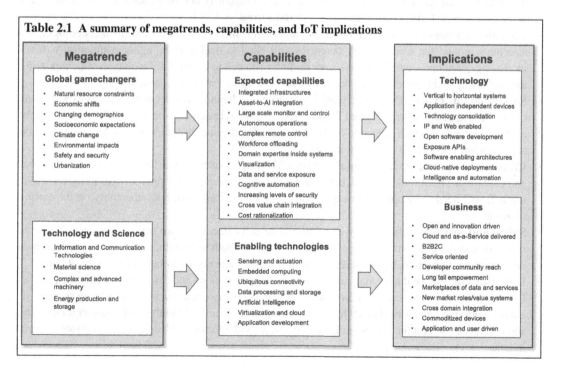

Table 2.1 A summary of megatrends, capabilities, and IoT implications

Megatrends	Capabilities	Implications
Global gamechangers • Natural resource constraints • Economic shifts • Changing demographics • Socioeconomic expectations • Climate change • Environmental impacts • Safety and security • Urbanization	**Expected capabilities** • Integrated infrastructures • Asset-to-AI integration • Large scale monitor and control • Autonomous operations • Complex remote control • Workforce offloading • Domain expertise inside systems • Visualization • Data and service exposure • Cognitive automation • Increasing levels of security • Cross value chain integration • Cost rationalization	**Technology** • Vertical to horizontal systems • Application independent devices • Technology consolidation • IP and Web enabled • Open software development • Exposure APIs • Software enabling architectures • Cloud-native deployments • Intelligence and automation
Technology and Science • Information and Communication Technologies • Material science • Complex and advanced machinery • Energy production and storage	**Enabling technologies** • Sensing and actuation • Embedded computing • Ubiquitous connectivity • Data processing and storage • Artificial Intelligence • Virtualization and cloud • Application development	**Business** • Open and innovation driven • Cloud and as-a-Service delivered • B2B2C • Service oriented • Developer community reach • Long tail empowerment • Marketplaces of data and services • New market roles/value systems • Cross domain integration • Commoditized devices • Application and user driven

2.3.1 GAME CHANGERS

The game changers come from a set of social, economic, and environmental shifts that create pressure for solutions to address issues and problems, but also opportunities to reformulate the manner in which our world faces them. There is an extremely strong emerging demand for monitoring, controlling, and understanding the physical world, and the game changers are working in conjunction with technological and scientific advances. The adoption of IoT is one of the important contributions of the technology evolution required to face these challenges.

We outline some of these more globally significant game changers below, and their relationship to IoT:

- **Natural resource constraints.** The world needs to increasingly do more with less, from raw materials to energy, water or food, the growing global population and associated economic growth demands put increasing constraints on the use of resources, including the introduction of regenerative systems as in the circular economy. The use of IoT to increase yields, improve productivity, and decrease loss across global supply chains is therefore escalating.
- **Economic shifts.** The overall economy is in a state of flux as it moves from the postindustrial era to a digital economy. One example of this is found in the move from product-oriented to service-oriented economies. This implies a lifetime responsibility of the product used in the service offering, and it will in many cases require the products to be connected and contain embedded technologies for gathering data and information. At the same time, there are fluctuations in global economic leadership. Economies across the globe must handle the evolving nature of these forces. As technology becomes increasingly embedded and more tasks are automated, countries need to manage this shift and ensure that IoT also creates new jobs and industries.
- **Changing demographics.** With increased prosperity, there will be a shift in the demographic structures around the world. Many countries will need to deal with an aging population without increasing economic expenditure. As a result, IoT will need to be used, for example, to help provide assisted living and reduce costs in healthcare and emerging "wellcare" systems.
- **Socioeconomic expectations.** The global emerging middle class results in increasing expectations on well-being and Corporate Social Responsibility. Lifestyle and convenience will be increasingly enabled by technology as the same disruption and efficiency practices evident in industries will be applied within people's lives and homes as well.
- **Climate change and environmental impacts.** The impact of human activities on the environment and climate has been long debated, but is now in essence scientifically proven. Technology, including IoT, will need to be applied to aggressively reduce the impact of human activity on the earth's systems.
- **Safety and security.** Public safety and national security become more urgent as society becomes more advanced, but also more vulnerable. This has to do both with reducing fatalities and health as well as crime prevention, and different technologies can address a number of the issues at hand.
- **Urbanization.** We see the dramatic increase in urban populations and discussions about megacities. Urbanization creates an entirely new level of demands on city infrastructures in order to support increasing urban populations. IoT technologies will play a central role in the optimization for citizens and enterprises within the urban realm, as well as providing increased support for decision makers in cities.

2.3.2 GENERAL TECHNOLOGY AND SCIENTIFIC TRENDS

Technological and scientific advances and breakthroughs are occurring across a number of disciplines at an increasing pace. Below is a brief description of the science and technology advances that have a direct relevance to IoT. The trends in the ICT sector are described separately in the subsequent section, as they are central for this book.

Material Science has a large impact across a vast range of industries, from pharmaceutical and cosmetics to electronics. Microelectromechanical Systems (MEMSs) and Nanoelectromechanical Systems (NEMSs) can be used to build advanced micro-sized motors and sensors like accelerometers and gyroscopes. Emerging flexible and printable electronics will enable a new range of innovations for embedding technology in the real world. New materials provide different methods to develop and manufacture a large range of different sensors and actuators, as well as being used in applications for environmental control, water purification, etc. Additionally, we will see other innovative uses, such as smart textiles, that will provide the capability to produce the next generation of wearable technologies. From an IoT perspective, these advances in material science will see an increasing range of applications and also a broader definition of what is meant by a sensor.

Complex and Advanced Machinery refers to OT that is autonomous or semiautonomous. Today they are used in a number of different industries; for example, robots and very advanced machinery are used in different harsh environments, such as deep-sea exploration, or in the mining industry in solutions such as the aforementioned Mine of the Future from Rio Tinto. Advanced machines have many modalities and operate with a combination of local autonomous capabilities as well as remote control. Sensing and actuation are key technologies, and local monitor-control loops for routine tasks are required in addition to reliable communications for remote operations. Often such solutions require real-time characteristics. These systems will continue to evolve and automate tasks today performed by humans – the ongoing boom around self-driving cars is a prime example.

Energy Production and Storage is relevant to IoT for two reasons. Firstly, it relates to the global interest of securing the availability of electricity while reducing climate and environmental impacts. Smart Grids, for example, imply microgeneration of electricity using, e.g., affordable Photovoltaic Panels. In addition, smart grids also require new types of energy storage, both for the grid itself and for emerging technologies, such as Electric Vehicles (EVs), that rely on increasingly efficient battery technologies. Secondly, powering embedded devices in WSNs will increasingly rely on different energy harvesting technologies and also rely on new miniaturized battery technologies and ultra-capacitors. As these technologies improve, IoT will be applicable in a broad range of scenarios that need long battery life.

2.3.3 TRENDS IN INFORMATION AND COMMUNICATIONS TECHNOLOGIES

While significant advances in the fields of Material Science, Advanced and Complex Machinery, and Energy Production and Storage will have an impact on IoT, first and foremost, ICT advances will drive the manner in which these solutions are provided as they are the core enabling factors behind IoT. Ever since the development of integrated circuits during the late 1950s and early 1960s, these technologies have had a growing impact on enterprises and society. The increasing rates of change have led to a situation where it is now cheap enough to "sensor the planet".

Today, **sensors, actuators, and tags function as the digital interfaces** to the physical world. Small-scale and cheap sensors and actuators provide the bridge between the physical realm and ICT systems. Tags using technologies such as RFID provide the means to put electronic identities on any object and can be cheaply produced.

Embedded processing is evolving, not only towards higher capabilities and processing speeds, but also extending towards the smallest of applications. There is a growing market for small-scale embedded processing such as 8-, 16-, and 32-bit microcontrollers with on-chip RAM and flash memory, I/O capabilities, and networking interfaces such as IEEE 802.15.4 and Bluetooth Low Energy that are integrated on tiny System-on-a-Chip (SoC) solutions. These enable very constrained devices with a small footprint of a few mm^2 and very low power consumption (in the milli- to micro-Watt range), yet are still capable of hosting an entire TCP/IP stack including a small web server.

Instant access to the Internet is available virtually everywhere today, mainly thanks to wireless and cellular technologies and the rapid deployment of cellular 3G, 4G/LTE, and coming 5G systems on a global scale. These systems provide ubiquitous and relatively cheap connectivity with the right characteristics for many applications, including low latency and the capacity to handle large amounts of data with high reliability. These systems can be further complemented with last-hop technologies such as IEEE 802.11, IEEE 802.15.4, Bluetooth Low Energy, and Power Line Communication (PLC) solutions to reach even the most cost-sensitive deployments and tiniest devices. Technologies like 6LoWPAN allow IP connectivity to be provided end-to-end, stretching into the capillary network domain, and legacy and proprietary protocols like ZigBee PRO can be avoided with the benefit of IP and the web anywhere. 3GPP are also extending LTE towards the lower end of the scale, providing very low-power extensions like NB-IoT targeting specific IoT applications, also for more constrained devices.

Software architectures have undergone several evolutions over the past decades, in particular with the increasing dominance of the web paradigm. A description of the evolution of software is beyond the scope of this book; here we instead look at those angles of software that are critical for successful IoT solutions. From a simplistic perspective, we can view software development techniques from what were originally closed environments towards platforms, where Open APIs provide a simple mechanism for developers to access the functionality of the platform in question (e.g., Android APIs for mobile devices). Over time, these platforms, due to the increasing use and power of the Internet, have become open platforms – ones that do not depend on certain programming languages or lock-in between platform developers and platform owners.

Software development has started applying the **web paradigm using a Service-Oriented Architecture (SOA)** and lately the microservices[3] approach. By extending the web paradigm to IoT devices, they can become a natural component of building any application and facilitate an easy integration of IoT device services into any enterprise system that is based on the SOA (e.g., that uses web services or RESTful interfaces). IoT applications can then become technology- and programming language-independent. This helps boost the IoT application development market as application development for IoT is not different than building applications for the web, and web developers are easily available. A key component in establishing the application development market is Open APIs.

Open APIs, in the same way that they have been critical to the development of the web, will be just as important to the creation of a successful IoT market, and we can already see developments in this space. Put simply, Open APIs relate to a common need to create a market between many companies, as is the case in the IoT market. Open APIs permit the creation of a fluid industrial platform, allowing

---

[3]https://www.martinfowler.com/articles/microservices.html

components to be combined in multiple different ways by multiple developers with little to no interaction with those who developed the platform or installed the devices. As we will discuss in more detail in Chapter 3, it is impossible for one company to guess what will be successful or liked by all of the customer segments associated with IoT. Open APIs are the market's response to this uncertainty; the choice of how to combine components is left to developers who are able to merely pick up the technical description and combine them together.

Without Open APIs, a developer would need to create contracts with several different companies in order to get access to the correct data to develop the application. The transaction costs associated with establishing such a service would be prohibitively expensive for most small development companies; they would need to establish contracts with each company for the data required and spend time and money on legal fees and business development with each individual company. Open APIs remove the need to create such contracts, allowing companies to establish "contracts" for sharing small amounts of data with one another and with developers dynamically, without legal teams, without negotiating contracts, and without even meeting one another. Open APIs therefore reduce the transaction costs associated with establishing a new market boundary [12], mitigating the risk of development, and help to establish a market for innovative capacity, which encourages creativity and application development.

Meanwhile, within ICT, **virtualization** has many different facets and has gained a lot of attention in the past few years, even though it has been around for a rather long time. The **cloud computing** paradigm, with different *as a Service* models, is one of the greatest aspects of the evolution of ICT for IoT as it allows virtualized and independent Execution Environments for multiple applications to reside in isolation on the same hardware platform, in large Data Centers and in distributed infrastructures alike.

Cloud computing allows elasticity in deployment of services and enables reaching long-tail applications in a viable fashion through "pay-as-you-grow", i.e., only paying for the resources that are actually used. It can be used to avoid in-house installations of server farms and associated dedicated IT service operations staff inside companies, thus enabling them to focus on their core business. Cloud computing also has the benefit of easing different businesses to interconnect if they are executing on the same platform. Handling of, for example, Service Level Agreements (SLAs) is easily facilitated with a high degree of control in a common virtualized environment. Cloud computing is also a key enabler when moving from a product-oriented offering to a service-oriented offering – instead of selling software, companies can sell the Software as a Service.

Somewhat related to the topic of Data Centers is the perspective of processing large amounts of data for gathering insights. The use of **Analytics, Machine Learning (ML), and Artificial Intelligence (AI)** in IoT solutions is rapidly gaining momentum. Dedicated hardware accelerators built on Graphics Processing Units (GPUs) and Field-Programmable Gate Arrays (FPGAs) are being embedded in Data Centers to more efficiently execute different ML and AI algorithms. Combined, these technologies are therefore key enablers for IoT, as they allow the collation and aggregation of the massive datasets that devices and sensors are likely to produce, as well as generating desired insights and automating operations. IoT shall not be confused as "yet another Big Data" scenario though. Even a single bit can be important in an industrial automated control application. It all depends on the type of IoT application at hand, e.g., the industrial automated control scenario, or a trend forecasting scenario. And this implies that IoT data can be treated both off-line and in mission critical real-time applications.

Therefore, IoT data typically also involves numerous and very different and heterogeneous sources, but also numerous and very different usages of the data. The analysis of IoT data may therefore be viewed as a complex set of interactions related to time (i.e., when the data is received) and relevance (i.e., the overall relevance of the piece of data to the question in hand). Managing these interactions is critical to the success of IoT solutions. Decision support or even decision making systems will therefore become very important in different application domains for IoT, as will the set of tools required to process data, aggregate information, and create knowledge. Knowledge representation across domains and heterogeneous systems are also important, as are semantics and linked-data tools. As a result, we can expect to see an increased usage of cognitive technologies and self-learning systems.

A fundamental addition to the data aspect of IoT is the dimension represented by actionable services as realized by actuators. There is a duality in sensing and actuation in terms of fusion and aggregation. Where data analytics is employed to find insights basically by aggregation, one can consider complex multimodal actuation services that need to be resolved down to the level of individual atomic actuation tasks. IoT also calls for intelligence in the form of closed control loops of sensing and actuation, which can be simple or very complex. This duality and the closed control aspect will put new requirements on technologies that stretch the boundaries from what can be achieved based on data-only-oriented technologies used in the prevalent approach of "Analytics".

As is illustrated in this chapter, the IoT market holds incredible promise for solving big problems for industry, society, and even individuals. One key thing to note, however, is the tremendous complexity that such systems need to handle in order to function efficiently and effectively. Partnerships and alliances are therefore critical – no single company will be able to produce all the technology and software required to deliver IoT solutions. Moreover, no single company will be able to house the innovative capacity required to develop new solutions for this market. IoT solutions bring together devices, networks, applications, software platforms, cloud computing platforms, business processing systems, and different aspects of control systems and AI techniques. This is quite simply not possible at scale without significant levels of **open technology development in standardization, open source, and alliance partnerships**.

This section discussed the global trends associated with ICT technology. The following section contains a discussion of the capabilities that are expected to be supported by IoT in order to address the challenges.

2.3.4 EXPECTED CAPABILITIES

As illustrated in previous sections, there are several recurring expected needs of ICT required to be delivered by IoT solutions. These needs address several aspects such as cost efficiency, effectiveness, and convenience; being lean and reducing environmental impact; encouraging innovation; and in general applying technology to create more intelligent systems, enterprises, and societies. The aforementioned ICT developments provide us with a rich toolbox to address these different aspects in general, and as part of that, IoT in particular. In the following paragraphs we exemplify how these expected capabilities, driven by global megatrends, can be met through the use of the enabling technologies.

While IoT to date primarily has targeted specific problems with tailored, siloed solutions, it is clear that emerging IoT applications will address the much more complex scenarios of large-scale distributed monitor and control applications. IoT systems are generally becoming multimodal in terms of sensing and control, complex in management, and distributed across large geographical areas. For example, the new requirements on electricity Smart Grids[4] involve end-to-end management of energy production, distribution, and consumption, taking into consideration needs from Demand Response, microgeneration, energy storage, and load balancing. Industrialized agriculture involving automated irrigation, fertilization, and monitoring of crops, soil or livestock at large scale is another example. We see clearly here heterogeneity across sensor data types, actuation services, and underlying communication systems and the need to apply intelligent software to reach various business objectives and Key Performance Indicators (KPI).

Take, for example, Smart City solutions: Here there is a clear need for integration of multiple disparate infrastructures such as utilities, including district heating and cooling, water, waste, and energy, as well as transportation such as road and rail. Each of these infrastructures has multiple stakeholders and separate ownership even though they operate in the same physical spaces of buildings, road networks, and so on. The optimization of entire cities requires the opening up of data and information, business processes, and services at different levels of the disjoint silos, creating a common fabric of services and data relating to the different infrastructures. This integration of multiple infrastructures will drive the need for a horizontal approach at the various levels of the system, for instance, at the resource level where data and information are captured by devices, the use of common network infrastructures, via the information and business process levels, up to an orchestrated optimization of all processes on the city level. We cover these issues in more detail in Chapter 14.

Meanwhile, advanced remotely operated machinery, such as drilling equipment in mines or deep sea exploration vessels, will require real-time control of complex operations, including various degrees of autonomous control systems. This places new requirements on the execution of distributed application software and real-time characteristics on both the network itself, as well as a need for flexibility where application logic is deployed and executed.

IoT will allow more assets of enterprises and organizations to be connected, thus allowing a tighter and more prompt integration of the assets into business processes and domain knowledge systems. Simple things can be used in a more controlled and intelligent manner, often called "Smart Objects", for instance power tools used at a construction site. These connected assets will generate more data and information, and will expose more service capabilities to ICT systems. Managing the complexity of information and services becomes a growing barrier for the workforce and places a high focus on using analytics tools of various kinds to gain insights. These insights, combined with domain-specific knowledge, can help the decision process of humans as individuals or professionals via decision support systems and visualization software and also assist workforce in the field.

As society operations involve a large number of actors taking on different roles in providing services, and as enterprises and industries increasingly rely on efficient operations across ecosystems, cross value chain and value network integration is a growing need. This requires technologies and business mechanisms that enable operations and information sharing across organizations and domains. Even industry segments that have been entirely unconnected will connect due to new needs; an example is the introduction of EVs. EVs are enabled by the new battery and energy storage technologies, but also require three separate elements to be connected, i.e., cars, road infrastructure via charging poles, and the electricity grid. In addition, there are new charging requirements that are created by the use of EVs that need new means for billing, in turn placing new requirements on the electricity grid itself for distribution and storage of electricity.

---

[4]https://www.nist.gov/engineering-laboratory/smart-grid

These sorts of collaboration scenarios will become increasingly important as industries, individuals, and government organizations work together to solve complex problems involving multiple stakeholders. This places an emphasis on the openness and exposure of services and information at different levels. What is important is to be able to share information and services across organizations in the horizontal dimension, as well as being able to aggregate and combine services and information to reach higher degrees of refinement and values in the vertical dimension. The open and collaborative nature of IoT means methods are required to publish and discover data and services, as well as means to achieve information interoperability, but also that care needs to be given to trust, security, and privacy. Information interoperability requires semantics of data so that data can be shared across systems without humans interpreting the data from one system to another. What is also needed is provenance of information so that the information can be trusted and the source can be identified for the sake of any liability issues. It also dramatically increases the required capability of system integration and the management of large-scale complex systems across multiple stakeholders and multiple organizational boundaries. As we come to increasingly rely on ICT solutions to monitor and control assets and physical properties of the real world, we not only require increased levels of cybersecurity, but what can be referred to as cyber-physical security. In the use of the Internet today, it is possible to exact financial damage via breaking into Information Technology (IT) systems of companies or bank accounts of individuals. Individuals, meanwhile, can face social damages from people hacking social media accounts. In an IoT, where it is possible to control assets (e.g., vehicles or moveable bridges), severe damage to property, or even loss of life, is possible. It is one thing to secure the IoT system itself, but it is another thing to secure that the IoT system does not cause physical damages and that it operates in a safe way. This raises requirements for trust and security to be correctly implemented in IoT systems.

2.3.5 IMPLICATIONS FOR IOT

Having gained a better understanding of the needed capabilities, as well as how technology evolution can support these needs, we can identify implications on both the technology and business perspectives. Many of these implications also have proof points of what is already ongoing.

The transition from vertically oriented systems, or application-specific silos, towards a horizontal systems approach is an already confirmed trend through the abundance of different IoT platforms on the market. These platforms are to the absolute majority cloud-based with a focus on offering IoT Platform as a Service, see Chapter 5. The horizontal approach with IoT is also confirmed through work in various industry alliances, where the Industrial Internet Consortium (IIC) is a very prominent example. In IIC, a large number of actors from different parts of the ecosystem and industry sectors come together to define common best practices on how to build industrial IoT systems.

The consolidation of technology fragmentation is also well under way. Legacy technologies for, e.g., building automation have been plentiful, e.g., BACnet, Lonworks, KNX, Z-Wave, and ZigBee. Now we see new alliances forming that take the Internet and Web set of technologies from the IETF as a baseline for smart buildings, for instance in the FairHair Alliance[5], and there one can also note the participation of some of the legacy standards organizations. Other examples of the adoption by various alliances of this same set of standards from the IETF as a common baseline across industries are

---

[5]https://www.fairhair-alliance.org

the Open Connectivity Foundation (OCF)[6], OMA SpecWorks[7] LwM2M solution, and Arm's Mbed[8] community. What is also seen from many of these efforts is the focus on and importance of open source software. What is happening here on the device technology side is also enabling devices to be more application-independent, thus making it easier to reuse devices for many different applications.

Further, where the innovation in the past has been very much about the possibility to connect things, e.g., with M2M solutions, the focus has now rapidly shifted to making use of the data from the things. Where the past focus was to be able to connect, the innovation today is about applying ML and AI to IoT.

As focus is shifting to the importance of data, and also for reasons of achieving interoperability by technology consolidation and making devices more application agnostic, solutions for semantic interoperability are gaining interest. Semantic interoperability helps extending IoT solutions with different data sources and also allows different IoT solutions to integrate, i.e., towards a "System of Systems". This will also be an enabler to achieve digital marketplaces of data and services related to IoT.

2.3.6 BARRIERS AND CONCERNS

With the outlined evolution of IoT, which involves many opportunities, we should not forget that some new concerns and barriers will also be raised. With the IoT, the first concern that likely comes to mind is the compromise of privacy and the protection of personal integrity. The use of RFID tags for tracing people is a raised concern. With a massive deployment of sensors in various environments, including smartphones, explicit data and information about people can be collected, and using analytics tools, users could potentially be profiled and identified even from anonymized data. Legislatory bodies are increasingly addressing these concerns, and the General Data Protection Regulation (GDPR)[9] from the European Union is one prominent example.

The reliability and accuracy of data and information when building solutions that integrate data from a large number of data sources that can come from different providers that are beyond one's own control is another concern. As there is a risk of relying on inaccurate or even faulty information in a decision process, the issue of accountability, and even liability, becomes an interest. Concepts like Provenance of Data and Quality of Information (QoI) become important, especially considering when data from various sources are aggregated to extract insights.

And what can be reiterated is also the topic of security that has one added dimension or level of concern. Not only are today's economical or social damages possible on the Internet, but with real assets connected and controllable over the Internet, damage of property as well as people's safety and even lives become an issue, and one can talk about cyber-physical security.

Not a concern, but a perceived barrier for large-scale adoption of IoT is in costs for massive deployment of IoT devices and embedded technologies. This is not only a matter of Capital Expenditure (CAPEX), but likely more importantly a matter of Operational Expenditure (OPEX). From a technical perspective, what is desired is a high degree of automated provisioning towards zero-configuration. Not only does this involve configuration of system parameters and data, but also contextual information

---

[6]https://openconnectivity.org

[7]https://www.omaspecworks.org

[8]https://www.mbed.com/

[9]https://www.eugdpr.org

such as location (e.g., based on Geographic Information System (GIS) coordinates or room/building information).

Concerns and barriers have consequences not only on finding technical solutions, but are more importantly having consequences also on business and socioeconomic aspects as well as on legislation and regulation. For instance, what is the long-term impact on jobs as more and more labor, both manual and intellectual, is being done by machines and software?

2.4 A USE CASE EXAMPLE

By now, the main features of working out IoT solutions should be understood, e.g., the approach towards open systems, the importance of data, multipurpose devices and application independence, information from various sources, and the role of analytics. And that in contrast to the mindset of "this is my problem, and to solve it, I deploy a device that I connect to get some data that I visualize", i.e., the "point problem – point solution" approach.

In order to illustrate this, we provide a fictitious illustrative example taking two different approaches to the solution. The first approach is the more traditional one, which is the way many M2M solutions have been built. The second approach is the Internet of Things approach to solving the problem. We want to highlight the potential and benefits of an IoT-oriented approach over the traditional approach like with M2M, but also indicate some key capabilities that will be needed going beyond what can be achieved with M2M. Our example is taken from personal well-being.

Studies from the US Department of Health and Human Services have shown that close to 50% of the health risks of the enterprise workforce are stress-related and that stress was the single highest risk contributor in a group of factors that also included such risks as high cholesterol, overweight issues, and high alcohol consumption. As stress can be a root cause for many direct negative health conditions, there are big potential savings in human quality of life, as well as national costs and productivity losses, if the factors contributing to stress can be identified and the right preventive measures taken. By performing the steps of stressor diagnosis, stress reliever recommendations, and logging and measuring the impacts of stress relievers for making a stress assessment, all in an iterative approach, there is an opportunity to significantly reduce the negative effects of stress.

Measuring human stress can be done using sensors. Two common stress measurements are heart rate and Galvanic Skin Response (GSR), and there are products on the market in the form of wearables that can do such measurements. These sensors can only provide the intensity of the heart rate and GSR, and they do not provide an answer to the cause of the intensity. A higher intensity can be the cause of stress, but this can also be due to exercise. In order to analyze whether the stress is positive or negative, more information is needed. The typical M2M solution would be based on getting sensor input from the person by equipping him or her with the appropriate device, in our case the aforementioned wearable, and using a smartphone as a mobile gateway to send measurements to an application server hosted by, e.g., a health service provider. In addition to the heart rate and GSR measurements, an accelerometer in the wearable measures the movement of the person, thus providing the ability to correlate any physical activity to the excitement measurements. The application server hosts the necessary functionality to analyze the collected data and, based on experience and domain knowledge, provides an indication of the stress level. The stress information can then be made available to the person or a caregiver via a smartphone application or a web interface on a computer. The M2M system solution and measured

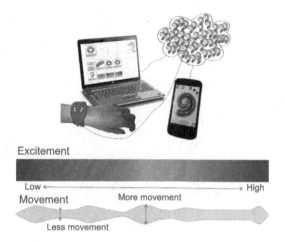

Excitement

Low ←————————————————————→ High

Movement   More movement

Less movement

FIGURE 2.6

Stress measurement M2M solution. Source: SICS & Ericsson.

data are depicted in Figure 2.6. This is in fact the typical setup provided by the many manufacturers of wearables that have been very popular these days. The manufacturer of the wearable is today also the service provider of the derived insights.

As already pointed out, this type of solution that is limited to a few measurement modalities can only provide very limited (if any) information about what actually causes the stress or excitement. Causes of stress in daily life, such as family situation, work situation, and other activities, cannot be identified. A combination of the stress measurement log over time, and a caregiver interviewing the person about any specific events at high levels of measured stress, could provide more insights, but this is a costly, labor-intensive, and subjective method. If additional contextual information could be added to the analysis process, a much more accurate stress situation analysis could potentially be performed.

Approaching the same problem situation from an IoT perspective would be to add data that provide much deeper and richer information of the person's contextual situation throughout the day. The prospect is that the more data is available, the more data can be analyzed and correlated in order to find patterns and dependencies. What is then required is to capture as much data about the daily activities and environment of the person as possible. The data sources of relevance are of many different types and can be openly available information as well as highly personal information. A resulting illustrative IoT solution is shown in Figure 2.7, where we see examples of a wide variety of data sources that have an impact on the personal situation. Depicted is also the importance of having access to domain expertise that can mine the available information and that can also provide proposed actions to avoid stressful situations or environments.

The contextual aspects include the physical properties of the specific environment, and these can be air quality and noise levels of the work environment, or the nighttime temperature of the bedroom, all having impacts on the person's well-being. Work activities can include the amount of emails in the inbox or calendar appointments, all potentially having a negative impact on stress. Leisure activities, on the other hand, can have a very positive impact on the level of excitement and stress and can have a more healing effect. Such different negative and positive factors need to be separated and filtered

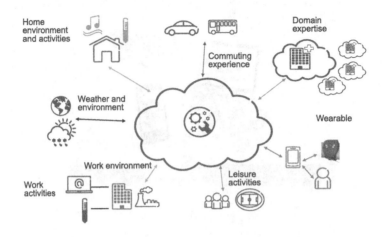

FIGURE 2.7

IoT-oriented stress analysis solution. Source: Ericsson.

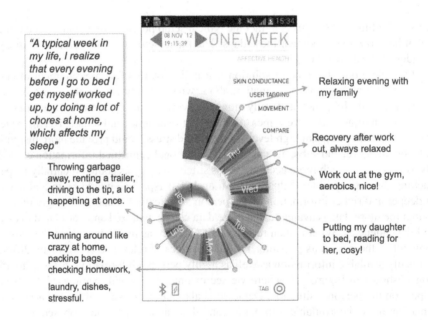

FIGURE 2.8

Stress analysis visualization. Source: SICS & Ericsson.

out; see Figure 2.8 for an example smartphone application that provides stress analysis feedback. The wearable is in this scenario just one component out of many. It should also be noted that the majority of the actual information sources are fully agnostic the specific application in mind, i.e., the measurement and prevention of negative stress.

By having the access to domain expertise, analytics can be proactive and preventive. And it should be noted that domain expertise does not imply that experts need to be consulted – domain expertise can rather be provided by an AI-based knowledge system. And the actual results from the stress analysis can be fed back to the knowledge base itself to further train the stress model. By understanding what factors cause negative stress, the system can propose actions to be taken, or even initiate actions automatically. They could be very elementary, such as suggesting to lower the nighttime bedroom temperature a few degrees, but also be more complex, such as having to deal with an entire workplace environment. Here, again, AI comes into play to make the necessary recommendations or even automatically taking the action, like lowering the room temperature, but still after consent by the person. As can be seen, this is in fact not far from a self-learning system that is adaptive.

As this simple example illustrates, an IoT-oriented solution to solving a particular problem could provide much more precision in achieving the desired results. We also observe some of the key features of an IoT solution; in other words, taking many different data sources into account, relying on sensor-originated data sources, but also other sources that have to do with the physical environment, and then also relying on both openly available data and data that is private and personal. The data sources, such as sensor nodes, should also focus on providing the information and should to the greatest extent be application-independent so that their reuse can be maximized. We also see the central role of analytics and knowledge extraction, as well as taking knowledge of actionable services that can involve controlling the physical environment using actuators. The increased complexity also comes at a cost. The solutions must ensure security and protection of privacy, and the need to deal with data and information of different degrees of accuracy and quality needs to be addressed in order to provide dependable solutions in the end. Also, another concern is how to handle the user interaction. Applications should be nonintrusive as information and decision overload is another factor contributing to human stress. In the IoT consumer space, this can also be noted as another barrier for IoT adoption.

2.5 A SHIFT IN MINDSET

To summarize, IoT solutions of the future compared to solutions of the past have differing characteristics, and the success of those solutions benefit from a shift of mindset. First of all, IoT solutions to date have generally been focusing on solving a very specific problem in isolation, and mostly for a single stakeholder. Another approach would be to take a broader perspective on solving a larger set of issues or ones that could involve several stakeholders. As a result of today's narrow view, most existing IoT devices are special purpose devices that are application-specific, often down to the device protocols. IoT solutions are hence also often effectively vertical silos with no horizontal integration or connection to adjacent use cases, and such an extension tends to be hard. IoT applications are also traditionally built by very specialized developers and are deployed inside enterprises. Two reasons are, firstly, the fact that existing legacy technologies are very industry-specific and, especially on the device side, technology use is highly fragmented with little or no standards across industries and, secondly, that state-of-the-art software engineering practices have not yet been widely adopted.

A key future approach to IoT solutions is to move away from the mentioned closed-silo deployments towards something that is characterized by openness, multipurpose, and innovation. This

Table 2.2 Characteristics and approaches of past and future IoT practices

Aspect	Past practices	IoT practices
Applications and services	Point problem driven	Innovation driven
	Single application - single device	Multiple applications - multiple devices
	Communication and device centric	Information and service centric
	Asset management driven	Insight and automation driven
Business	Closed business operations	Open market places
	Business objective driven	Collaborative and community driven
	B2B	B2B, B2B2B, B2C
	Established value chains	Ecosystems and value networks
	Consultancy and Systems Integration enabled	Open Web and as-a-Service enabled
	In-house deployment	Cloud deployment
Technology	Vertical system solution approach	Horizontal enabler approach
	Specialized device solutions	Generic commodity devices
	De facto and proprietary	Standards and open source
	Specific closed data formats and service descriptions	Open APIs and data specifications
	Closed specialized software development	Open software development
	Enterprise integration	Open APIs and web development

transition consists of a few main steps, namely: moving away from isolated solutions to an open environment; the use of IP and Web as a technology toolbox and the use of the current Internet as a foundation for enterprise and government operations; multimodal sensing, actuation, and data sources; insight and automation technologies; and the general move towards a horizontal layering of both technology and business. The main differing characteristics and shift of mindset between past and future IoT practices with respect to IoT applications, business, and technology are summarized in Table 2.2.

IOT – A BUSINESS PERSPECTIVE 3

CHAPTER OUTLINE

3.1 INTRODUCTION

Over the past few years, IoT uptake has increased with a variety of solutions being offered on a commercial scale. The global Internet of Things (IoT) market reached 598.2 billion US dollar in 2015 and the market is expected to reach 724.2 billion US dollar by 2023. Further, the market was projected at a compound annual growth rate (CAGR) of 13.2% during the forecast period 2016–2023 globally [13]. Infrastructure manufacturer Juniper predicts that the number of connected IoT devices, sensors and actuators will reach over 46 billion in 2021.

The market development for IoT, however, is intricately linked to the technology implemented today and to how this will evolve to provide new economic benefits and value creation opportunities. Chapter 2 outlined these processes from a global technology perspective; this chapter presents an overview of the market drivers active within the IoT market. We discuss firstly the concepts behind the drive towards information-driven Value Chains [14] and then turn our attention to business model innovation.

3.1.1 INFORMATION MARKETPLACES

A key aspect to note between Machine-to-Machine (M2M) and IoT is that the technology used for these solutions may be very similar – they may even use the same base components – but the manner in which the data is managed will be different. In an M2M solution, data remains within strict boundaries – most often within the boundaries of one company and solely within the scope of one project. It is used solely for the purpose that it was originally developed for. With IoT, however, data may be used and reused for many different purposes, perhaps beyond the original intended design, thanks to web-based technologies. In theory, data can be shared between companies and value chains in so-called information marketplaces. Alternatively, data could be publicly exchanged on a public information marketplace.

While public information marketplaces are generally the vision around IoT, in particular for Smart Cities as discussed in Chapter 14, it is unlikely such marketplaces will become commonplace before trust, risk, security, and insurance for data exchanges are able to be fully managed appropriately. In the following sections, we therefore focus on the business drivers for delivering IoT solutions and marketplaces that span multiple value chains, rather than publicly traded IoT marketplaces.

3.2 DEFINITIONS

We do not intend to go into detailed economic theory within this chapter, but we provide some basic working definitions that will provide a working understanding of the market dynamics driving the move from M2M towards IoT and what business and economic enablers must be put in place in order to drive the overall market.

3.2.1 GLOBAL VALUE CHAINS

A value chain describes the full range of activities that firms and workers perform to bring a product from its conception to end use and beyond, including design, production, marketing, distribution, and support to the final consumer [15]. A simplified value chain is illustrated in Figure 3.1; it is comprised of five separate activities that work together to create a finalized product.

These activities may be contained within a single firm or divided among different firms [16]. Analyzing an industry from a Global Value Chain (GVC) perspective permits understanding of the context of globalization on the activities contained within them by "focusing on the sequences of tangible and intangible value-adding activities, from conception and production to end use. GVC analysis therefore provides a holistic view of global industries – both from the top down and from the bottom up" [15].

Within the context of the technology industries, GVC analysis is particularly useful as such an analysis can help identify the boundaries between existing industrial structures such as M2M solutions and emerging industrial structures, as seen within the IoT market.

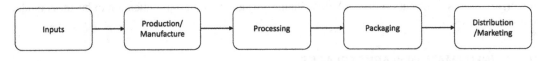

FIGURE 3.1

Example of SLA Security Integration.

3.2.2 ECOSYSTEMS VS. VALUE CHAINS

Business Ecosystems, defined by James Moore [17], refer to "an economic community supported by a foundation of interacting organizations and individuals ... The economic community produces goods and services of value to customers, who are themselves members of the ecosystem. The member organisms also include suppliers, lead producers, competitors, and other stakeholders. Over time, they co-evolve their capabilities and roles, and tend to align themselves with the directions set by one or more central companies. Those companies holding leadership roles may change over time, but the community values the function of ecosystem leader because it enables members to move toward shared visions to align their investments, and to find mutually supportive roles."

Many people discuss the IoT market as an "ecosystem", with multiple companies establishing loose relationships with one another that then may "piggy back" on larger companies in the ecosystem to deliver products and services to end-users and customers. While this is a useful description to begin with, a value chain is associated with the creation of value – it is the instantiation of exchange by a certain set of companies within an ecosystem. This is an important distinction when we are talking about market creation. A value chain is a useful model to explain how markets create value and how they evolve over time. While a market space composed of only competing value chains will eventually see the overall market value decrease (as they will compete only on price), in an ecosystem, the value chains will complement one another. In this chapter, we are interested in the creation of a marketplace for IoT data, and we therefore use a GVC analysis; an ecosystem analysis is out of the scope of this book.

3.2.3 INDUSTRIAL STRUCTURE

Industrial structure refers to the procedures and associations within a given industrial sector. It is the structure that is purposed towards the achievement of the goals of a particular industry. This is one of the key differences between the M2M and IoT markets – how the industrial structures will be formed around these solutions, despite very similar technology implementations. This is covered in more detail in the following sections.

3.3 VALUE CHAINS OVERVIEW

For the purposes of this section, we take a simplified view of a value chain for data where the ultimate product being developed is information that is used for decision making – or in some cases for selling to others. As illustrated in Figure 3.2, there are several inputs and outputs in such a value chain.

Inputs: Inputs are the base raw ingredients that are turned into a product. In the physical world this could be something like cocoa beans for the manufacture of chocolate. In the case of IoT this will be a piece of raw data that will ultimately be turned into information.

Production/Manufacture: Production/Manufacture refers to the process that the raw inputs are put through to become part of a value chain. In the physical world for example, cocoa beans may be dried and separated before being transported to overseas markets. Data from an M2M solution, meanwhile, need to be verified and tagged for provenance. This is where newly emerging technologies such as blockchain or distributed ledger technology can prove useful. These technologies are covered in more detail in Chapter 5.

Processing: Processing refers to the process whereby a product is prepared for sale. For example, cocoa beans may now be made into cocoa powder, ready for use in chocolate bars. For an IoT solution,

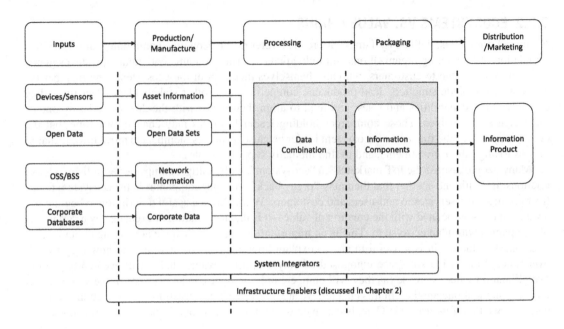

FIGURE 3.2

A simplified information value chain.

this refers to the aggregation of multiple data sources to create an information component – something that is ready to be combined with other datasets to make it useful for corporate decision making.

Packaging: Packaging refers to the process whereby a product can be branded as would be recognizable to end-user consumers. For example, a chocolate bar would now be ready to eat and have a red wrapper with the words "KitKat™" on it. For IoT, the data will have to be combined with other information from internal corporate databases, for example, to see whether the data received require any action. This data would be recognizable to the end-users that need to use the information, either in the form of visualizations or an Excel spreadsheet.

Distribution/Marketing: This process refers to the channels to market for products. For example, a chocolate bar may be sold at a supermarket, a kiosk, or even online. An IoT solution, however, will have produced an Information Product that can be used to create new knowledge within a corporate environment – examples include more detailed scheduling of maintenance based on real-world information or improved product design due to feedback from the IoT solution.

3.4 IOT VALUE CHAIN EXAMPLE

Meanwhile, the move towards IoT – from a value creation perspective – comes with the desire to make some of the data from sensors publicly available as part of an "information marketplace" or other data exchange that allows the data to be used by a broader range of actors rather than just the company that the system was originally designed for. It should be noted that such a marketplace could still be internal to a company or strictly protected between the value chains of several companies. Another alternative is a public marketplace, where data may be treated as a derivative, but such public trading of data is probably a long way from real-world market realization, although with the increasing use of tokenization techniques it is increasingly possible to think of such a market.

IoT value chains based on data are to some extent enabled by Open APIs and the other open web-based technologies discussed in Chapter 2. Open APIs allow for the knowledge contained within different technical systems to become unembedded, creating the possibility for many different economic entities to combine and share their data as long as they have a well-defined interface and description of how the data is formatted. Open APIs in conjunction with the Internet technologies described in Chapter 2 mean that knowledge is no longer tied to one digital system. The cognitive and conceptual human skills that were first embedded in semiconductors during the 1950s and 1960s are now decoupled from the specific technological system that was developed to house them. It is this decoupling of technology systems that allows for the creation of information marketplaces. This can initially make the value chain of an IoT solution look significantly more complex than one for a traditional product such as chocolate, but the principles remain the same. Let us take a closer look at a possible IoT value chain, including an Information Marketplace.

Both the processing and packaging sections of the Information-Driven Global Value Chain (I-GVC) are where Information Marketplaces will be developed. At this point, datasets with appropriate data tagging and traceability could be exchanged with other economic actors for feeding into their own information product development processes. Alternatively, a company may instead select to exchange information components, which represent a higher level of data abstraction of their corporate information.

3.5 AN EMERGING INDUSTRIAL STRUCTURE FOR IOT

Where the technologies of the industrial revolution integrated physical components together much more rapidly, M2M and IoT are about rapidly integrating data and workflows that form the basis of the global economy at increasing speed and precision. In contrast to fixed broadband technologies, which are limited to implementation in households mainly in the developed world, mobile places consumer electronic goods into the hands of over 4 billion end-users across the globe and connects billions of new devices into the mobile broadband platform. Concepts such as cloud computing, meanwhile, have the ability to provide low-cost access to computational capacity for these billions of end-users via these mobile devices. Combined, these two technologies create a platform that will rapidly redefine the global economy. A new form of value chain is actually emerging as a result – one driven by the creation of information, rather than physical products.

The adoption of the mobile broadband platform is therefore different from previous incarnations of Information and Communication Technology (ICT) industrial platforms as it reshapes how economic actors within a value chain interact not just with one another, but also with employees and the wider economic environment in a manner similar to the technology of the industrial revolution. More importantly perhaps, it changes fundamentally the manner in which individuals interact with economic actors in a digital world.

As mentioned in Chapter 2, the need for System Integrators in the communications industries has increased over the decades. With each generation of platform, a new type of system integrator has emerged. For IoT, however, new sets of system integrator capacity are required for two main reasons:

- Technical: The factors driving the technical revolution of these industries mean that the complexity of the devices in question require massive amounts of R&D, as do semiconductors with large amounts of functionality built into the silicon. Services will require multiple devices, sensors, and actuators from suppliers to be integrated and exposed to developers. Only those companies with sufficient scale to understand the huge number of technologies well enough to integrate them fully on behalf of a customer can handle this technical complexity. While niche integrators will continue to exist, full solutions will be integrated and managed by large companies, or partnerships between vendors.
- Financial: Only those companies that are able to capture the added value created in the emerging industrial structure will recoup enough money to reinvest in the R&D required to participate in the systems integration market. It is highly likely that the participants that do not capture part of the integration market will be relegated to "lower" ends of the value chain, producing components as input for other system integrators.

There is in fact a new type of value chain emerging – one where the data gathered from sensors and Radio Frequency Identification (RFID) are combined with information from smartphones that directly identify a specific individual, their activities, their purchases, and their preferred method of communication. This information can be combined in a number of ways to create tailored services of direct relevance to the individual or corporation in question. Search queries can be localized based on where a person is, and advertising can be targeted directly to the end-user in question based on personalized information about their age, level of education, employment, and tastes. While it is perhaps questionable whether the world really needs new methods to advertise goods and services, beneath this development lies a fundamental change in some aspects of the global economy.

Firstly, information about individuals is now captured, stored, processed, and reused across many different systems that sit on top of the mobile broadband platform. This data has always existed, but with the increasingly low cost of computing capacity in the form of cloud computing platforms, it is now cheap enough to store this data for an extremely long length of time. It is now possible, therefore, for information about individuals and digital systems to be packaged, bundled, and exchanged between economic entities with an ease that previously was impossible. Value is no longer solely measured through "value-in-use" or "value-in-exchange", but there is now also a "value-in-reuse", specifically because the commodity, data, is not consumed within the processes of production as with previous generations of commodity creation.

Actors that perform this data collection, storage, and processing are forming the basis of what may be viewed as an I-GVC, a value chain where the product is information itself. As an example, a difference in value can be identified in knowing my location when I step off a train in a new city and am looking for a decent cup of coffee. I may choose to activate my smartphone and perform a localized search using my phone's GPS and browser features. Alternatively, I may be happy enough to just walk around until I find a place that I think looks OK. In this case, the value that I as an individual place on my phone knowing my location and assisting me to find a local coffee store is relatively low – personally, I might not value this very highly. In comparison, however, there is a great deal of value for a coffee company to know that a few hundred women have stepped off a train in search of an espresso.

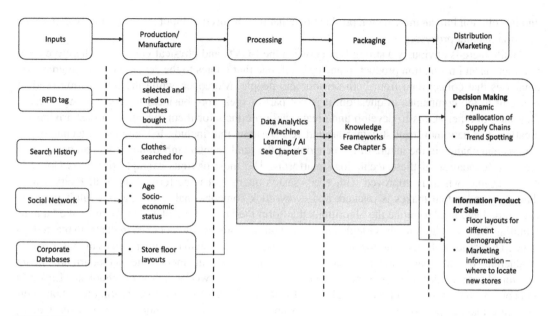

FIGURE 3.3

An Information-Driven Value Chain for retail.

A coffee shop chain would know that it is potentially quite profitable to open a new store there. In addition, understanding the age group of those women, their level of education, and their general tastes would allow the chain to tailor the coffee store to their target market with much greater precision.

Similarly, if I was in a clothing store searching for a new outfit for work, through a combination of information about myself and the RFID tags on the different clothes, I could be guided to the correct clothing selection for my age group, my education level, and also my current employer. Information about what path I take through the store during my search for the clothes could be fed back into an information system that would allow the store to reorganize their floor layouts more effectively and track the clothes that I was interested in and those which I actually selected to try on and purchase. This information can be used to streamline the supply chains of corporations even further than is possible today, and this represents the next phase of the impact of communication technologies on the boundaries of the firm within the global economy: companies that share this type of information would be more deeply embedded in one another's workflow, leading to highly concatenated supply chains and a further blurring of the boundaries of the firm within the digital economy. This is illustrated in Figure 3.3.

This streamlining could also be extended into the processes of production, changing orders based on consumer interest in products, not just their purchasing patterns. This would result in less wasted stock and a much closer understanding of seasonal trends and an increased level of control for those companies working as system integrators. The integration of these data streams allows for concatenation of supply chains not just internally to one company, therefore, but across industrial boundaries [14]. The level of analysis described above requires aggregation of data from many different people and its collation into an information product, one that may be used as input into corporate and end-user decision making processes.

While there is obviously a strong link between the I-GVC and physical goods, it is therefore clear that there is an information product in and of itself, one that relates to the development of aggregation databases that collect data from both sensors and people. Moreover, while the information product is useful to the companies in question, it is not part of their core business to create it. As a result, they look to other actors to develop and create these products for them, which is further driving the creation of a new industrial structure around ICT systems that include IoT and cloud technologies. The second change in the nature of the economy is the fundamental embedding of human beings into the very foundation of these technology platforms. The most obvious example of this is Google's search engine, which is improved with every search query that is performed using it. Every search that every individual makes is tracked, and every click someone makes through Google's products is recorded and used to refine the algorithms that form the basis of the platform. Without the humans inputting their searches into the Google platform, it quite simply would not exist in the form that it does today. The broad-scale consumerization of technology, combined with the cheap means of "information production" due to cloud computing, has led to information management systems that are now being developed for end-user consumers, not just enterprises. Social networks such as Facebook and LinkedIn and content sharing sites such as YouTube or Blogger allow end-users to store information about their lives in a manner previously not possible. Consumers now store their photos, contact lists, videos, documents, and financial data online. Ostensibly, this is provided for "free"; end-users receive access to the websites merely through creating an account, logging in, and uploading their data.

Within the capitalist economy, however, no service is ever really free. Companies must pay for the costs of computing resources, even those that are housed in the cloud. While the cost to end-users appears to be zero, they are in fact being charged on a daily basis through the use of their profile data for targeted advertising. In the early days of social networks, for example, this targeted advertising was no worse than the "traditional" direct advertising methods: based on the data provided by the end-user, they would receive an age-appropriate advertisement for their demographic. With the mobile broadband platform, however, the level of data that can be gathered about end-users is orders of magnitude larger than previously imaginable. My location, level of education, employment status, health records, tax data, credit rating, purchasing patterns, search history, social networks (both private and professional), relationship status, and even how often I call my mother are recorded, stored, and interconnected in a vast array of disparate systems that are now linked together through the platform of a converged communications industry.

With the addition of new levels of data that it is possible to retrieve about individuals through mobile devices and sensor networks, the emerging ICT platforms are forcing a redefinition of our established understandings of the notion of value both within the communications industries and even beyond these industrial boundaries to all companies and individuals that use these platforms on a daily basis [14]. The following section outlines the emerging value chain and the roles that must be filled in order to create a flourishing IoT marketplace.

3.5.1 THE INFORMATION-DRIVEN GLOBAL VALUE CHAIN

There are five fundamental roles within the I-GVC that companies and other actors are forming around, illustrated in Figure 3.4:

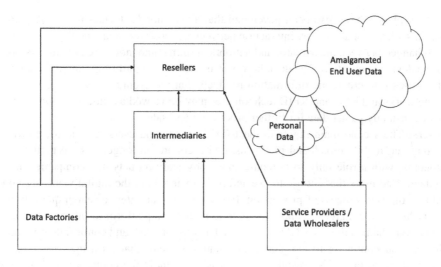

FIGURE 3.4

The Information-Driven Global Value Chain.

- Inputs:
    - Sensors, RFID, and other devices.
    - End-users.
- Data factories.
- Service providers/data wholesalers.
- Intermediaries.
- Resellers.

3.5.1.1 Inputs to the information-driven global commodity chain

There are two main inputs into the I-GVC:

- Sensors and other devices (e.g., RFID and NFC).
- End-users.

Both of these information sources input tiny amounts of data into the I-GVC chain, which are then aggregated, analyzed, repackaged, and exchanged between the different economic actors that form the value chain. As a result, sensor devices and networks, RFIDs, mobile and consumer devices, Wi-Fi hotspots, and end-users all form part of a network of "subcontractors" in the value chain, all contributing to the increased value of the information products, which were discussed in Section 3.4.

Sensors and Radio Frequency Identification: Sensors and RFID are already found in a multitude of different applications worldwide (as discussed in Chapter 2), helping to smooth supply and demand in various supply chains worldwide and gathering climate and other localized data that is then transmitted back to a centralized information processing system. These devices are working as inputs to the I-GVC through the capture and transmission of data necessary for the development of information products. Smartphones have also been developed that allow mobile devices to interact with sensors and RFID. This allows for a two-way interaction between a mobile terminal and the sensor technology. The data exchanged between the actuator and a mobile terminal may not be readily understood or even useful for the device in question. The data, however, is used as one part of the input to the commodity chain, which uses it to create the information products that are eventually exchanged. In this sense, the sensor networks and NFC and RFID technologies may be viewed as subcontractors to the I-GVC, workers that constantly gather data for further processing and sale.

End-users: The second main inputs to the I-GVC are the end-users. Due to the convergence of the computing and mobile broadband platforms, end-users are no longer passive participants in the digital economy, with a role only to purchase those physical products that companies develop and market to them. End-users that choose to use and participate within the digital world are now deeply embedded into the very process of production. Every human that enters a search query into a search engine, every human that agrees to allow the mobile broadband platform to inform a service of their location, and every human that uses NFC to allow a bank to establish and confirm their identity is also functioning as a subcontractor to the global information systems that form the basis of the I-GVC. In fact, the creation of the I-GVC would not be possible without the contribution of many millions of individuals worldwide. This is perhaps the most unique aspect of the I-GVC – there is no national boundary for the contribution of humans to the I-GVC, the data about individuals can be collected from any person in any language, in almost any data format. Each individual's data can be treated as unique within this value chain; in fact, it is the ability to capture the uniqueness of every person that is a key aspect of the I-GVC in comparison to the other commodity chains that are at work within the global economy. Every person worldwide that has to use digital technologies to do their banking, their taxes, and their information searches and to communicate with friends and colleagues is constantly working on behalf of the I-GVC, contributing their individual profile data and knowledge to the value chain. In the same manner as the actuators constantly gather localized data, humans are now contributing to the development of information products within the I-GVC nearly 24 hours a day.

3.5.1.2 Production processes of the information-driven global value chain

Data factories: Data factories are those entities that produce data in digital forms for use in other parts of the I-GVC. Many of these companies existed in the predigital era; for example, Ordnance Survey (OS) in the UK has always collected map information from the field and collated and produced maps for purchase. Previously, such data factories would create paper-based products and sell them to end-users via retailers. With the move to the digital era, however, these companies now also provide this data via digital means; for example, OS now makes maps and associated data available in digital format. Essentially, its business model has not changed significantly – it still produces maps – but its means of delivery of products has changed. Moreover, its products can now be combined, reused, and bundled together with other products by actors in the commodity chain as the foundation of other services. For example, maps from OS can be combined with other data from travel services such as TFL to provide detailed travel applications on mobile devices. A more complex example is Sveriges Meterologiska och Hydrologiska Institut (SMHI), which provides weather and climate data throughout Sweden. SMHI has a large number of weather stations across Sweden through which it collects weather and environmental information. In addition, it purchases its data from Yr.no, its Norwegian equivalent. SMHI therefore produces raw data, but it also processes the data and bundles it in different ways based on customer requests and requirements. SMHI functions not only as a data factory, therefore, but also a reseller, which is described separately below.

Service providers/data wholesalers: Service providers and data wholesalers are those entities that collect data from various sources worldwide and, through the creation of massive databases, use it to either improve their own information products or sell information products in various forms. Many examples exist; several well-known ones are Twitter, Facebook, and Google. Google "sells" its data assets through the development of extremely accurate, targeted, search-based advertising mechanisms that it is able to sell to companies wishing to reach a particular market. Twitter, meanwhile, through collating streams of "Tweets" from people worldwide, is able to collate customer sentiment about different products and world events, from service at a restaurant to election processes across the globe; through what Twitter refers to as a "data hose", companies and developers can access 50% of end-user Tweets for 360,000 US dollar per annum. A new set of data wholesalers is starting to emerge, however: those companies that handle the massive amount of data that is produced by sensor networks and mobile devices worldwide. These companies are collating those transactions that are made by the millions of devices worldwide that utilize communication networks to transmit data. The sheer quantity of data that is being transmitted via actuators and mobile devices will be orders of magnitude higher than previously imagined within the supply chains of multinational companies alone. We cover the technologies associated with these, IoT and Big Data Analytics, in Chapter 5.

Intermediaries: In the emerging industrial structure of the I-GVC, there is a need for intermediaries that handle several aspects of the production of information products. As mentioned above, there are many privacy and regional issues associated with the collection of personal information. In Europe, the manner in which Facebook collects and uses the data of the individuals that participate in its service may actually be in contravention of European privacy law. The development of databases such as the ones created by Google, Facebook, and Twitter may therefore require the creation of entities that are able to "anonymize" data sufficiently to protect individuals' privacy rights in relevant regional settings. These corporations will provide protection for the consumer that their data is being used in an appropriate manner, i.e., the manner in which the consumer has approved its usage. For example, I may happily share my personalized information about my tastes with a clothing company or music store in order to receive better service, while I may not be happy for my credit rating or tax data to be shared freely with different companies. I would therefore allow an intermediary to act on my behalf, tagging the relevant information in some form to ensure that it was not used in a manner that I had not previously agreed to. Another reason for an intermediary of this nature is to reduce transaction costs associated with the establishment of a market for many different companies to participate in. As discussed in the previous section on service providers/data wholesalers, the amount of data that is being produced is also problematic – even with cloud computing, it is difficult to process the amounts of data produced in the I-GVC. The different types of information products that are to be produced are only of interest to certain types of companies – for example the marketing division of a company may be interested to understand customer sentiment about a particular product within a certain age group. Another company may want to understand what searches are being performed in their local area, while a local authority may wish to use sensor data to obtain real-time data about pollution from local factories. The type of data and style of analysis for each of these information products is fundamentally different, and each requires unique skills – it is highly improbable that one company will be able to handle all of these types of data in one place. It is therefore more likely that different companies, intermediaries, will develop that target their information products to different niche markets. These companies will be able to focus on certain datasets and become specialists within that particular information product field. The quantity and nature of data being developed into information products also require a completely new type of intermediary, one that is able to handle the scalability issues and the associated security and privacy questions raised by the use of this data to build products. This is perhaps the most obvious role for operators and network vendors with global services operations to take, as they have many decades of experience in developing, operating, and maintaining secure systems that scale to millions of users. The average operator network is designed to scale to approximately 100 million end-users. With the advent of data networks for devices (not just human subscribers), operators are now investigating systems that can scale to at least ten-fold that size. Systems that can handle this number of devices and end-users require a huge level of cooperation across industrial structure, and the development of this scale of intermediary will therefore require closer cooperation between equipment manufacturers and service providers at all levels. The I-GVC may therefore also be seen to further blur the boundaries of the firm between the high-technology companies that form its basis.

Resellers: Resellers are those entities that combine input from several different intermediaries, combine it, analyze it, and sell it to either end-users or to corporate entities. These resellers are currently rather limited in terms of the data that they are able to easily access via the converged communications platform, but they are indicative of the types of corporate entities that are forming within this space. One example is BlueKai, which tracks the online shopping behavior of Internet users and mines the data gathered for "purchasing intent" in order to allow advertisers to target buyers more accurately. BlueKai combines data from several sources, including Amazon, Ebay, and Alibaba. Through this data, it is able to identify regional trends, helping companies to identify not just which consumer group to target their goods to, but also which part of the country. As an example, BlueKai is able to identify all those end-users in West Virginia currently searching for a washing machine in a certain price bracket.

3.6 THE INTERNATIONAL-DRIVEN GLOBAL VALUE CHAIN AND GLOBAL INFORMATION MONOPOLIES

Currently, within the industrial structure of the converged communications industry, there is a large regional disparity between those companies that produce the infrastructure for the I-GVC and those that make a significant profit from it. Through positioning themselves within the correct part of the GVC, these companies are able to take the lion's share of the profit. Through the breakdown of regional boundaries for collection of data by the development and implementation of a global converged communication infrastructure, these companies are able to enlist every person using a mobile device worldwide as a contributor to the development of their information products – in effect, every person worldwide is working for these corporations so that they are able to sell aggregated data for a huge profit. Despite this data being collected from people in every corner of the globe, from the UK, Thailand, Australia, China, and Africa to even the remotest parts of Kashmir, the surplus value of the mobile broadband platform is currently being overwhelmingly captured, developed, and molded into information products, mostly by U.S. companies. Through being able to collect and analyze data without being restricted by the same level of privacy regulation as in Europe, for example, they are able to create a much better information product. Companies in Europe, Asia, and other parts of the globe are therefore dependent on these companies in order to gain the most appropriate knowledge for their companies' needs. In the same way that the use of IT became a critical success factor for enterprises in the late

1980s and 1990s, the use of information products developed within the I-GVC is becoming critical to securing a competitive advantage in a global market. Companies are therefore compelled to use the most effective information product for their needs. In effect, the I-GVC, rather than breaking down the digital divide, as many have predicted, is in fact leading to a new form of digital discrimination and a new sort of dependency relationship between large multinationals and those participants, or "workers", within the I-GVC. While there may appear to be huge differences between the industrial revolution and the birth of the digital planet in the nature of how workers are treated, in particular with so much being advertised as "free" for end-users, there are in fact many similar parallels in the aggregation of human endeavor in the processes of the accumulation of capital. A multitude of workers contribute to the information products developed, but only a few large corporations capture the surplus value. This has in fact led to a few interesting discussions within industry about who actually owns this data – is it the company that provides the service, the service provider that delivers the connectivity, or the end-users themselves? An end-user might potentially be able to receive money from the use of his or her data, a nominal contribution for each time that data is used for creation of an information product. Profit sharing arrangements might even be possible between companies that develop the platform and those that collate the data into product form. The fact remains, however, that it is only those companies with the R&D budgets, the scale, and the global reach necessary to exploit the aggregation of this data that will be able to make significant profits from it.

3.7 BUSINESS MODEL INNOVATION IN IOT

For those companies interested in IoT development, a useful hands-on approach to developing business models is through the lens of business model innovation. This is most effectively done through assessing business cases from the perspective of transactions – how exchanges happen across the industry and how these exchanges are controlled. Through this approach, companies are able to identify partners and work more broadly across the IoT ecosystem. There are three main ways that companies should approach transactions in business model innovation.

Content refers to what sort of activities are being performed and – from a transaction perspective – what is being exchanged. **Structure** refers to how activities are linked and the exchange mechanisms used to link them. **Governance** refers to who performs what activities and to issues of control.

The company within the ecosystem that has most control generally captures the majority of the value creation. Within IoT systems, however, this will often need to be approached from a revenue sharing perspective.

The technical developments emerging in IoT enable many companies to fundamentally change the way they do business – in particular how they organize and conduct exchanges and activities across firm and industry boundaries with customers, vendors, partners, and other stakeholders. IoT solutions can often force companies to rethink the structure, content, and governance of transactions within their activity system – or business model. IoT will, however, also provide an increased number of options for how a company can structure its networks – allowing boundary spanning exchanges and activities to be developed in a manner that allows operators to create sustainable performance advantage.

3.7.1 CURRENT EXAMPLES

So far across the industry, multiple IoT applications have been deemed to be promising candidates and trials or Proofs of Concept (PoC) have been started. Examples include environmental, security (e.g., crowd management, port management/logistics), utilities (e.g., metering), and other "Smart Cities" applications (e.g., street lighting, parking, waste management). Other industries – many of which have long experience in M2M solutions and high operational expenditure (e.g., oil & gas) – are actively investing in IoT installations due to their annual spend on monitoring and maintenance that often runs into tens of billions of dollars.

The benefits of these systems are obvious to many of the stakeholders involved, but a key issue in the development of new markets is the long-term viability of involvement of all stakeholders outside of the "pilot" or prototype stage. Taking these solutions to real-world scale requires significantly more than small- – or even large- – scale test beds. It requires the development of appropriate and viable business models that provide and reward the value creation activities of all those involved.

Furthermore, stakeholders are unlikely to accept high connectivity or maintenance costs for embedded devices on their own without significant payback from the initiative. IoT therefore requires new partnerships, ecosystems, and revenue sharing schemes between technology providers (e.g., chip manufacturers, systems integrators, network operations, to ensure that the split is equitable). It is reasonable to suggest that the net benefit is impossible to achieve in isolation; therefore IoT is a perfect candidate for business model innovation, rather than merely product or process innovation. This may occur outside of the MNO community, relegating operators to connectivity providers only, which is a risk to maximizing potential revenues and justifying managing IoT network costs.

The following sections outline several potential business model innovations to provide in-depth illustrations of a few use cases in order to assist the reader understand how to develop/build them.

3.7.2 BUSINESS TO BUSINESS MODELS

One of the main use cases that IoT can help enable is Business to Business models – where companies are able to share data between one another more easily and in an automated manner, rather than relying on paper-driven processes or, e.g., spreadsheets. Often this means IoT solutions require a new type of pricing model. Often, the business models require collaboration across a much broader ecosystem of actors. See Figure 3.5.

3.7.3 DATA ANALYTICS BUSINESS MODELS

Another possible business model innovation is the creation of data analytic services for specific industry verticals. Through working together with other industry players who have invested in data analytics platforms, new offerings can be created that fully leverage the potential of IoT data. See Figure 3.6.

3.7.4 NEW DATA MARKETPLACES MODELS

Through deploying existing resources within the IoT solutions but linking them in new ways and new governance models, it is possible for mobile operators to play a key role in the delivery of new marketplaces for data. Through combining this approach with innovative technologies such as blockchain, it can also enable micropayments for IoT data. See Figure 3.7.

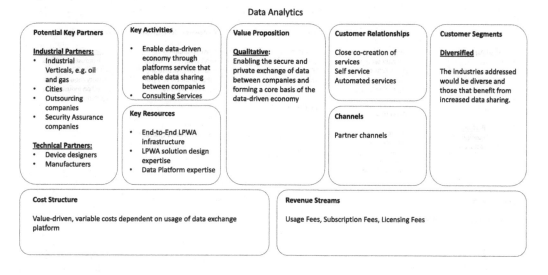

B2B Business Models

Potential Key Partners	Key Activities	Value Proposition	Customer Relationships	Customer Segments
Industrial Partners: • Oil and Gas Verticals • Water Verticals • City Leaders • Outsourcing companies **Technical Partners:** • Device designers • Manufacturers	• Provide standards based communications networks • Consulting Services **Key Resources** • Network infrastructure • LPWA solution design experts • MNO branding	**Quantitative:** Solutions built on standards-based communications provide cheaper overall solutions to customers **Qualitative:** MNO branding and experience provides better overall experience and service delivery for industry verticals	Close co-creation of services and delivery of final infrastructure and other products **Channels** Enterprise sales channels	**Niche markets** LPWA solutions in this market will need to be tailored to the long-tail in order to create sufficient coverage of devices **Segmented** Different industry verticals will require different solutions. Each vertical will have multiple segments that need to be targeted through

Cost Structure	Revenue Streams
Cost-driven, Economies of Scale and Economies of Scope	Usage Fees and Subscription Fees

FIGURE 3.5

Example of Business to Business models for IoT.

Data Analytics

Potential Key Partners	Key Activities	Value Proposition	Customer Relationships	Customer Segments
Industrial Partners: • Industrial Verticals, e.g. oil and gas • Cities • Outsourcing companies • Security Assurance companies **Technical Partners:** • Device designers • Manufacturers	• Enable data-driven economy through platforms service that enable data sharing between companies • Consulting Services **Key Resources** • End-to-End LPWA infrastructure • LPWA solution design expertise • Data Platform expertise	**Qualitative:** Enabling the secure and private exchange of data between companies and forming a core basis of the data-driven economy	Close co-creation of services Self service Automated services **Channels** Partner channels	**Diversified** The industries addressed would be diverse and those that benefit from increased data sharing.

Cost Structure	Revenue Streams
Value-driven, variable costs dependent on usage of data exchange platform	Usage Fees, Subscription Fees, Licensing Fees

FIGURE 3.6

Example of data analytics business models.

3.7.5 SLA SECURITY INTEGRATION

Probably one of the most important new business models will be the increasing requirements for security integration from an end-to-end perspective. This includes ensuring the security and integrity of the data produced, transmitted, and used within such IoT solutions. This is a critical area for the cre-

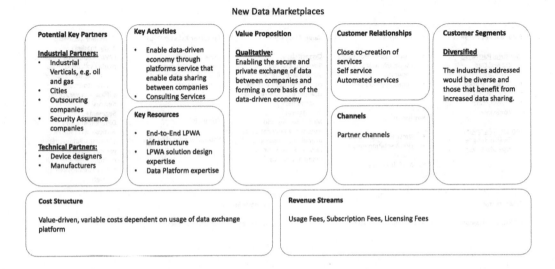

FIGURE 3.7

Example of data marketplaces models.

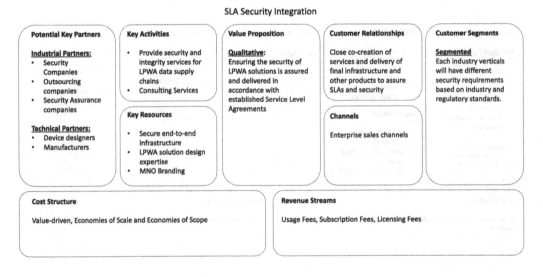

FIGURE 3.8

Example of SLA security integration.

ation of trust within the IoT-enabled digital economy. If upstream users of the data produced by IoT solutions cannot be assured of the integrity of data (i.e., that the data has come from the sensor that it says it has, that the device has not been tampered with, and that the data itself has not been tampered within in transit), then the decisions made upon such data can be of poor quality – and in some cases dangerous. Through applying business model innovation, however, mobile operators are able to create a consortium approach to delivering a new solution to the market that provides critical assurances to the upstream users of IoT data. See Figure 3.8.

3.8 CONCLUSIONS

This chapter outlined IoT from a business perspective. Thanks to open, web-based technologies, IoT solutions will drive the creation of Information Marketplaces that allow the exchange of data between different economic entities within an information value chain. In Chapter 4, we turn to the architectural implementation for IoT solutions.

AN ARCHITECTURE PERSPECTIVE

4

CHAPTER OUTLINE

4.1 BUILDING AN ARCHITECTURE

This chapter provides an introduction to how we use the term "architecture" in this book and, secondly, how it relates to problems, applications of interest, and actual Internet of Things (IoT) solutions. The term *architecture* has many interpretations. Traditionally, *architecture* is both the process and the product of planning, designing, and the construction of buildings. In computer systems, an architecture is a set of rules and methods that describe the functionality, structure, and implementation of computer systems.

Within this book, architecture refers to the description of the main conceptual elements, as well as the actual elements of a target system, how they relate to each other, and principles for the design of the architecture. A conceptual element refers to an intended function, a piece of data, or a service. An actual element, meanwhile, refers to a technology building block or a protocol. The term "reference architecture" in this book relates to a generalized model that contains the richest set of elements and relations that are of relevance to the domain "Internet of Things". A detailed description of the theory and philosophy of architectures is beyond the scope of this book; interested readers will find a good example in Rozanski and Woods [18].

Whenever embarking on building something, a house, a car, or an IoT system, it is customary to first do a description of the desired construction, i.e., a blueprint. Nobody starts building a house at random from construction material like bricks and wood. In order to create a blueprint, systems engineers rely on a range of methodologies in order to give a structured approach to solving the problem at hand. This is the role of an architecture within IoT, i.e., giving the right guidelines for how to build a system, the typical elements to use, and how are they composed.

When looking at solving a particular problem or designing a target application, the reference architecture is to be used as an aid to design an *applied* architecture, i.e., an instance created out of a subset of the reference architecture. The applied architecture is then the blueprint used to develop the actual system solution (Figure 4.1). The approach taken here is inspired by the work in the EU FP7 project IoT-A [19].

An architecture can be described from several different views to capture specific properties that are of relevance to the model, and the views chosen in this book are the functional view, deployment view, process view, and information view. The topic of architecture is the subject of Chapter 8 of this book, where more details of the definitions and purpose of an architecture are given. State-of-the-art examples of IoT architectures are also briefly covered in Chapter 7 and Chapter 8.

Internet of Things. https://doi.org/10.1016/B978-0-12-814435-0.00015-8

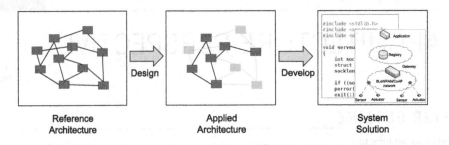

FIGURE 4.1

From a reference architecture to a system solution.

When creating a model for the reference architecture, one first needs to establish overall objectives for the architecture as well as design principles that come from understanding some of the desired major features of the resulting system solution. For instance, an overall objective might be to decouple application logic from communication mechanisms, and typical design principles might then be to design for protocol interoperability and to design for encapsulated service descriptions. These objectives and principles have to be derived from a deeper understanding of the actual problem domain, and this is typically done by identifying recurring problems or solution types, thus extracting common design patterns. The problem domain establishes the foundation for the subsequent solutions. It is common to partition the architecture work and solution work into two domains, each focusing on specific issues of relevance at the different levels of abstraction (Figure 4.2).

The top level of the pyramid is referred to here as the "problem domain" ("domain model" in software engineering). The problem domain is about understanding the applications of interest, for example, developed through scenario building and use case analysis in order to derive requirements. In addition, constraints are typically identified as well. These constraints can be technical, like limited power availability in wireless sensor nodes, or nontechnical, like constraints coming from legislation or business considerations. Real-world design constraints are covered in more detail in Chapter 9.

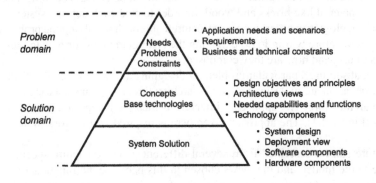

FIGURE 4.2

Problem and solution domain partitioning.

The lower level of the pyramid is referred to as the solution domain. This is where design objectives and principles are established, conceptual views are refined, and required functions are identified and where logical partitions of functionality and information are described. Often this is where a logical architecture is defined, or network architecture in the form of a network topology diagram is produced. It is also common to identify suitable technology components such as Operating Systems and protocols or protocol stacks at this level. The actual system solution is finally captured by a system design that typically results in actual software and hardware components, as well as information on how these are to be configured, deployed, and provisioned. The next section outlines the design objectives and principles for IoT, and the final section outlines the main capability domains of an IoT architecture.

4.2 REQUIREMENTS AND MAIN DESIGN PRINCIPLES

Having introduced some basic concepts in architecture design and how that facilitates the subsequent development of an actual system solution, we now turn to the main design principles that underpin architecture design for IoT. Within existing work for deriving requirements and creating architectures or reference models for IoT, two primary sources can be identified as both represent early and seminal work with a lasting impact. Both of them come from two larger European 7th Framework Program research projects, SENSEI [20] and IoT-A [19]. Another early work towards an architecture specification by primarily the telecommunication industry was initiated by ETSI in 2009 that resulted in the ETSI M2M (Appendix A) specifications and that later carried over to oneM2M[1]; see Section 7.10.1. ETSI also has moved on from using the term M2M to the term IoT[2]. Following the earlier work by these projects and organizations, several other IoT architecture activities have been initiated in the standardization and industry alliance communities; see Chapter 7 for an overview. The important point of all of these architectures – and any others that you may come across – is the drive towards horizontalization.

The approach taken in SENSEI [20] was to develop an architecture and a set of technology building blocks that enable a "Real World integration in a future Internet". The concept of a "Real-World Internet" was about creating digital representations of physical objects, what today is popularly referred to as Digital Twins[3]. Key features include the definition of a real-world services interface and the integration of numerous Wireless Sensor and Actuator Network (WSAN) deployments into a common services infrastructure on a global scale. The service infrastructure provides a set of services that are common to a vast range of applications and is separated from any underlying communication network for which the only assumption made was that it should be based on the Internet Protocol (IP) stack. The architecture relies on the separation of resources providing sensing and actuation from the actual devices, a set of contextual and real-world entity-centric services, and the users of the services. SENSEI [20] further relies on an open-ended constellation of providers and users and also provides a reference model for different business roles. Here we clearly see horizontal points in the architecture: the use of IP; a common service layer; and exposure of those services. A number of design principles and guidelines are identified, and so is a set of requirements. Finally, the architecture itself contains a set of key functional capabilities.

---

[1] http://www.onem2m.org/

[2] http://www.etsi.org/technologies-clusters/technologies/internet-of-things

[3] https://en.wikipedia.org/wiki/Digital_twin

The telecommunications industry, meanwhile, focused on defining a common service core for supporting various M2M applications that is agnostic to underlying networks in ETSI M2M as well as oneM2M. The approach taken has been to analyze a set of M2M use cases, derive a set of M2M service requirements, and then to specify an architecture as well as a set of supporting system interfaces. Similar to SENSEI [20], there was a clear approach towards a horizontal system with separation of devices, gateways, and communications networks and the creation of a common service core and a set of applications, all separated by defined reference points.

Finally, the approach taken in IoT-A differs from the two approaches above in the sense that instead of defining a single architecture, a reference architecture was created, captured in what the IoT-A refers to as the Architectural Reference Model (ARM). The vision of IoT-A is, via the ARM, to establish a means to achieve a high degree of interoperability between various IoT solutions at the different system levels of communication, service, and information. IoT-A provides a set of different architectural views and establishes a proposed terminology and a set of Unified Requirements [21,22]. Furthermore, IoT-A proposes a methodology for how to arrive at a concrete architecture based on use cases and requirements. The IoT-A architecture approach and propositions are covered in more detail in Chapter 8.

Comparing these different approaches, and as already mentioned, a common feature is the focus on a high-level horizontal system approach:

- There is a clear separation of the underlying communication networks and related technologies from capabilities that enable services.
- There is a clear desire to define uniform interfaces towards the devices that provide sensing and actuation, including abstraction of services the devices provide.
- There is also a clear desire to separate logic that is highly application-specific from logic that is common across a large set of applications.

Returning to our previous discussion on understanding trends, needed capabilities, and implications for IoT, we also clearly identified the need for a horizontal approach from a set of different perspectives – the need for horizontal integration across value chains at different levels; the need to integrate multiple infrastructures; and the need to reuse existing deployments, to name a few. Taking other key identified features into consideration as well, such as being able to support open-ended service development and providing security reliability, we can formulate what can be seen as an overall IoT architecture objective:

The overall design objective of an IoT architecture shall be to target a horizontal system of real-world centric services that are open, service-oriented, and secure and offer trust.

Further analyzing both the referenced existing work as well as our conclusions in Chapter 2 on both needed capabilities and direct IoT implications, we can also derive a set of supporting design principles that target different means to fulfill the overall architecture objective. These design principles have a set of interpretations and further expectations on needed technology solutions that we now describe.

Design for reuse of deployed IoT resources across application domains. Deployed IoT resources shall be able to be used in a vast range of different applications. This implies that devices shall be application-independent and that the basic and atomic services they expose in terms of sensing and actuation shall be done in a (to the greatest extent possible) uniform way. A system design will benefit from providing an abstracted view of these basic underlying services that also are decoupled from the devices that provide the services.

Design for a set of support services that provide open service-oriented capabilities and can be used for application development and execution. What we already have seen in the service layer-oriented M2M standardization in ETSI M2M and oneM2M is the definition of a set of common application-independent service capabilities. These support services shall in general cater to the typical environment of a stakeholder where IoT applications are to be built, such as an open environment, and shall in particular provide support for a few key service capabilities that are central from an IoT perspective. The open environment of IoT will, for instance, require mechanisms for authorized and authenticated access to and usage of services, resources, and IoT data.

The key support services that are required from an IoT perspective include the means to access IoT resources, how to publish and discover resources, tools for modeling contextual information and information related to the real-world entities that are of interest, and capabilities that provide different levels of abstracted and complex services. The latter can include data and event filtering and analytics, as well as dynamic service composition and automation involving control loops of sensing and actuation. Furthermore, well-defined service interfaces and Application Programming Interfaces (APIs) are required to facilitate application development, as are the appropriate Software Development Kits (SDKs). As we will also see later, service orientation increasingly implies cloud-native implementations and a microservices approach.

Design for different abstraction levels that hide underlying complexities and heterogeneities to facilitate interoperability. As we have already seen, typical IoT solutions can involve a large number of different devices and associated sensor modalities, and they can involve a large set of different actors providing services and information that need to be composed and accessed with different levels of aggregation. A system design will greatly benefit from providing the necessary abstractions both of underlying technologies and of data and service representations, as well as granularity of information and services. This is needed to ensure interoperability between different components within a particular IoT system, as well as between different IoT systems. This will ease the burden of both system integrators and application developers. Again, hiding device-side technologies and providing simple abstractions of the sensing and actuation services is one aspect. Another is the means to perform aggregation of information or knowledge. A third is to provide different levels of automation, and at different time scales ranging from real-time robot control to long-term forecasting or planning.

Design for stakeholder actors taking on different roles of providing and using services across different business domains and value chains. As discussed in Chapter 3 on the IoT Market aspects, as well as in Chapter 2, there are different levels of openness of the business context in which IoT solutions are deployed and running. IoT solutions can be run across a set of departments within an enterprise, or across a set of enterprises in a value system, or even be provided in a truly open environment. The business contexts can then be viewed as no market (entirely intraorganizational), as closed markets (finite and predetermined set of business actors in a specific value system or value chain), or as open markets (undefined and open-ended number of participants). In these different setups there are varying degrees of needed capabilities that address the multistakeholder perspective and are of both a technical and a business nature. The first thing that needs to be provided is a set of mechanisms that ensure security and trust. Trust and identity management that refer to the different stakeholders is a fundamental requirement. Authentication and authorization of access to use services as well as to be able to provide services is then a second requirement. The third requirement is the capability to be able to do auditing and to provide accountability so that stakeholders can enforce liability if the need occurs. The next fundamental requirement is to ensure interoperability. This is needed at different levels across the interaction points between the stakeholders. Primary examples are to ensure data and information interoperability on the semantic level and means to connect business processes across organizational and administrative boundaries. The third fundamental requirement is related to the market perspective, whether the markets are closed or fully open. Mechanisms that provide compensation for used services or data between service users and service providers are needed. As an IoT market can involve everything from trading individual sensor data to aggregated insights and knowledge, compensation and billing mechanisms are needed that can operate on the microlevel as well as on more traditional macrolevels. An open market environment also calls for means to publish or advertise services, as well as a means for finding services. These different main requirements also provide opportunities for new market roles, such as aggregator roles, broker roles, and clearing houses, all well known in other existing markets.

Design for ensuring trust, security, and privacy. Trust within IoT often implies reliability, which can be both ensuring the availability of services as well as how dependable the services are and that data is only used for the purposes the end-user has agreed to. One important aspect of dependability is the accuracy of data or information, as you can have multiple sources of IoT data. Concepts like Quality of Information become important, especially considering that a piece of information can very well be accurate enough for one application, but not for another. As has been already mentioned, security and privacy are potential barriers for IoT adoption and represent key areas to address when building solutions. Privacy needs to be ensured by, for example, anonymization of data, seeing that profiling of individuals is not easily done or even made undoable. Still, it is foreseen that authorities and agencies will require support to get access to data and information for the purpose of national security or public safety.

Design for scalability, performance, and effectiveness. IoT deployments will happen on a global scale and are foreseen to involve billions of deployed nodes. Sensor data will be provided with a wide range of different characteristics. Data may be very infrequent (e.g., alarms or detected abnormal events) or may be coming as real-time data streams, all dependent on the type of data needed or based on application needs. Scalability aspects of importance include the large number of devices and amounts of data produced that need to be processed or stored. Performance includes consideration of mission-critical applications such as Supervisory Control and Data Acquisition (SCADA) systems with extreme requirements on latency, for example.

Design for evolvability, heterogeneity, and simplicity of integration. Technology is constantly changing, and given the nature of IoT deployments where especially devices and sensor nodes are expected to be operational and in the field for many years, sometimes with lifecycles of over 15 years (e.g., smart meters), IoT solutions must be able to withstand and cater to introduction and use of new technologies as well as handle legacy deployments. Handling heterogeneity is also important since in particular device-oriented technologies used across industries are very different. Means to integrate legacy devices using many different protocols becomes a necessity, and gateways of different types and with different capabilities are essential to expose capabilities of legacy devices in a uniform manner.

Design for simplicity of management. Again going back to one of the potential barriers for IoT adaptation, simplicity of management is an important capability that needs to be properly taken care of when designing IoT solutions. Considering the prospect of IoT device deployments in the range of

10–100 billion units, autoconfiguration, autoprovisioning, and automated management are key capabilities for viable IoT solutions in order to lower Operating Expenditures (OPEX).

Design for different service delivery models. We already know about the clear trends to move from product offerings to a more combined product and service delivery model in a number of industries, for instance, the automotive industry delivering applications to connected vehicles, and the software industry delivering Software as a Service (SaaS). IoT with the wide span of possible applications clearly benefits from elasticity in deployment of solutions, all to meet the long-tail aspect. Cloud and virtualization technologies play a key enabler role in delivering future IoT services, and cloud-native software is an increasing fundamental requirement.

Design for lifecycle support. The lifecycle phases are: planning, development, deployment, operation, and decommissioning. Management aspects include deployment efficiency, design time tools, and run-time management. Again, automation is key from the perspectives of both scale and complexity, as is a lifecycle support of security including identity management. Especially ensuring that devices are reliable throughout the lifecycle from manufacturing to decommissioning is important considering that they in general will be unattended and out of reach by humans.

From these design principles, and taking into consideration detailed use cases and target applications, it is possible to identify both functional and nonfunctional requirements that form the basis for a more detailed architecture design. Different sets of requirements have been identified in the already referenced work, and the reader interested in the more detailed requirements is referred to, e.g., [21–23].

4.3 AN IOT ARCHITECTURE OUTLINE

We have now arrived at a better understanding of the design objectives and principles that capture the main desired characteristics of an IoT solution, and we have also identified some high-level capabilities that generally are needed. As indicated above, there is an increasing common understanding of how a typical IoT solution looks. However, there is no generally accepted IoT system architecture or universally agreed set of standards that define the solution components. What is state-of-the-art today is mainly coming from a group of standardization bodies and industry alliances that have specified either protocols as systems components, or system and functional architectures for various parts of a more complete end-to-end IoT architecture. Chapter 7 gives an overview of the main architecture activities, and prominent examples include work from IIC[4] and IETF[5,6].

When it comes to more mid- to long-term IoT technology research on both solution components and various architectural aspects, there are a number of research activities and projects on the European level to be found in the Internet of Things European Research Cluster (IERC)[7].

A fundamental consideration to keep in mind when evaluating both IoT system architecture patterns and the choice of appropriate technologies is diversity. Diversity comes both from the range of possible

---

[4] https://www.iiconsortium.org/

[5] http://ietf.org

[6] http://ietf.org/topics/iot/

[7] http://www.internet-of-things-research.eu

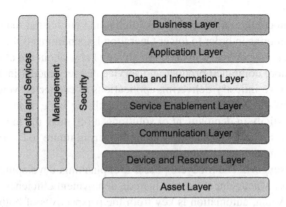

FIGURE 4.3

Functional layers and capabilities of an IoT solution.

and sough-after applications and from the diversity in deployment scenarios and situations. These diversities together produce a large aggregate set of different requirements and constraints. And these in turn also imply that there is not one single IoT system architecture, nor a single narrow set of system components that can realize all the imaginable IoT solutions. Again, it is the purpose of a reference architecture to provide the methodology and commonly recurring design patterns to guide the definition of a particular target IoT system solution in the end; see Figure 4.1. Attempting to produce a single reference architecture consequently results in a number of optional and conditional requirements, all depending on the particular problem at hand or application in focus.

Nevertheless, the identified key features that are needed when building an IoT solution can now be put together into a larger context by proposing a single view of the main functional capabilities (see Figure 4.3). This is not a strict and formal functional architecture, but provides a conceptual overview. It also follows the approach of looking at the system capabilities from a layered point of view, as well as key functional domains that go across the layers.

Other common approaches when describing an architecture are the software approach and network approach that are more focused on how functions are realized and implemented, including how they are distributed. These are formalized into the different architecture views mentioned above and covered in more detail in Chapter 8. Here the different proposed functional layers and cross-cutting capabilities provided are discussed.

At the lowest level is the **Asset Layer**. This layer is, strictly speaking, not providing any functionality within a target solution, but represents the *raison d'être* for any IoT application. The assets of interest are the real-world objects and entities that are subject to being monitored and controlled, as well as having digital representations and identities. The typical examples include vehicles and machinery and fixed infrastructures such as buildings and utility systems, homes, and people themselves – thus being inanimate as well as animate objects. Assets can also be of a more virtual character, being subjective representations of parts of the real world that are of interest to a person or an organization. A typical example of the latter is a set of particular routes used by trucks in a logistics use case. Information of interest may then be traffic intensity, roadwork, or road conditions based on the actual weather situation.

The **Device and Resource Layer** provides the main functional capabilities of sensing, actuation, and embedded identities and is hence the bridge between the digital realm with the physical world that enables assets and physical infrastructures to be monitored, controlled, and identified. The sensors and actuators can be in various devices that may be smartphones or WSANs, smart meters, or other sensor/actuator nodes. Sensors and actuators are the resources that can be abstracted and integrated into higher-level data processing capabilities, and hence they represent the sources and sinks of monitor and control data. This layer is also where gateways of different types are placed that can provide aggregation or other capabilities that are closely related to these basic resources. Identification of assets can be provided by different types of tags; for instance, Radio Frequency Identification (RFID) as in ISO/IEC 18000[8] family of standards or optical codes like bar codes or Quick Response (QR)[9] codes. The topic of devices and gateways is further dealt with in Section 5.1.

The purpose of the **Communication Layer** is to provide the means for connectivity between the devices and resources on one end and the different computing infrastructures that host and execute service support logic and application logic on the other end. Different types of networks realize the connectivity, and it is customary to differentiate between the notion of a Local Area Network (LAN) and a Wide Area Network (WAN). WANs can be realized by different wired or wireless technologies, for instance, fiber, Digital Subscriber Line (DSL) for the former and cellular mobile networks, satellite, or microwave links for the latter. WANs can also be provided by different actors, where some networks can be regarded as public (i.e., offered as commercial services for the general public) or as private (i.e., dedicated networks that provide services in a more closed business or entirely company internal environment). Particularly in the mobile network industry, there are different models for how the communications services are provided that include wholesale of access and dedicated virtual network operators that focus on managed M2M connectivity offerings without owning licensed mobile spectrum or actual network resources.

When it comes to LANs, there are many examples of different types, and there is also no stringent definition of what might be considered a LAN or a WAN. Prime examples of LANs include Wireless Personal Area Networks (WPANs; also known as Body Area Networks, BANs) for fitness or healthcare applications, Home or Building Area Networks (HANs and BANs, respectively) used in automation and control applications, and Neighborhood Area Networks (NANs) or Field Area Networks (FANs) which are used in the Distribution Grid of a Smart Electricity Grid. Communication can also be used in more ad hoc scenarios. Vehicle-to-Vehicle (V2V) is one example that can target safety applications like collision avoidance or car platooning. As opposed to the situation for WANs, the interface technologies used within LANs are characterized by being very industry segment-specific, being supported by a plethora of different standards, or even being proprietary or at best de facto standards. LANs use both wired and wireless technologies. General examples of wired LANs include Ethernet, Power Line Communication (PLC), and the family of different fieldbus[10] technologies used in industrial real-time control applications. Prominent examples of wireless LAN networking technologies include the IEEE 802.11[11] and IEEE 802.15.4[12] families, as well as Bluetooth, which has a protocol addition called

---

[8]https://www.iso.org/ics/35.040.50/x
[9]https://en.wikipedia.org/wiki/QR_code
[10]https://en.wikipedia.org/wiki/Fieldbus
[11]http://www.ieee802.org/11
[12]http://www.ieee802.org/15/pub/TG4.html

Bluetooth Low Energy (BTLE or BLE) that targets typical IoT applications. IEEE 802.15.4 is the basis for protocol stacks that target different IoT applications in different sectors, for instance, the ZigBee specifications[13], the proprietary Z-Wave protocol stack[14] for home automation, Wi-SUN Alliance[15], and ISA100.11a[16]. Many of the legacy industry-specific LAN protocol stacks do not use IP as the networking protocol, but there is a growing number of examples where the legacy protocol stacks are migrated towards IP, for instance, ZigBee IP and IPv6 over Bluetooth under the IETF 6lo work group (see Section 7.3). To provide an end-to-end communication service that bridges a LAN and a WAN, gateways are used. From a communication layer perspective, gateways are primarily used to do interworking or protocol translation at different levels of the protocol stack. This can involve the physical and link layers, but it can also involve interworking on the communications or messaging level, for example, to do interworking between a legacy protocol like ZigBee to exchanging service operations using HTTP as the means for communication. Section 5.2 deals with the various LAN and WAN aspects and technologies in detail.

As described earlier, IoT applications benefit from simplification by relying on support services that perform common and routine tasks. These enabling services are provided by the **Service Enablement Layer** and are typically executing in Data Centers or server farms inside organizations or in a cloud environment. These enabling services can provide uniform handling of the underlying devices and networks, thus hiding complexities in the corresponding layers. Examples include remote Device Management that can do remote software upgrades, do remote diagnostics or recovery, and dynamically reconfigure application processing such as setting event filters, machine learning algorithms or control rules. Communication-related functions include selection of communication channels if different networks can be used in parallel, for example, for reliability purposes, and integrate different means for transferring data and data ingestion, such as RESTful protocols like HTTP or CoAP, and different publish–subscribe and message queue mechanisms like MQTT and AMQP. Location-Based Service (LBS) capabilities and various Geographic Information System (GIS) services are also important for many IoT applications. Of more specific relevance for IoT are services that relate to sensor originating data and actuation services and services that relate to different tags like RFID. A directory that holds information of available resources and associated service capabilities that can function as a rendezvous mechanism is one example. In such a directory, nodes in WSANs can publish themselves with service descriptions and how to be reached. Applications then perform look-ups to find which device can provide the sensor reading of interest. Another directory service example is the GS1 Object Naming Service (ONS)[17] (see Section 7.9) that can resolve an RFID code to a URL where information about the tagged object can be found. Examples of different enabling services are provided in Chapters 7 and 8.

Where the Resource, Communication, and Service Enablement Layers have concrete realizations in terms of devices and tags, networks and network nodes, and computer servers, the **Data and Information Layer** provides a more abstract set of functions as its main purposes are to extract insights, capture knowledge, and provide advanced automation in different IoT applications. Key concepts here

---

[13] http://www.zigbee.org

[14] http://www.z-wave.com

[15] https://www.wi-sun.org/

[16] https://www.isa.org/isa100/

[17] https://www.gs1.org/epcis/epcis-ons/2-0-1

include data and information models and knowledge representation in general, and the focus is on the representation and organization of information. Different tools are required to process data, extract insights, and automate processes, and they are very dependent on the use case at hand. We refer to Machine Intelligence as the collective term of the necessary diverse tools of this layer. The different capabilities include various types of analytics, Machine Learning algorithms, control systems logic, and AI technologies. This is the subject of Section 5.3.

The Application Layer in turn provides the specific IoT applications. There is an open-ended array of different applications from various sectors across the industry, enterprise, consumer, and society domains as laid out in Chapter 1. Part 3 is devoted to providing examples of different IoT applications.

The final layer in our architecture outline is the Business Layer, which focuses on supporting the core business or operations of any enterprise, organization, or individual that is interested in IoT applications. This is where any integration of the IoT applications into business processes and enterprise systems takes place. The enterprise systems can, for example, be Customer Relationship Management (CRM), Enterprise Resource Planning (ERP), or other Business Support Systems (BSSs). The business layer also provides exposure via APIs for third parties to get access to data and information and can also contain support for direct access to applications by human users; for instance, city portal services for citizens in a Smart City context, or providing necessary data visualizations to the human workforce in a particular enterprise. The business layer relies on IoT applications as one set of enablers out of many (e.g., field force automation) and takes care of necessary orchestration and composition to support a business process workflow. A detailed discussion on business integration is provided in Section 5.6.

In addition to the functional layers, three functional domains span across the different layers, namely Management, Security, and IoT Data and Services. The former two are well-known functions of a system solution, whereas the latter one is more specific to IoT.

Management, as the name implies, deals with management of various parts of the system solution related to its operation, maintenance, administration, and provisioning throughout the lifecycle of the system. This includes management of devices, communications networks, and the general Information Technology (IT) infrastructure, as well as configuration and provisioning of data, performance of services delivered, etc. IoT management aspects are in part covered in Chapter 8.

Security is about protection of the system, its information, and its services, from external threats or any other harm. Security measures are usually required across all layers, for instance, providing communication security and information security. Trust and identity management and authentication and authorization are key capabilities. From an IoT perspective, management of privacy via, for example, anonymization, is in many instances a specific requirement. Security is the topic of Chapter 6.

The final cross-functional domain of our outlined architecture is denoted **Data and Services**. Processing of IoT data and services (e.g., actuation) can be done in a very distributed fashion and at different levels of granularity, abstraction, and complexity. A fundamental and first step of processing is the collection and curation of data, i.e., preparing the data so that it is clean to be processed by application logic. Different basic data processing includes event filtering and simpler aggregation, such as data averaging, that can take place in individual sensor nodes in WSANs. Contextual metadata such as spatial and temporal information can be added to sensor readings, and further aggregation can take place higher up in the network topology. More advanced processing is, for instance, data analytics that can be done in near real-time or in batch. Different technologies are used to support the different levels of insight extraction, processing, reasoning, decision making, and automation. Data and service processing also include different types of sensing–actuation control loops that can be simple in nature based on rules, or be complex to automate large physical infrastructures. In general, data and service processing can be distributed all the way from the edge in a sensor node to a centralized data center, all based on the needs of the application. This set of functions thus represents the vertical flow of data into knowledge, the abstraction of data and services in different levels, and the process steps of extracting knowledge and providing automation at different time horizons. The aspects of Machine Intelligence, distributed computing, and data management are covered in Sections 5.3, 5.4, 5.5, respectively.

What is not reflected in the architecture outline is the IoT business solution lifecycle aspect. This includes the steps of identifying and analyzing the business needs, the design of a solution, and the subsequent steps of solution planning, design, deployment, and maintenance. These different steps are outside the scope of this book, but some aspects related to the deployment phase are covered in Chapter 8.

To summarize, we have now provided an overview of the main different functional domains that are required to build IoT solutions. For the reason of simplicity and to convey a general understanding, this is conceptual and not formal. The more detailed approach following a formal process of describing a reference architecture is provided in Chapter 8.

4.4 STANDARDS CONSIDERATIONS

The main purpose of standardization can be summarized as being a means to achieve technical interoperability and replicability. Standardization of different types result in technical standards that are agreements among the involved parties. Interoperability[18] implies that different systems, or subsystems, that are developed and used by different parties can work with each other. A benefit is that standardization facilitates market competition and is a move away from customized solutions from single suppliers, i.e., it reduces vendor lock-in. Interoperability can be achieved at different levels as well, both technical (e.g., protocol) syntactic (data structures) and semantic (meaning), but also beyond those.

Replicability means that solutions can be replicated and used in multiple contexts and that technologies such as a piece of software or hardware can be reused in different system solutions. Replicability of solutions has the benefit that best practices on solving a particular type of problem can be reused, and examples include a particular architecture pattern or a solution blueprint. Replicability of software or hardware implies that one can rely on readily available solution components that are proven instead of doing one's own costly development and testing. Standardization also has the benefits of facilitating commoditization and economy of scale.

The technical standards resulting from any standardization activity can be of different types and can be achieved through different processes, mainly through standards organizations, open source projects, and industry alliances. Standards Developing Organizations (SDOs) define formal standards that are mandated for use. The SDOs publish the standards mainly as system or interface specifications. SDOs can be global, regional or national. Prominent global SDOs include the International Organization for Standardization (ISO), the International Electrotechnical Commission (IEC), and the International Telecommunication Union (ITU). In addition, there is a group of independent international standards

---

[18]https://en.wikipedia.org/wiki/Interoperability

organizations that agree upon and publish standards that achieve widespread adoption. Examples include the Internet Engineering Task Force (IETF) and the World Wide Web Consortium (W3C). SDOs do not only develop their own standards, but many times adopt the successful and widespread standards produced by the independent standards organizations, some of them even being de facto standards.

The open source model is a decentralized development model that builds on open collaboration. The products of the open source model is typically source code, blueprints and documentation that is openly available. Open source projects are primarily targeting software, but hardware projects exist for IoT, and the Arduino[19] electronic prototyping platform is a notable example. A popular infrastructure for open source projects is provided by GitHub[20]. Open source projects for IoT are many[21] and a few prominent examples include projects under the Eclipse foundation[22] and the Linux foundation[23]. Typical examples of what open source projects target include operating systems for IoT devices, e.g., RIOT[24] and Zephyr[25], and different IoT protocol stacks, like Leshan[26] for LwM2M. When relying on open source, the licensing terms for using and contributing to open source vary depending on the project and are an important consideration when engaging either as a user or as a contributor. Open source activities and projects revolve rapidly and it is beyond the scope of this book to cover them properly.

Industry alliances and consortia have become of growing importance in IoT. This comes as a response to the notion that traditional standards organizations have a tendency to make slower progress than the pace at which the technology itself is evolving. These alliances can focus on a specific IoT technology, but can also form to agree on how to use existing standards in a specific applied industry sector or context, like providing different application-specific profiles, for instance, OMA SpecWorks adopting standards from the IETF to define a framework for device and service enablement in LwM2M (Section 7.4), or Wi-SUN Alliance defining conformance and interoperability test specifications for wireless IoT mesh networks based on the IEEE 802.15.4g standard for utility networks. Other alliances can work on identifying, agreeing on, and defining best practices on how to build IoT solutions. An example consortium that focuses on delivering best practices agreed upon by members from across ecosystems is the Industrial Internet Consortium[27]; see Section 8.9.1.

These different ways to do standardization are not mutually exclusive or separate. On the contrary, they are to be seen as different tools that combined achieve the goal of agreed technical standards. For example, the IETF has defined a set of standards on how to use the RESTful approach in resource-constrained environments typical to IoT (see Section 7.3.2) that has then been picked up by the consortium Open Connectivity Foundation (OCF)[28] that develops specifications and certifications based on the IETF standards. OCF also sponsors an open source reference implementation of the OCF

---

[19]https://www.arduino.cc
[20]https://github.com
[21]https://www.postscapes.com/internet-of-things-award/open-source/
[22]https://iot.eclipse.org
[23]https://www.linuxfoundation.org/projects/
[24]https://www.riot-os.org
[25]https://www.zephyrproject.org
[26]http://www.eclipse.org/leshan/
[27]https://www.iiconsortium.org
[28]https://openconnectivity.org

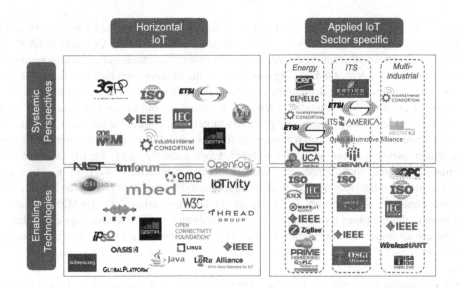

FIGURE 4.4

A schematic overview of the IoT standardization landscape.

specifications called IoTivity[29], which is a Linux Foundation collaborative project. This combined approach of orchestrated efforts across developing standards, engaging in open source and participating in industry alliances, is a common practice in the IoT ecosystem. In this book we hence use the word "standardization" as a collective term to include these three main tools to achieve the desired levels of interoperability and replicability.

The landscape of different concrete IoT standardization activities is very diverse and wide. The purpose here is to provide an overview of the topology of this landscape and to highlight some prominent standardization organizations and activities rather than trying to list all relevant standards. This will serve as an illustration that the standardization around IoT is rather rich and multidimensional; see Figure 4.4 for a schematic overview of the structure and highlighted organizations. To find more information, one overview is provided by ETSI ([24]), and another online compilation is provided by Postscapes.com[30].

The first top-level dimension is that standards are developed either as horizontal IoT standards or as applied IoT standards. The applied IoT standards target a particular industry vertical or sector, such as buildings, utilities, energy, manufacturing, transportation or the health sectors. The horizontal technical standards, on the other hand, can be generally applied across a number of different industries. For historical reasons, there are many standardization activities across the different industry verticals. As these applied industries have a long history of providing their own industry-specific standards, there is an inheritance and legacy of practices and technologies that continue to develop, but as we see more

---

[29]https://iotivity.org

[30]https://www.postscapes.com/internet-of-things-protocols/

and more converging interests across industries, and due to the need for replicability mentioned above, standards will increasingly cater to reducing technology fragmentation by adoption of horizontal IoT standards.

The second top-level dimension is that some standardization activities define entire systems or parts of systems, and other standards organizations target the development of specific pieces of technologies, for instance, specific protocols or software. System standards can address a mobile communication network as defined within the 3rd Generation Partnership Project (3GPP) or IoT solution best practices for multiple industry verticals, like IIC. Organizations like the IETF, on the other hand, focus on developing the protocol suite of the Internet without any effort to specify a system standard beyond what is already in existence in a few key IETF Request For Comments (RFCs) such as RFC1958[31] that establishes the architectural principles of the Internet. The natural observation is that systemic standards rely on the enabling technology components as the foundation, but as there generally are many competing technology components (e.g., protocol stacks), the adoption into a system standard is not a straightforward route.

Another important consideration is about the lifecycle process of standards. Many times, standards are emerging as a result of collaborative research involving both academia and industry. In other situations, technology selection for standardization can happen as part of regulatory or legislative processes. Within the European Union, the European Commission has issued so called Mandates that can have a direct impact on the choice of technology, which hence precedes any subsequent standardization activity. An example of this is the European Mandate M490 [25] on the Smart Grid that was issued by the European Commission to the European Standardization Organizations to come together to develop and update a set of consistent standards within a common European framework that integrates various ICT and electrical architectures and processes to achieve interoperability for the European Smart Grid. As a conclusion, technology selection does not only happen in the process of standardization.

Some specific references to different standardization examples were provided above, but the relevant standards for IoT are covered in the respective chapters and sections of Parts 2 and 3 of the book.

---

[31] https://doi.org/10.17487/rfc1958

IOT TECHNOLOGIES AND ARCHITECTURES *2*

Part 1 provided an overview of the vision and market conditions for the Internet of Things (IoT). In the following chapters, we turn our attention to the technology building blocks that form the basis of IoT solutions. During the last few years the technical building blocks have grown exponentially in breadth and depth. We look at the security, privacy, and trust issues of IoT in the next few chapters. We then delve into an overview of the architectural foundations of IoT, more specifically the IoT Architecture (IoT-A) and the Industrial Internet Consortium (IIC) standard in detail. Part 3 then outlines the application of these technologies within real-world use case contexts, illustrating how the IoT vision combines with the reference architectures to create real-world value.

PART 2

IoT TECHNOLOGIES AND ARCHITECTURES

TECHNOLOGY FUNDAMENTALS

CHAPTER 5

Internet of Things. https://doi.org/10.1016/B978-0-12-814435-0.00017-1
Copyright © 2019 Elsevier Ltd. All rights reserved.

5.1 DEVICES AND GATEWAYS
5.1.1 INTRODUCTION

As discussed in Chapter 1, embedded processing is evolving, not only towards higher capabilities and processing speeds, but also towards allowing a multitude of applications to run autonomously. There is a growing market for small-scale embedded processors; 8-, 16-, and 32-bit microcontrollers with on-chip RAM and flash memory, I/O capabilities, and networking interfaces such as IEEE 802.15.4, Bluetooth, and Wi-Fi, which are increasingly integrated as tiny System-on-a-Chip (SoC) solutions. SoCs enable the design of devices with small physical footprints of a few mm^2 with very low power consumption, e.g., in the milli- to microwatt range, but which are capable of hosting complete communication protocol stacks, including small web servers.

There is an increasingly broad spectrum of IoT devices, and to avoid confusion, it is worth explaining what is referred to as a *device* here. Typically, a *device* is an embedded computer, which can be characterized as having several properties, including some or all of the following:

- Computational capability: typically 8-, 16-, or 32-bit working memory and storage.
- Power supply: wired, battery, energy harvester, or hybrid.
- Sensors and/or actuators: used to sample an environmental variable and/or exert control (e.g., flicking a switch, tuning a motor).
- Communications interfaces: wireless or wired technologies are used to connect devices to one another, the Internet, remote servers, etc.
- Operating System (OS): main-loop, event-based, real-time, or full-featured OS.
- User Interface: display, buttons, or other functions for user interaction.
- Device Management (DM): provisioning, firmware, bootstrapping, and monitoring.
- Execution Environment (EE): Application Lifecycle Management and Application Programming Interface (API).

For several reasons, one or more of these functions may often be hosted on gateway-types of devices instead. This can be to reduce energy consumption, for example, by letting the gateway handle computationally expensive functions, such as Wide Area Network (WAN) connectivity and application-level processing that require more powerful processors. This may also lead to reduced costs as more expensive components would be needed otherwise. Another reason is to reduce complexity by letting a central node (the gateway, or an application server) handle functionality such as DM and advanced applications (e.g., implementing Machine Learning (ML) algorithms), while letting the devices focus on sensing and actuating. Moreover, reallocating functions on a gateway type of device may also lead to reduced communication overheads and thus improved energy performance for *basic* devices.

5.1.1.1 Device types

There are no clear criteria today for categorizing IoT devices. The closest reference on device classification in several dimensions is the Internet Engineering Task Force (IETF) Request For Comments (RFC) or simply RFC7228[1] titled "Terminology for Constrained-node Networks". At the time of writ-

---

[1] https://doi.org/10.17487/rfc7228

ing of this book, RFC7228 is updated as an individual draft (draft-bormann-lwig-7228bis[2]) under the IETF Working Group (WG) `lwig` (see Section 7.3). The classification from the first edition of this book was based on datasheets from ST Microelectronics (a chip manufacturer); however, there are similarities with the definitions in RFC7228 and its recent update because device capabilities are often dictated by manufacturers' portfolios which are mainly estimates of the current market needs or customer requirements.

The purpose of RFC7228 is to provide a common terminology and a common reference for the different IoT-related protocols in development within IETF. While the current Internet technologies assume implicitly a set of device, host, and machine capabilities, the IoT-related protocols are mainly supposed to operate nontypical Internet devices, hosts, and machines. RFC7228 defines several terms such as constrained nodes, constrained networks, challenged networks, and constrained-node networks because of the different combinations of computing, communication, and energy supply capabilities for nodes and networks. The node and network terms are defined loosely and in relation to the current Internet device, hosts, and machines. A constrained node, for example, is defined as a "node where some of the characteristics that are otherwise pretty much taken for granted for Internet nodes at the time of writing are not attainable, often due to cost constraints and/or physical constraints on characteristics such as size, weight, and available power and energy". A constrained node typically has constraints on device capabilities such as memory (read-only or ROM/Flash and random-access RAM), processing power, energy resources, and User Interface (e.g., an IoT node may not have a screen or buttons to get input from a user). With respect to the communication capabilities, constrained networks can exhibit low bit rate, high packet loss, high communication delay, availability of an Internet Protocol (IP) stack, etc. RFC7228 classifies devices in three classes (C0, C1, C2) with respect to memory (ROM/Flash and RAM) sizes and in four classes (E0, E1, E2, E9, allowing future updates of Ex classes) depending on the availability of energy resources on the nodes; for example E9 is the mains-powered devices with theoretically unlimited energy supply.

The update of RFC7228 (RFC7228bis) goes into more detail with respect to these device classes and also extends the definitions towards network capabilities. RFC7228bis first defines two rough groups of classes that correspond to the two types from the first edition of the book (Table 5.1).

- Basic Devices or Microcontroller-class devices, Group-M (RFC7228bis): Devices that only perform simple tasks, like creating sensor readings and/or executing actuation commands, and in some cases have limited support for user interaction. These devices are likely to communicate locally, require a gateway device for Internet connectivity, have memory on-chip, have Analog to Digital (ADC) and Digital to Analog (DAC) converters and have the capabilities to go into different sleep modes to achieve low-power operations.
- Advanced Devices or General purpose-class devices, Group-J: These may host application-level logic and complex communications protocol stacks. They may also feature DM capabilities and provide an EE for hosting multiple applications. Gateway devices that provide physical interfaces to multiple media almost always fall into this category. These classes of devices typically have large amounts of ROM/Flash and RAM off-chip, include a Memory Management Unit (MMU), interface with other chips that perform specialized operations (e.g., video playout) and not directly

---

[2]https://www.ietf.org/archive/id/draft-bormann-lwig-7228bis-02.txt

Table 5.1 Example of basic and advanced devices

	CPU	Memory	Power	Comm	OS, EE
Basic	8-bit PIC, 8-bit 8051, 32-bit Cortex-M	Kilobytes	Battery	802.15.4, 802.11, Z-Wave	Main-loop, Contiki, RTOS
Advanced	32-bit ARM9, Intel Atom, more powerful Intel, AMD etc processors	Megabytes, Gigabytes	Fixed	802.11, LTE, 3G, GPRS, wired	Linux, Java, Python

the physical world as Group-M, and include fewer Low-Power Modes as they are expected to be connected to mains power.

It is interesting to note that the RFC7228 device classes (C0–C2) fall under the Group-M and the device classes now include the Group-J (general purpose) devices, representative examples of which are embedded routers, the RaspberryPi[3], smartphones, laptops, and servers.

5.1.1.2 Deployment scenarios for devices

The deployment scenario differs for basic and advanced application scenarios. Example deployment scenarios for **basic** devices include:

- Home alarms: Such devices typically include motion detectors, magnetic sensors, and smoke detectors. A central unit takes care of the application logic that calls security and sounds an alarm if a sensor is activated when the alarm is armed. The central unit also handles the WAN connection towards the alarm central. These systems are currently often based on proprietary radio protocols.
- Smart meters: The meters are installed in households and measure consumption of, for example, electricity and gas. A concentrator gateway collects data from the meters, performs aggregation, and periodically transmits the aggregated data to an application server over a cellular connection. Using a capillary network technology (e.g., IEEE 802.15.4), it is possible to extend the range of the concentrator gateway by allowing meters in the periphery to use other meters as extenders and interface with handheld devices on the Home Area Network (HAN) side.
- Building Automation Systems (BASs): Such devices include thermostats, fans, motion detectors, air quality sensors, and boilers, which are controlled by local facilities, but can also be remotely operated.
- Standalone Smart Thermostats: These use Wi-Fi and typical routers to communicate with web services.

Examples for **advanced** devices, meanwhile, include:

---

[3]https://www.raspberrypi.org

- Onboard units in cars that perform remote monitoring and configuration over a cellular connection.
- Robots and autonomous vehicles such as Unmanned Aerial Vehicles that can work both autonomously or by remote control using a cellular connection.
- Video cameras for remote monitoring connecting using 4G/LTE, etc.
- Oil well monitoring and collection of data points from remote devices.
- Connected printers that can be upgraded and serviced remotely.

The devices and gateways of today often use legacy technologies such as KNX, Z-Wave, and ZigBee on the capillary side, but the vision for the future is that every device can have an IP address and be (in)directly connected to the Internet, by implementing the IETF protocol stacks (see Section 7.3) leveraging the Timeslotted Channel Hopping (TSCH) mode of IEEE 802.15.4e, 6LoWPAN, RPL, and CoAP for example.

Some of the examples listed above, BAS for example, require some form of autonomous mode, where the system operates even in the absence of a WAN connection. Also, in these cases it is possible to use IoT technologies to form an "Intranet of Things". The term Intranet by itself means an isolated communication network typically in a corporate network which is not accessible from the Internet. Intranet hosts typically use Internet technologies (i.e., a TCP/IP-based stack) and may access Internet hosts but not vice versa. An Intranet of Things refers to an isolated snapshot of an IoT network without any IoT nodes being accessible by Internet hosts.

5.1.2 BASIC DEVICES

Basic devices are often intended for a single purpose, such as measuring air pressure or closing a valve. In some cases several functions (via corresponding sensors and actuators) are deployed on the same device, such as monitoring humidity, temperature, and light level. The requirements on hardware for this kind of devices are low, both in terms of processing power and memory. The main focus is on keeping the Bill of Materials (BOM) as low as possible by using inexpensive microcontrollers with built-in memory and storage, often on an SoC-integrated circuit with all main components on one single chip (Figure 5.1). While earlier SoCs included just one microprocessor for general purpose computing and wireless connectivity (e.g. IEEE 802.15.4), modern models (e.g., the STM32WB family) are typically SoCs with at least two internal microprocessors, one for general purpose applications and one acting as a radio processor. In this way the SoC could contain multiple wireless communication stacks such as Bluetooth and IEEE 802.15.4 and they are implemented mainly in software. This SoC architecture allows SoC manufacturers to diversify in terms of wireless connectivity since currently there is no single communication standard that dominates the market. Another common and equally important goal for basic devices is their energy consumption. Typical solutions include the use of a battery or energy harvester as a power source, extending device lifespans towards decades in the most aggressive cases, thus reducing the need for maintenance and associated operating costs [26–28].

Microcontrollers typically provide a number of ports (physically accessible pins) that allow interaction with sensors and actuators. These include General Purpose I/O (GPIO), support for digital sensors that communicate using standard interfaces such as Serial Peripheral Interface (SPI) and Inter-Integrated Circuit (I^2C) and ADCs for supporting analog input. For certain actuators, such as motors, Pulse-Width Modulation (PWM) can be easily implemented. As low-power operation is paramount to battery-powered devices, the microcontroller hosts functions that facilitate sleeping and operations in different sleep modes and interrupts that can wake up the device on

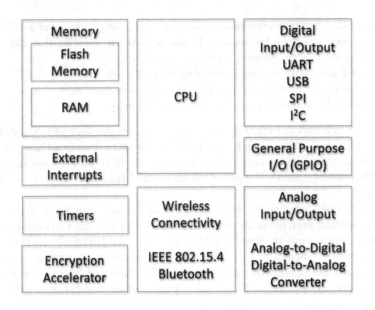

FIGURE 5.1

Typical microcontroller/microprocessor internal architecture.

external and internal events, e.g., when there is activity on a GPIO port or the radio (hardware interrupt), as well as timer-based wake-ups. Some devices even go as far as harvesting energy from their environment, e.g., solar and thermal energy. Where the application allows, specifically in the case of periodic monitoring as opposed to continuous monitoring of phenomena, devices tend to be highly *duty-cycled*, remaining in ultra-Low-Power Mode (LPM) for as long as possible.

In order to interact with peripherals such as external on-board storage or display, it is common to use serial interfaces such as SPI, I²C, or Universal Asynchronous Receiver Transmitter (UART). These interfaces can also be used to communicate with another microcontroller on the device (e.g., a separate controller managing the energy subsystem, performing power point tracking). This is common when there is a need for offloading certain tasks, or when in some cases the entire application logic is put on a separate host processor. It is not unusual for the microcontroller, Radio Frequency Integrated Circuit (RFIC) or SoC to host hardware security processors, e.g., to accelerate the Advanced Encryption Standard (AES) operations. This is necessary to allow encrypted communication over the radio link without the need for a host processor and is far more energy-efficient than using software implementations of cryptographic ciphers [29].

Because basic devices lack WAN interfaces, according to our working definition, a gateway of some form is necessary in order to provide WAN connectivity. The gateway, together with the connected devices, forms a capillary network. The microcontroller implements in software most of the functions needed for communicating with the gateway and other devices in the same capillary network. Antennas and associated front-end RF circuitry are always needed for wireless communication, where the radio chip, or radio block of the SoC, implements the necessary filtering and signal processing in compliance with the wireless communications standard in use, e.g., IEEE 802.15.4.

Owing to limited computational resources, basic devices do not use an OS in the traditional sense. Something as simple as a single-threaded main-loop or a lightweight OS such as FreeRTOS, Atomthreads, AVIX-RT, ChibiOS/RT, ERIKA Enterprise, TinyOS, or Thingsquare Mist/Contiki are most often implemented. These lightweight OSs offer significant functionality, including memory and concurrency model management, sensor, actuator, and radio drivers, (multi-) threading, TCP/IP, and higher-level protocol stacks.

The application logic at the device level is usually implemented on top of the OS, calling existing functions provided, as a main application, or within a main loop for simpler cases. A typical task for the application logic might be to read values from sensors and to provide these over the radio interface to a listening gateway or sink node at a predefined frequency. This may or may not be done in a semantically correct manner with the correct units. In the case of raw data (e.g., only numeric values with no units, no location or time of sensor data), the gateway device or a web-hosted application may be needed to perform a transformation on the data such that it becomes semantically correct, appends relevant metadata, etc.

For the basic class of devices, the use of constrained hardware and nonstandard (often specialized and community-developed) software limits third-party development and has traditionally made development relatively cost-intensive.

5.1.3 GATEWAYS

A gateway typically includes multiple networking technologies and serves as a translator between different physical interfaces and protocols, e.g., IEEE 802.15.4 or IEEE 802.11, to Ethernet or cellular. Gateways also often include a local or capillary networking technology and a WAN interface. A gateway refers to a device that performs translation or conversion on different levels of the stack (Physical, Link, Network, Transport Layers) between different network interfaces, but Application Layer Gateways (ALGs) are also common. The latter is preferably avoided because it adds complexity and is a common source of error in deployments.

Some examples of ALGs include the ZigBee Gateway Device[4], which translates from ZigBee to Simple Object Access Protocol (SOAP) and IP, or gateways that translate from Constrained Application Protocol (CoAP) to Hypertext Transfer Protocol/Representational State Transfer (HTTP/REST). For some Local Area Network (LAN) technologies, such as Wi-Fi and Z-Wave, the gateway is used for inclusion and exclusion of devices. This typically works by putting the gateway into an inclusion or exclusion mode and by pressing a button on the device to be added or removed from the network. We cover network technologies in more detail in Section 5.2.

For very basic gateways, the hardware is typically focused on simplicity and low cost, but frequently the gateway device is also used for many other tasks, such as Data Management, DM, and local applications. In these cases, more powerful hardware with GNU/Linux is commonly used. The following sections describe these additional tasks in more detail.

---

[4]http://www.zigbee.org/zigbee-for-developers/zigbee-gateway

5.1.3.1 Data Management

Typical Data Management functions include sensor data collection, data visualization, local storage & synchronization with remote storage (e.g., cloud), processing of sensor readings, and caching of processed data, as well as filtering, concentrating, and aggregating the data before transmitting it to backend or cloud servers. Data Management is discussed in more detail in Section 5.5.

5.1.3.2 Local applications

Examples of local applications that can be hosted on a gateway include event aggregation and closed loops (more specifically home alarm logic and ventilation control), and/or the Data Management functions as above and Section 5.5. The benefit of hosting this logic on the gateway instead of in the network is to avoid downtime in case of WAN connection failure, minimize usage of costly cellular data, and reduce latency. This is also becoming more popular in the context of "edge computing" [30].

To facilitate efficient management of applications on the gateway, it is necessary to include an EE. The EE is responsible for the Lifecycle Management of the running applications, including installation, pausing, stopping, configuration, and uninstallation of the applications. A common example of an EE for embedded devices like gateways is the Open Service Gateway initiative (OSGi), which is based on Java. Bosch ProSyst and Eurotech are two examples of companies providing OSGi-based IoT gateways. Java-based OSGi Gateway Applications are built as one or more Bundles, which are packaged as Java JAR files and installed using a so-called Management Agent. The Management Agent can be controlled from, for example, a terminal shell or via a protocol such as Customer Premise Equipment (CPE) WAN Management Protocol (CWMP).

OSGi Bundle packages can be retrieved from the local file system or the Internet (HTTP), for example. OSGi also provides security and versioning for Bundles, which means that communication between Bundles is controlled, and several Bundle versions can exist. The benefit of versioning and the Lifecycle Management functions is that the OSGi environment never needs to be shut down when upgrading, thus avoiding downtime in the system. Increasingly, stripped back Linux kernels are being used as EEs, where cheap and popular devices like Raspberry Pi[5] may be used as gateways. Another typical software architecture for IoT gateways includes a Linux-based OS, a scripting language (e.g., Javascript, Python, Lua) platform such as Node.js[6] and a local Web server (e.g., Apache[7], nginx[8]) for visualization and serving simple local requests. The concept of microservices and their deployment in the edge is getting increasingly popular however there is no single definition of microservices. Examples of technologies that are associated with the concept of microservices are Linux containers, OSGi bundles, scripts in scripting platforms, and actors in an actor models (e.g., akka[9]).

5.1.3.3 Device Management

DM is an essential element of IoT and provides efficient means to perform many of the management tasks for devices:

---

[5]https://www.raspberrypi.org
[6]https://nodejs.org/en
[7]https://httpd.apache.org
[8]https://www.nginx.com
[9]https://akka.io

- Provisioning: Initialization (or activation) of devices in regard to configuration and features to be enabled.
- Device configuration: Management of device settings and parameters.
- Software upgrades: Installation of firmware, system software, and applications on the device.
- Fault management: Enables error reporting and access to device status.

Examples of DM standards include Broadband Forum (BBF[10]) TR-069, Open Mobile Alliance-Device Management (OMA-DM[11]), and OMA Lightweight M2M (OMA-LwM2M, Section 7.4). Recently OMA has merged with the IPSO Alliance and changed names to OMA SpecWorks. OMA has grown in importance with respect to IoT protocols with OMA LwM2M; however, OMA LwM2M is used more on the lower-end devices without excluding IoT gateways.

In the simplest deployments, devices communicate directly with the DM server. This is, however, not always optimal or even possible due to network or protocol constraints, e.g., due to a firewall or mismatching protocols. In these cases, the gateway functions as mediator between the server and the devices and can operate in three different ways:

- If devices are visible to the DM server, the gateway can simply forward messages between the device and the server, and is not a visible participant in the session.
- In the case where devices are not visible but can process the DM protocol in use, the gateway can act as a proxy, essentially acting as a DM server towards the device and a DM client towards the cloud DM server.
- For deployments where devices use a different DM protocol than the one used by the gateway, the gateway can abstract/represent the devices in the gateway DM protocol and translate between different protocols (e.g., TR-069, OMA-DM, or OMA LwM2M). The devices can be represented either as virtual devices or as part of the gateway (which is typically also a device that is managed by the DM server).

5.1.4 ADVANCED DEVICES

As mentioned earlier, the distinction between basic devices, gateways, and advanced devices is well defined, but some features that can characterize an advanced device are the following:

- A powerful CPU or microprocessor with enough memory and storage to host advanced applications, such as a printer offering functions for copying, faxing, printing, and remote management.
- A more advanced User Interface with, for example, display and advanced user input in the form of a keypad or touch screen.
- Video or other high-bandwidth functions.
- Unlimited supply of energy to power the aforementioned advanced capabilities.

It is not unusual for an advanced device to also function as a gateway for local devices on the same LAN or capillary network. For these more computationally capable devices, the OS can be,

---

[10]https://www.broadband-forum.org
[11]http://www.openmobilealliance.org/wp/overviews/dm_overview.html

for example, GNU/Linux or a commercial RTOS, such as ENEA OSE[12] or WindRiver VxWorks[13]. This class of device comes with optimized and high-performance IP stacks, thus making networking a nonissue. By offering a more common and open OS, along with community-standardized APIs, software libraries, programming languages, and development tools, the number of potential developers grows significantly.

5.1.5 SUMMARY AND VISION

This section covered different device classes and the role of the gateway in an IoT deployment. There are also other aspects that must be taken into account in regard to devices. The most important of these is security, both in terms of physical security as well as software and network security. As this is a very extensive topic, in-depth coverage is broadly beyond the scope of this book, but is addressed in Chapter 6. Another aspect that must be managed is the matter of external environmental factors that can affect the operation of devices, such as rain, wind, chemicals, and electromagnetic influences. These elements are slowly being understood and studied in the context of Cyber-Physical Systems, as IoT applications move from the laboratory into real-world deployments. Fundamentally, these external factors necessitate adaptability and situational awareness capabilities as features of the devices in the field, which are typically unaccounted for during the software engineering phase of development.

A major effect that the IoT will have on devices is to disrupt current value chains, where one actor controls everything from device to service. This will happen due to standardization and consolidation of technologies, such as protocols, OSs, software and programming languages, and the business drivers discussed in Chapter 2. New types of actors will be able to enter the market, e.g., specialized device vendors, cloud solution providers, and service providers. Standardization will improve interoperability between devices, as well as between devices and services, resulting in commoditization of both.

Another potential outcome of improved interoperability is the possibility to reuse the same device for multiple services; for example, a motion detector can be used both for security purposes as well as for reducing energy consumption in the absence of activity in a room. The barrier for new developers is currently and will be further reduced thanks to the consolidation of software and interfaces, e.g., it is possible to interact with a device using simple HTTP/RESTful protocols and to easily install a Java application on a device, resulting in an increased number of developers. Linked cloud-hosted applications are already leveraging data streams coming from heterogeneous deployments sensors, actuators and other devices. Amazon's Alexa controlling lighting (Phillips), music (Spotify), and ambient conditions (Hive) in many homes around the world is a prime example of how heterogeneous systems, companies, and technologies can converge in this way.

Thanks to developments in hardware and network technologies, entirely new device classes and features are expected, such as:

- Battery-powered devices with ultra-low-power cellular connections (e.g., NB-IoT).
- Devices that harvest energy from their environment [31].

---

[12]https://www.enea.com/products/operating-systems/enea-ose
[13]https://www.windriver.com/products/vxworks/

- Smart bandwidth management and protocol switching, i.e., using adaptive RF mechanisms to swap between, for example, Bluetooth Low Energy (BLE) and IEEE 802.15.4, such as the Texas Instruments Sensor Tag[14].
- Multiradio/multirate to switch between bands or bit rates (slower bit rate implies better sensitivity at longer range).
- Microcontrollers with multicore processors.
- Novel software architectures for better handling of concurrency.
- The possibility to automate the design of integrated circuits based on business-level logic and use case.

All these improvements are expected to accelerate the adoption of IoT in the future. In the next section, we cover LANs and WANs – the technological building blocks that allow devices to communicate with Information and Communications Technology (ICT) systems and the wider world.

5.2 LOCAL AND WIDE AREA NETWORKING
5.2.1 THE NEED FOR NETWORKING

A network is created when two or more computing devices exchange data or information. The ability to exchange pieces of information using telecommunication technologies has changed the world and will continue to do so for the foreseeable future, with applications emerging in nearly all contexts of contemporary and future living. Typically, devices are known as *nodes* of the network, and they communicate over *links*.

In modern computing, nodes range from Personal Computers, servers, and dedicated packet switching hardware to smartphones, game consoles, television sets, and, increasingly, heterogeneous devices that are generally characterized by limited resources and functionalities. Limitations typically include computation, energy, memory, communication (range, bandwidth, reliability, etc.), and application specificity (e.g., specific sensors, actuators, tasks). Such devices are typically dedicated to specific tasks, such as sensing, monitoring, and control (discussed later).

Network links rely upon a physical medium, such as electrical wires, air, and optical fibers, over which data can be sent from one network node to the next. It is not uncommon for these media to be grouped either as wired or wireless.

A selected physical medium determines a number of technical and economic considerations. Technically, the medium selected, or more accurately, the technological solution designed and implemented to communicate over that medium, is the primary enabler of bandwidth – without which certain applications are infeasible. Simultaneously, different technological solutions require certain economic considerations, such as the cost of deployment and maintenance of the networking infrastructure. For example, consider the cost of embedding wires across a metropolitan, or larger, geographic region (e.g., electricity and legacy telephone networks).

When direct communication between two nodes over a physical medium is not possible, networking can allow for these devices to communicate over a number of *hops*. To achieve this, nodes of the

---

[14]http://www.ti.com/ww/en/wireless_connectivity/sensortag

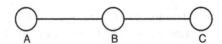

FIGURE 5.2

Simplest possible network.

network must have an awareness of all nodes in the network with which they can (in)directly communicate. This can be a direct connection over one link (edge, the transition or communication between two nodes over a link) or knowledge of a route to the desired (destination) node by communicating through cooperating nodes, over multiple edges (consider Figure 5.2).

This is the simplest form of network that requires knowledge of a route to communicate between nodes that do not have direct physical links. Therefore, if node A wishes to transfer data to node C, it must do so through node B. Thus, node B must be capable of the following: communicating with both node A and node C and advertising to node A and node C that it can act as an intermediary, i.e., advertising to node A that node C can be reached through node B and vice versa.

It is essential to uniquely identify each node in the network, and it is necessary to have cooperating nodes capable of linking nodes between which physical links do not exist. In modern computer networking, this equates to IP addresses and routing tables. Thanks to standardization, IP in particular, physical media (links) no longer need to be the same across the network, and the nodes need not have the same capabilities and/or mission.

Beyond the basic ability to transfer data, the speed and accuracy with which data can be transferred is of critical importance to the application. Irrespective of the ability to link devices, without the necessary bandwidth some applications are rendered impossible. Consider the differences between streaming video from a surveillance camera, for example, and an intrusion detection system based on a passive sensor. Simplistically, streaming video requires high bandwidth, whereas transmitting a small amount of information about the detection of an intruder requires a tiny amount of bandwidth, but a high degree of reliability with respect to both the communications link and the accuracy of the detection.

Today, we have complex, heterogeneous networks. The simple example above is useful for explaining the basics of networking at a very high level, but is also useful when abstracting the types of nodes that A, B, and C might be, the different physical links between them, and their methods of interaction.

Summarizing, consider that node A is a device that can only communicate over a particular wireless channel of limited range (e.g., Channel 11 in the 2.4 GHz ISM band, over less than 200 meters). Node B is capable of communicating with node A, but also with an application server with service capabilities (node C, with which it can connect using wired Ethernet, e.g., over a complex link using a standardized protocol and/or web service such as REST at the application layer) over the Internet. Now consider that node B may be connected to a subnetwork (of child nodes, similar to node A) of up to thousands of similarly constrained devices ($A_1 \ldots A_n$). These thousands of devices may be equipped with sensors, deployed specifically to monitor some physical phenomenon. They may only communicate with one another and node B, and they may communicate with each other over single or multiple hops (thus increasing the range of the sensing field, not all nodes requiring direct connectivity with B). This is representative of a Wireless Sensor Network (WSN) in the traditional sense.

Consider that the owner of the WSN wishes to obtain the data from each of the $(A_1 \ldots A_n)$ devices in the WSN. However, the preferred way to read the data is through a web browser, or application on a smartphone/tablet, via node C. In this case, a networking solution is required to transfer all of the WSN data from nodes $A_1 \ldots A_n$ to node C, through node B. This is now a complex networking infrastructure, and this is representative of many potential embodiments of sensor networks and IoT technologies. This concept maps directly to the IoT Reference Architecture with devices and IoT Gateways in capillary networks (described later in the book) and cloud-based IoT services where nodes $A_1 \ldots A_n$ constitute the capillary network, node B is an IoT Gateway, and node C is representative of an IoT cloud service.

A LAN was traditionally distinguishable from a WAN based on the geographic coverage requirements of the network and the need for third-party, or leased, communication infrastructure. In the case of the LAN, a smaller geographic region is covered, such as a commercial building, an office block, or a home, and does not require any leased communications infrastructure.

WANs provide communication links that cover longer distances, such as across metropolitan, regional, or by textbook definition, global geographic areas. In practice, WANs are often used to link LANs and Metropolitan Area Networks (MANs) – where LAN technologies cannot provide the communications ranges to otherwise interconnect – and commonly to link LANs and devices (including smartphones, Wi-Fi routers that support LANs, tablets, and IoT devices) to the Internet. Quantitatively, LANs tended to cover distances of tens to hundreds of meters, whereas WAN links spanned tens to hundreds of kilometers. Low-Power WAN (LP-WAN) technologies in the licensed and the unlicensed spectrum have recently emerged to provide an extended range for applications expected to require metropolitan-scale communication ranges. Examples in the unlicensed spectrum include Sigfox, Long Range (LoRa) and Wireless Smart Ubiquitous Networks (Wi-SUNs), whereas Extended Coverage-GSM-IoT (EC-GSM-IoT), Long Term Evolution Category M1 (LTE-M), and Narrowband IoT (NB-IoT) have been standardized for licensed-spectrum use. These technologies are described in more detail in Section 5.2.3.

There are differences between the technologies that enable LANs and WANs. In the simplest case for each, these can be grouped as wired or wireless. The most popular wired LAN technology is Ethernet. Wi-Fi is the most prevalent Wireless LAN (WLAN) technology. Wireless WAN (WWAN), as a descriptor, covers cellular mobile telecommunication networks, a significant departure from WLAN in terms of technology, coverage, network infrastructure, and architecture. The current generation of WWAN technology includes LTE (or 4G), while 5G is on the horizon. Acting as a link between LANs and Wireless Personal Area Networks (WPANs), Gateway Devices (Section 5.1) typically include cellular transceivers and allow seamless IP connectivity over heterogeneous physical media. An example of a Gateway's logical functionality is presented in Figure 7.7 in Chapter 7, which shows an IETF CoAP HTTP proxy.

An intuitive example of a similar device is the wireless access point commonly found in homes and offices. In the home, the "wireless router" typically behaves as a link between the Wi-Fi (WLAN, and thus connected laptops, tablets, smartphones, etc., commonly found in the home) and Digital Subscriber Line (DSL) broadband connectivity, traditionally arriving over telephone lines. "DSL" refers to Internet access carried over legacy (wired) telephone networks and encompasses numerous standards and variants. "Broadband" indicates the ability to carry multiple signals over a number of frequencies, with a typical minimum bandwidth of 256 Kbps. In the office, the Wi-Fi wireless access points are typically connected to the wired corporate (Ethernet) LAN, which is subsequently connected to a wider-area network and Internet backbone, typically provided by an Internet Service Provider (ISP). Many ISPs now offer fiber (high-speed optical) connections.

Considering the breadth of IoT applications, there are likely to exist a combination of traditional networking approaches. There is a need to interconnect devices (generally integrated microsystems) with central data processing and decision support systems, in addition to one another. The business logic and requirements for each embodiment tend to differ on a case-by-case basis.

Practically, these devices will not warrant individual connections to leased networking infrastructure (e.g., putting a SIM card in each device and using the cellular network for fast IP connectivity). This approach is thought to be prohibitive due to cost, among other factors. A more likely scenario is where, similar to WLAN technologies, a geographic region can be covered by a network of devices that connect to the Internet via a gateway device, which may use a leased network connection.

The potential complexity of these networks is enormous. For example, a gateway device can access the IP backbone over a WWAN (e.g., GPRS/UMTS/LTE) link, or over a WLAN link, where the leased infrastructure would be that of the ISP providing backbone connectivity to the WLAN in its own right, as above.

It is worth extending the consideration of WAN and LAN to encompass the idea of WPANs, which is the description used for the newer standards that govern low-power, low-rate networks suitable for IoT applications. Indeed, the standard upon which many popular recent networking technologies are built (including ZigBee, WirelessHART, ISA00.11a, and other IETF initiatives such as 6TiSCH, 6LoWPAN, RPL, and CoAP, all discussed later in this chapter) is "IEEE 802.15.4 – Wireless Medium Access Control (MAC) and Physical Layer (PHY) Specifications for Low-Rate Wireless Personal Area Networks (LR-WPANs)". This standard was first approved in 2003, and it has been subject to numerous amendments over the past decade. These amendments have related to modifying and/or extending PHY parameters to ensure global utility with regard to licensing, application suitability, and modifications to the MAC layer. This is similar to the evolution of Wi-Fi WLAN technology (e.g., IEEE 802.11, a, b, g, n, etc.). The naming convention is ultimately nonintuitive, as communication ranges for IEEE 802.15.4 technology may range from tens of meters to kilometers. It is probably more useful to think of these technologies as "low-rate, low-power" networks.

It is reasonable to suggest that the traditional boundaries between LAN and WAN technologies, and their working definitions, require updating to account for contemporary amendments to the standards and use cases to which they are applicable.

From an M2M perspective, ETSI (Appendix A) and similarly (albeit with a different terminology) oneM2M (Section 7.10.1) considers, as part of its functional architecture, M2M Area Networks (Figure A.1). Devices in an M2M Area Network connect to the IP backbone, or Network Domain, via an M2M Gateway device. Typically, a Gateway device is equipped with a cellular transceiver that is physically compatible with UMTS or LTE-Advanced, for example, WWAN. The same device will also be equipped with the necessary transceiver to communicate on the same physical medium as the M2M Area Network(s) in the M2M Device Domain. This is covered in more detail in Chapters 7–8 and Appendix A. The IoT Reference architecture and Model described in Chapter 8 also defines IoT devices including gateways and device networks also known as capillary networks (described later in the book). The topology is similar to the ETSI M2M topology; however the ETSI M2M terminology has been adapted in oneM2M and is not used widely anymore.

IoT networks may include a plethora of wired or wireless technologies, including: Bluetooth LE/Smart, IEEE 802.15.4 (LR-WPAN; e.g., ZigBee, IETF 6LoWPAN, RPL, CoAP, ISA100.11a, WirelessHART), M-BUS, Wireless M-BUS, KNX, and Power Line Communication (PLC). The "Internet of Things", as a term, originated from Radio Frequency Identification (RFID) research, wherein the original IoT concept was that any RFID-tagged "thing" could have a new type of Electronic Product Code (EPC) and a virtual presence on the Internet. In reality, there is little conceptual dissimilarity between RFID and bar codes, or more recently, Quick Response (QR) codes – they simply use different technological means to achieve the same result (i.e., an "object" has an online presence). For more information about RFID-related architectures and system standards please refer to Chapter 7.

The original concept has evolved from a reasonably simple idea, with immediate utility in logistics (i.e., track and trace, inventory management applications; see Chapter 17), to complex networks, functionalities, and interactions, without any satisfactory working definition(s). As IoT applications, networks, and systems evolve, it is necessary to understand the technologies, limitations, and implications of the networking infrastructure. Essentially, the ability to remotely communicate with devices, and resultant new capabilities, is what sets modern IoT thinking apart from the originating concepts.

The following subsections describe the technologies traditionally used to achieve WAN and LAN. We are further motivated to shift away from conventional thinking on how to describe networking and communication technologies based on simplistic concepts around geographic coverage or leased infrastructure.

5.2.2 WIDE AREA NETWORKING

WANs are typically required to bridge a capillary network to the backhaul network, thus providing a proxy that allows information (data, commands, etc.) to traverse heterogeneous networks. This is seen as a core requirement to provide communication services between cloud-based IoT services and the physical deployments of devices in the field. Thus, the WAN is capable of providing the bidirectional communication links between services and devices. This, however, must be achieved by means of physical and logical proxy.

The proxy is achieved using an IoT Gateway device. Depending on the situation, there are, in general, a number of candidate technologies to select from. As before, the IoT Gateway device is typically an integrated microsystem with multiple communication interfaces and computational capabilities. It is a critical component in the functional architecture, as it must be capable of handling all of the necessary interfacing to the cloud-based IoT services.

By way of example, consider a device that incorporates both an IEEE 802.15.4-compliant transceiver (a popular example of which is the Texas Instruments CC2520[15]), capable of communicating with a capillary network of similarly equipped devices, and a cellular transceiver (a popular example of which is a Telit LN940A9) that connects to the Internet using the LTE network. This assumes the handover to the backbone IP network is handled according to the 3GPP specifications. Transceivers (sometimes referred to as modems) are typically available as hardware modules with which the central intelligence of the device (gateway or cell phone) interacts by means of standardized (sometimes vendor-specific) AT Commands. This device is now capable of acting as a physical proxy between the Low-Rate Wireless Personal Area Network (LR-WPAN) and the cloud.

The ETSI M2M Functional Architecture is shown in Figure 5.3 and described in more detail in Appendix A. The ETSI M2M architecture is obsolete since the ETSI M2M specification was incorporated

---

[15] http://www.ti.com/product/cc2520

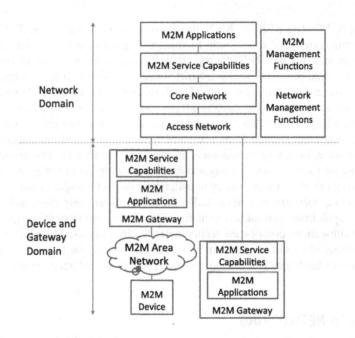

FIGURE 5.3

ETSI M2M High-Level Architecture (redrawn from [32]). Copyright European Telecommunications Standards Institute 2013. Further use, modification, copy and/or distribution is strictly prohibited.

into the oneM2M (Section 7.10.1) specifications in 2012 and the ETSI M2M Technical Committee concluded its work. The architecture is only used here for the purpose of illustrating an example WAN since the ETSI M2M architecture focuses more on communication and networking than Data Management and services which are the focus of more recent IoT architectures and models. A reader interested in the state-of-the-art of IoT architectures and an IoT Architecture Reference and Model (ARM) should refer to Chapter 7 and Chapter 8, respectively.

Device types were discussed in more detail in Section 5.1. The Access and Core Network in the ETSI M2M Functional Architecture are foreseen to be operated by a Mobile Network Operator (MNO) and can be thought of simply as the "WAN" for the purposes of interconnecting devices and backhaul networks (Internet), thus, M2M Applications, Service Capabilities, Management Functions, and Network Management Functions. The WAN covers larger geographic regions using wireless (licensed and unlicensed spectra) as well as wire-based access. WAN technologies include cellular networks (using several generations of technologies), DSL, Wi-Fi, Ethernet, and Satellite. The WAN delivers a packet-based service using IP as default. However, circuit-based services can also be used in certain situations.

In the M2M context, important functions of the WAN include the following:

- The main function of the WAN is to establish connectivity between capillary networks, hosting sensors, and actuators and the M2M service enablement. The default connectivity mode is packet-based using the IP family of technologies. Many different types of messages can be sent and received. These include messages originating as, for example, a message sent from a sensor in an M2M Area Network and resulting in an SMS received from the M2M Gateway or Application (e.g., by a rel-

evant stakeholder with SMS notifications configured for when sensor readings breach particular sensing thresholds).
- Use of identity management techniques (primarily of M2M devices) in cellular and noncellular domains to grant right-of-use of the WAN resource. The following techniques are used for these purposes:
  - A Trusted Environment (TRE) on the M2M devices for remote provisioning of a Subscriber Identity Module (SIM) targeting M2M devices.
  - xSIM (x-Subscription Identity Module), like SIM, Universal Subscriber Identity Module (USIM), IP Multimedia Subscriber Identity Module (ISIM).
  - Interface identifiers, an example of which is the MAC address of the device, typically stored in the hardware.
  - Authentication/registration type of functions (device focused).
  - Authentication, Authorization, and Accounting (AAA), such as Remote Authentication Dial-In User Service (RADIUS) services.
  - Dynamic Host Configuration Protocol (DHCP), e.g., employing deployment-specific configuration parameters specified by device, user, or application-specific parameters residing in a directory.
  - Subscription services (device-focused).
  - Directory services, e.g., containing user profiles and various device(s), parameter(s), setting(s), and combinations thereof.

M2M-specific considerations include, in particular:

- An M2M/IoT identification module such as the old concept of the Machine Communications Identity Module (MCIM) in the 3GPP SA3 work.
- User Data Management (e.g., subscription management).
- Network optimizations (cf. 3GPP SA2 work).

There may be many suppliers of WAN functionality in a complete M2M solution. It follows that an important function in the M2M Service enablement domain will be to manage westbound Business-to-Business (B2B) relations between a number of WAN service providers.

5.2.2.1 3rd Generation Partnership Project technologies and Machine Type Communications

Machine Type Communications (MTC) is heavily referred to in the ETSI documentation. MTC, however, lacks a firm definition and is explained using a series of use cases. Generally speaking, MTC refers to small amounts of data that are communicated between machines (devices to backend services and vice versa) without the need for any human intervention. In the 3rd Generation Partnership Project (3GPP), MTC is used to refer to all M2M communication [33]. Thus, they are interchangeable terms.

5.2.3 LOW-POWER WIDE AREA NETWORKS

Recently specified Low-Power Wide Area (LPWA) wireless communications technologies, such as LTE-M, EC-GSM-IoT, LoRa, SigFox, and NB-IoT, have the potential to serve a range of hitherto unconnected devices. These devices are typically embedded sensors and actuators requiring long-range, low-data rate communication, and extreme operational energy efficiency and are deployed at the metropolitan scale within various vertical industries' applications.

A number of standards bodies, industrial consortia, and Special Interest Groups are developing competing LPWA technologies, including 3GPP, IEEE, IETF, ETSI, LoRa Alliance, Weightless SIG, and Wi-SUN Alliance. In the following subsections the LPWA technologies currently available are discussed with regard to those standardized by the 3GPP for use in licensed spectrum, proprietary LPWA technologies, and additional standards-driven LPWA technologies.

5.2.3.1 3GPP Licensed Spectrum LPWA Technologies

In June 2016, the 3GPP (Release 13[16]) completed the standardization Narrow-Band IoT (NB-IoT) to complement EC-GSM-IoT and LTE-M[17] [34]:

- **NB-IoT**: NB-IoT aims to enable deployment flexibility, long battery life, low device cost and complexity, and signal coverage extension. NB-IoT is not compatible with 3G but can coexist with GSM, GPRS, and LTE. It may be supported with a software upgrade on top of existing LTE infrastructure. It can be deployed inside a single GSM carrier of 200 KHz, inside a single LTE Physical Resource Block (PRB) of 180 KHz, or inside an LTE guard band. Compared to LTE-M (next), NB-IoT cuts cost and energy consumption further by reducing data rates and bandwidth requirements and simplifying protocol design and mobility support. Standalone deployment in a dedicated licensed spectrum is supported. NB-IoT aims for coverage that can serve up to 50k devices per cell, with the potential to scale up by adding more NB-IoT carriers.
- **LTE-M**: LTE end devices offer high data rate services at cost and power consumption levels unsuitable for IoT-type use cases. To reduce the cost while being compliant to LTE system requirements, 3GPP reduces the peak data rate from LTE Category 1 to LTE Category 0 and then to LTE Category M, the different stages in the LTE evolution process. To extend the battery lifetime for MTC, 3GPP implements "Power Saving Mode" and "extended Discontinuous Reception" or eDRX.
  These techniques enable devices to enter deep sleep modes for hours or days without losing network registration. End devices thus avoid monitoring downlink control channels for prolonged periods of time in order to save energy. The same power saving features are exploited in EC-GSM-IoT.
- **EC-GSM-IoT**: 3GPP is in the process of proposing the Extended Coverage GSM (EC-GSM) standard that aims to extend the GSM coverage by +20 dB using <GHz bands for better signal penetration in indoor environments. With a software upgrade to GSM networks, the legacy GPRS spectrum can use new channels to accommodate EC-GSM-IoT devices. It exploits repetitive transmissions and signal processing techniques to improve coverage and capacity of legacy GPRS. It can provide variable data rates up to 240 Kbps and aims to support 50k devices per base station with enhanced security features compared to existing GSM-based solutions.

5.2.3.2 Proprietary LPWA technologies

This subsection describes some of the recent proprietary LPWA technologies standardized by industrial consortia and Special Interest Groups (SIGs):

---

[16]http://www.3gpp.org/release-13
[17]EC-GSM-IoT and LTE-M are optimizations to 2nd (GPRS) and 4th (LTE) generation infrastructure to accommodate MTC suited to various IoT scenarios. NB-IoT is designed to work in the 2G and the 4G/LTE spectrum.

- **SigFox**[18] offers end-to-end LPWA connectivity based on its patented technologies, sometimes in partnership with other network operators. SigFox Network Operators (SNOs) deploy proprietary base stations with software-defined radios that connect to backend systems over IP networks. End devices connect to these base stations using Binary Phase Shift Keying (BPSK) modulation in an ultra-narrow (100 Hz) <GHz ISM band carrier. By exploiting Ultra Narrow Band (UNB), SigFox uses the bandwidth efficiently and experiences low noise levels that result in high receiver sensitivity, ultra-low power consumption, and inexpensive antenna design. These benefits come at the expense of a throughput limited to 100 bps.

- **LoRa**[19] is a Physical Layer technology that modulates signals in <GHz ISM bands using a proprietary spread spectrum technique developed by Semtech Corp[20]. Bidirectional communication is enabled by Chirp Spread Spectrum (CSS) techniques, spreading a narrowband input signal over a wide channel bandwidth. The resulting signal has noise-like properties, making it harder to detect or jam. Processing gain allows resilience to interference and noise. Data rates can range from 300 bps to 37.5 Kbps, depending on the spreading factor and channel bandwidth. Furthermore, multiple transmissions using different spreading factors can be received simultaneously by a LoRa base station. In [35], the authors point to a number of trials and comparisons of LoRa and SigFox which have been published in recent years.

- **Weightless**[21] proposed three LPWA standards, each providing different features, range, and power consumption, and each of which can operate in license-free as well as in licensed spectrum. Weightless-W leverages TV white spaces and supports several modulation schemes and a range of spreading factors. Depending on the link budget, packets with sizes upwards of 10 bytes can be transmitted at a rate between 1 Kbps and 10 Mbps. End devices transmit to base stations in a narrow band but at a lower power level than the base stations to save energy. Weightless-W has a one drawback in that shared access to TV white spaces is permitted only in a few regions, and therefore the Weightless SIG defines the other two standards in globally available ISM bands. Weightless-N is a UNB standard for one-way communication from end devices to a base station and achieves significant energy efficiency and lower cost than the other Weightless standards. One-way communication, however, limits the number of use cases for Weightless-N. Weightless-P blends two-way connectivity with two nonproprietary Physical Layers. Single 12.5 KHz narrow channels in the <GHz band allow data rates between 0.2 Kbps and 100 Kbps.

- **Ingenu**[22] (previously On-Ramp Wireless) uses proprietary LPWA that does not rely on better propagation properties of <GHz bands, but operates in the 2.4 GHz ISM band and leverages relaxed regulations on the spectrum use across different regions. Its patented physical access scheme is called Random Phase Multiple Access (RPMA) Direct Sequence Spread Spectrum (DSSS), which it uses for uplink communication only. Ingenu is heavily involved in efforts to standardize IEEE 802.15.4k (more later), making RPMA compliant with the IEEE specifications.

---

[18]https://www.sigfox.com/en
[19]https://www.lora-alliance.org/
[20]https://www.semtech.com
[21]http://www.weightless.org/
[22]https://www.ingenu.com/

- **Telensa**[23] provides end-to-end LPWA solutions for applications incorporating fully designed vertical network stacks with support for integration with third-party software. Telensa uses proprietary UNB modulation techniques operating in license-free <GHz ISM bands at low data rates. Little is known about the implementation of their wireless technology and Telensa is aiming to standardize its technology using ETSI Low Throughput Networks (LTNs) specifications for easy application integration.
- **Qowisio**[24] deploys dual-mode LPWA networks combining proprietary UNB technology with LoRa. It provides LPWA connectivity as a service to the end users: It offers end devices, deploys network infrastructure, develops custom applications, and hosts them at a backend cloud. Less is known about the Technical Specifications of their UNB technology and system components.

5.2.3.3 LPWA standards landscape

In addition to the 3GPP standards previously mentioned, it is worth mentioning in a little more detail the other standards emerging and relevant to the LPWA space. These specifically relate to those emerging from IEEE, IETF, ETSI, and LoRa Alliance.

- **IEEE** is extending range and reducing power consumption of IEEE 802.15.4 and 802.11 with a set of new specifications for the physical and the MAC layers. Specifically, IEEE 802.15.4k: Low Energy, Critical Infrastructure Monitoring Networks (TG4k)[25] proposes a standard for low-energy critical infrastructure monitoring applications and IEEE 802.15.4g: Low-Data-Rate, Wireless, Smart Metering Utility Networks (TG4g)[26] proposes amendments to address process control applications such as smart metering networks (typically comprising large numbers of fixed-end devices deployed over large geographic scales). These address the problem with earlier standards with insufficient range and device densities required for LPWA applications.
- **ETSI** is leading efforts to standardize a bidirectional low data rate LPWA standard called Low Throughput Network (LTN)[27]. A main goal is to reduce communication by exploiting short payload sizes and low data rates of M2M/IoT communication. Beyond recommendations on the Physical Layer, LTN defines interfaces and protocols for the cooperation between end-devices, base stations, network servers, and operational and business management systems.
- **IETF** set up a WG on LPWA Networks in April 2016 which identified challenges and the design space concerns for IPv6 connectivity for LPWA technologies. Future efforts will likely result in multiple standards defining a full IPv6 stack for LPWA that can connect devices with each other and the external ecosystem[28].
- **LoRa Alliance** defined the upper layers and system architecture in the LoRaWAN Specification, which was released in July 2015. This is built on LoRa[29].

---

[23]http://www.telensa.com/

[24]https://www.qowisio.com/en/

[25]http://www.ieee802.org/15/pub/TG4k.html

[26]http://www.ieee802.org/15/pub/TG4g.html

[27]http://www.etsi.org/technologies-clusters/technologies/low-throughput-networks

[28]https://datatracker.ietf.org/wg/lpwan/about/

[29]https://www.lora-alliance.org/What-Is-LoRa/Technology

5.2.4 LOCAL AREA NETWORKING

The term LAN originates mainly from the Internet topology and is often associated with a certain type of access technologies. An IoT-related term to represent a LAN is the term "Capillary Network". Capillary networks are typically autonomous, self-contained systems of IoT devices that may be connected to the cloud via an appropriate Gateway. They are often deployed in controlled environments such as vehicles, buildings, apartments, factories, and bodies (Figure 5.4) in order to collect sensor measurements, generate events should sensing thresholds be breached, and sometimes control specific features of interest (e.g., heart rate of a patient, environmental data on a factory floor, car speed, air conditioning appliances). There are and will exist numerous capillary networks that already are or will employ short-range wired and wireless communication and networking technologies.

For certain application areas, there is a need for autonomous local operation of the capillary network. That is, not everything needs to be sent to, or potentially be controlled via, the cloud. In the event that application-level logic is enforceable via the cloud, some capillary networks will still need to be managed locally. The complexity of the local application logic varies by application. For example, a building automation network may need local control loop functionality for autonomous operation, but can rely on external communication for the configuration of control schedules or parameters.

The IoT devices in a capillary network are typically thought to be low-capability nodes (e.g., battery operated, with limited security capabilities) for cost reasons and should operate autonomously. For this reason, a GW and/or Application Server (AS) will naturally also be part of the architected solution for capillary networks. More and more (currently closed) capillary networks open up for integration with the enterprise backend systems. For capillary networks that expose devices to the cloud/Internet, IP is envisioned to be the common protocol waist. IPv6 is to be the protocol of choice for IoT devices that

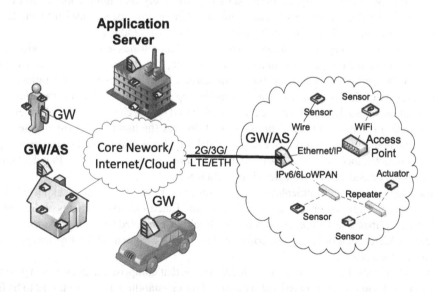

FIGURE 5.4

Capillary networks and their inside view.

operate a 6LoWPAN-based stack. IPv4 is used for capillary networks operating in non-6LoWPAN IP stacks (e.g., Wi-Fi capillary networks).

In terms of short-range communication technology convergence, an IPv6 stack with 6LoWPAN running above the physical medium is expected. The physical medium may be IEEE 802.15.4 (i.e., wireless), but can also be various PLC or other wired solutions (e.g., Homeplug). Legacy ZigBee application profiles are expected to be used in the future in addition to newer ZigBee IP and IEEE 802.15.4 6LoWPAN/RPL/CoAP networks. It is expected that the binary versions of the application profiles will be used for efficiency reasons (e.g., an automation profile device may be a temperature sensor not necessarily connected to the mains power).

Currently ZigBee addresses the Smart Home, Utility, Lighting, and Retails market segments which overlap somewhat with BLE markets (Smart Home, Smart Buildings, Smart Industry, Smart City, among others). However Bluetooth includes other market areas such as audio equipment and it is already on mobile phones and laptops comanufactured with Wi-Fi chipsets. There are no indications that KNX and ZigBee will be consolidated, as there are big players that support these for the smart grid application area. The situation is further exacerbated by the development of the IEEE 802.15.4g standard, a Physical Layer amendment to support Smart Utility Networks (SUNs) – Smart Grid in particular – designed to operate over much larger geographic distances (wireless links spanning tens of kilometers) and specifically designed for minimal infrastructure, low power, many-device networks.

5.2.4.1 Deployment considerations

The nature of the intended application plays a significant role in determining the appropriate technological solution. Typically, these are defined by the business logic that motivates initial deployment. There are increasing numbers of innovative IoT applications (hardware and software) marketed as consumer products. These range from intelligent thermostats for effectively managing comfort and energy use in the home to precision gardening tools (sampling weather conditions, soil moisture, etc.). At scale, similar solutions are, and will continue to be, applied in and across industry.

Scaling up for industrial applications and moving from laboratories into the real world creates significant challenges that are not yet fully understood. Low-rate, low-power communication technologies are known to be "lossy". The reasons for this are numerous. They can relate to environmental factors, which impact upon radio performance (such as time-varying stochastic wireless propagation characteristics), technical factors such as performance trade-offs based on the characteristics of Medium Access Control and routing protocols, physical limitations of devices (including software architectures, runtime and EEs, computational capabilities, energy availability, local storage, and so forth), and practical factors such as maintenance opportunities (scheduled, remote, accessibility, etc.). Practical constraints for deployment are considered in more detail in Section 8.7.

Numerous deployment environments (factories, buildings, roads, vehicles) are expected in addition to wildly varying application scenarios and operational and functional requirements of the systems. For example ETSI describes a set of use cases, namely eHealth, Connected Consumer, Automotive, Smart Grid, and Smart Meter, that only capture some of the breadth of potential deployment scenarios and environments that are possible.

Section 5.1 describes the various hardware technologies that comprise the devices and gateways that make up the current and future art in IoT technologies. Notwithstanding, there continues to be fragmentation at the Physical Layer in terms of communication technologies. Assuming that IP connectivity can be the fundamental mechanism to bridge heterogeneous physical and link layer technologies, it stands to reason that fragmentation can continue such that appropriate technologies are available for the breadth of potential application scenarios.

5.2.4.2 Key technologies

This section details a number of the standards and technologies currently in use and under development that enable ad hoc connectivity between the devices that will form the basis of the IoT. These are the communication technologies that are considered to be critical to the realization of massively distributed M2M applications and the IoT at large.

PLC refers to communicating over power (or phone, coax, etc.) lines. This amounts to pulsing, with various degrees of power and frequency, the electrical lines used for power distribution. PLC comes in numerous flavors. At low frequencies (tens to hundreds of Hertz) it is possible to communicate over kilometers with low bit rates (hundreds of bits per second). Typically, this type of communication was used for remote metering and was seen as potentially useful for the smart grid. Enhancements to allow higher bit rates have led to the possibility of delivering broadband connectivity over power lines. There have been a number of attempts to standardize PLC in recent years. NIST included IEEE 1901 (in 2011) and ITU-T G.hn (G.9960-PHY in 2009 and G.9961-Data Link Layer in 2010) as standards for further review for potential use in the smart grid in the United States. In 2011, the ITU referenced G.9903 that specifies the use of IPv6 over PLC, borrowing techniques (specifically 6LoWPAN, below) originally developed in the wireless community.

LAN (and WLAN) continues to be important technology for IoT applications. This is due to the high bandwidth, reliability, and legacy of these technologies. Where power is not a limiting factor, and high bandwidth is required, devices may connect seamlessly to the Internet via Ethernet (IEEE 802.3) or Wi-Fi (IEEE 802.11). The utility of existing (W)LAN infrastructure is evident in a number of early IoT applications targeted at the consumer market, particularly where integration and control with smartphones is required (irrespective of the actual technical architecture or optimality of the solution).

The IEEE 802.11 (Wi-Fi) standards continue to evolve in various directions to improve certain operational characteristics depending on usage scenario. A widely adopted release was IEEE 802.11n, which was specifically designed to enhance throughput (typically useful for streaming multimedia). Moreover IEEE 802.11ac (adopted in 2013) aims at an even higher throughput version to replace this, focusing efforts in the 5 GHz band.

IEEE 802.11ah was adopted in 2017 targeting an evolution of the 2007 standard that will allow a number of networked devices to cooperate in the <1 GHz (ISM) band. The idea is to exploit collaboration (relaying, or networking in other words) to extend range and to improve energy efficiency (by cycling the active periods of the radio transceiver). The standard aims to facilitate the rapid development of IoT applications that could exploit burst-like transmissions, such as in metering applications. This type of thinking is very similar to traditional WSN theory and practice, which foreran the development of technologies like 6LoWPAN, RPL, and CoAP, below.

BLE ("Bluetooth Smart") is an integration of Nokia's Wibree (2006) standard with the main Bluetooth standard (originally developed and maintained as IEEE 802.15.1 and Bluetooth SIG). It is designed for short-range (<50 m) applications in healthcare, fitness, security, etc., where high data rates (millions of bits per second) are required to enable application functionality. It is deliberately low-cost and energy-efficient by design, and it has been integrated into the majority of recent smartphones. Apart from the typical star topology (1-to-1 or 1-to-many communication patterns), BLE currently supports the full mesh topology (many-to-many, Bluetooth Mesh) directly competing with IEEE802.15.4 types of communication stacks.

Low-Rate, Low-Power Networks are among the key technologies that form the basis of the IoT. For example, the IEEE 802.15.4 family of standards was one of the first used in practical research and experimentation in the field of WSNs. It was originally presented in 2003 as Part 15.4: Low-Rate Wireless Personal Area Networks (LR-WPAN). The original release covered the Physical and Medium Access Control layers, specifying use in the ISM bands at frequencies around 433 MHz, 868/915 MHz, and 2.4 GHz. This supported data rates of between 20 Kbps up to 256 Kbps, depending on the selected band, over distances ranging from tens of meters to kilometers (also depending on transmission power level). Typically, radio transceivers developed in compliance with this standard consume power in the tens of milli-Watts range in active modes. This means that they are still insufficiently energy-efficient to provide long-lasting battery life (i.e., >10 years) for continuous operation, or energy harvested operation, without aggressively duty cycling the active period of the transceiver. Radio duty cycling refers to managing the active periods of the Radio Frequency Integrated Circuit (RFIC) during transmission and listening to the medium. It is typically quantified as a percentage with respect to active time.

Notwithstanding, IEEE 802.15.4 defines the PHY layer, and in some instances the MAC layer, upon which a number of low-energy communication specifications have been built. Namely, ZigBee, and its derivatives ZigBee IP and ZigBee RF4CE, WirelessHART, ISA100.a, and others use this technology at the most basic level. While the IEEE 802.15.4 standard refers to WPANs in its title, the reality is much different. Recent developments, such as the PHY Amendment for Smart Utility Networks (SUNs), IEEE 802.15.4g, seek to extend the operational coverage of these networks up to tens of kilometers in order to provide extremely wide geographic coverage with minimal infrastructure. As the name of the WG suggests, an intuitive use case for this amendment is in the future smart grid. With IPv6 Networking, attention is paid to the ongoing work to facilitate the use of IP to enable interoperability irrespective of the physical and link layers (i.e., making the fact that devices are networked, with or without wires, with various capabilities in terms of range and bandwidth, essentially seamless). It is foreseeable that the only hard requirement for an embedded device will be that it can somehow connect with a compatible gateway device (assuming it is a capillary device that does not have in-built WAN connectivity, i.e., a cellular modem). The advances in this space have been driven by the IETF from a standardization perspective. As explained in Section 7.3.2, the IETF produces specifications called Request for Comments (RFC) documents[30].

6LoWPAN (IPv6 Over Low Power Wireless Personal Area Networks) was developed initially by the 6LoWPAN WG of the IETF as a mechanism to transport IPv6 over IEEE 802.15.4-2003 networks. Specifically, methods to handle fragmentation, reassembly, and Header Compression were the primary objectives. This is due to the large size of IPv6 packets (1280 octets) and the limited space in the protocol data unit (81 octets in the worst case, after security) as a result of the maximum physical packet size of 127 octets, respectively, specified by IEEE 802.15.4-2003. The WG also developed methods to handle address autoconfiguration, the hooks for mesh networking, and network management. We refer the interested reader to the following IETF (open access) RFCs for detailed information: RFC 4919 (Problem statement), RFC 6282 (Header Compression), and RFC 6775 (Neighbor Discovery).

RPL (IPv6 Routing Protocol for Low Power and Lossy Networks) was developed by the IETF Routing over Low Power and Lossy Networks (RoLL) WG. They defined Low Power Lossy Networks

---

[30]The general reference for an IETF RFC with a serial number "XXXX" is https://doi.org/10.17487/rfcXXXX.

as those typically characterized by high data loss rates, low data rates, and general instability. No specific physical or Medium Access Control technologies were specified, but typical links considered include PLC, IEEE 802.15.4, and low-power Wi-Fi. The logic behind the development of the protocol was founded on the traffic flow characteristics of such networks, where typical use cases involve the collection of data from many (for example) sensing points, nodes towards a sink, or alternatively, flooding information from a sink to many nodes in the network. Thus, the well-known concept of a Directed Acyclic Graph (DAG) structure was concentrated to a Destination Oriented DAG (DODAG) for the purposes of initial development. The group defined a new ICMPv6 message, with three possible types, specific for RPL networks. These include a DAG Information Object (DIO), that allows a node to discover an RPL instance, configuration parameters and parents, a DAG Information Solicitation (DIS) to allow requests for DIOs from RPL nodes, and a Destination Advertisement Object (DAO), used to propagate destination information upwards (i.e., towards the root) along the DODAG (specific RPL details are available in RFC 6550 and related RFCs). The Trickle Algorithm, standardized in 2011 (RFC 6206) and well known in the WSN community, is an important enabler of RPL message exchange.

CoAP (Constrained Application Protocol) was developed by the IETF Constrained RESTful Environments (CoRE) WG as a specialized web transfer protocol for use with severe computational and communication constraints typically characteristic of IoT applications. Essentially, CoAP elaborates a simple request/response interaction model between application endpoints (e.g., from an IoT Application to an IoT Device). REST is essentially a simplification of the ubiquitous HTTP and thus allows for simple integration between them. This is especially powerful for integrating typical Internet computing applications with constrained devices.

For more information about the IoT-related IETF efforts the interested reader can refer to Section 7.3.2, which outlines the IETF WGs, the main protocols and IETF RFCs, and the relationships between the IETF results.

Having covered devices and networking for IoT solutions, we turn our attention to Data Management, which is a core enabling ICT function within IoT systems.

5.3 MACHINE INTELLIGENCE
5.3.1 THE ROLE OF MACHINE INTELLIGENCE IN IOT

As already mentioned, the foundation for any IoT solution is the IoT device. IoT devices are capable of interacting with the physical environment through their sensors and actuators. Sensors produce data about physical assets, processes involving the assets as well as capturing other contextual data of interest. Actuators on the other hand consume commands with the capability to control or affect the physical assets, e.g., opening a valve or controlling the heating element in a water boiler. The means to put the IoT data and control commands to meaningful work in IoT solutions is Machine Intelligence (MI), it is the capability that makes environments and physical assets appear to be smart.

Data on its own has limited value, but data is the raw material used to extract richer information and insights, all together so it can be visualized and exposed for human understanding, or turned into actions that automate processes or the control of an asset. In order to reap the benefits of data, the data first needs to be produced, and then the right strategy and technologies for making sense and business value of the data are needed. Even if data is available, many times it is in fact not used [4].

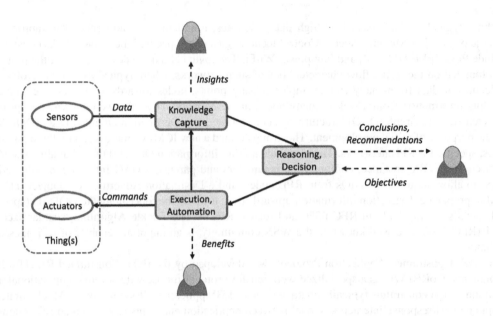

FIGURE 5.5

Applying MI to IoT.

The process can be simple involving a single data type and a single control command, e.g., based on the room temperature automatically adjusting a heater. The process can also be rich and complex and be multimodal in terms of different types of data and control commands, e.g., monitoring and controlling data about nutrition, soil acidity and moisture, irrigation, and ventilation and heating in a smart indoor farming scenario.

MI is the set of capabilities that turn data into information, knowledge, and actionable insights. MI is also capable of reasoning and making recommendations and decisions, as well as taking actions – it is hence capable of also controlling the physical environment. MI can also capture knowledge about the environment, i.e., information about the asset such as its state and behavior, and via inferencing creating new knowledge. In many scenarios, it is useful to have a proper digital representation of the physical reality, as pioneered in the SENSEI project [36], which today is popularly referred to as the Digital Twin. A simplistic flow view of MI applied to IoT is provided in Figure 5.5.

As can be intuitively understood, depending on the particular IoT application and usage scenario at hand, there is a difference in how data is used and consumed, such as extracting insights, capturing knowledge or control loops of sensing and actuation. Consequently, depending on the IoT application, different MI technologies need to be considered. Four examples follow to illustrate the diversity in needs, and more concrete use cases can be found in Part 3.

- **Massive monitoring**: Consider the monitoring of the climate or the environment. Example sensor data include temperature, humidity, air quality (gases and particles), and solar radiation. Data is typically collected on an hourly or daily basis and from a very large number of sensors that are geographically spread out even on a global level. The interests lie in understanding seasonal or yearly variations and longer-term trends, to detect abnormalities (e.g., harmful pollution) or even to make predictions about the future of climate change. MI provides the means to do the trend analysis, the foresight, and to understand when the situation is abnormal by analyzing the vast amount of data that has been collected. Key MI features include anomaly detection, clustering of data, and predictive analytics.

- **Asset management**: An example of managing an asset is building automation. The asset in question is the entire building and the purpose is to optimize its operation, e.g., energy and resource consumption, and the indoor climate, as well as control for safety and security. Sensor data of relevance include indoor and outdoor temperature, air quality, electricity and water consumption, and motion and smoke detectors. To optimize the resource consumption and indoor climate, different appliances are used such as heating, ventilation, and air conditioning (HVAC), water boilers, and lighting, and safety and security can be supported by smart locks, alarm bells, and sprinkler systems. As can be seen, this is an example where there are different sensor modalities as well as different actuator capabilities, and the Building Automation System relies on control loops for different purposes and of different types. It could be simple alarm triggering or more advanced regulated control loops involving heating and ventilation. Key MI features include simple triggering or anomaly detection, programmatic control, and tuned regulators involving multimodal sensing and actuation with fairly complex yet deterministic algorithms.

- **Logistics**: Distribution and transportation of goods involves a number of actors, primarily suppliers, distributors, and consumers. Important objectives include appropriate and timely delivery and cost optimization. Cost optimization includes reducing the overall cost of distribution, e.g. fuel consumption in trucks, and stock levels. Further, goods need to be tracked and many times monitored in how they are handled in order to avoid or detect damages. Keeping product stock levels low ensures avoiding tying up capital, but at the same time poses a risk in meeting unexpected increasing product demand. As can be seen, logistics involve conflicting objectives that require optimization to ensure reaching the right overall Key Performance Indicators (KPIs). Further, considering multiple suppliers, consumers, and a distribution network in between consisting of transporters, warehouses and logistics centers require complex planning to optimize supply and demand, as well as continuous replanning in the event of unforeseen disturbances. Sensors involved can monitor the handling of goods, e.g., temperature and vibrations for perishable goods, and identification and localization, e.g., via GPS, of goods are other needed capabilities.

- **Robots**: A last example is the general use case of Robots. They can be industrial robots in manufacturing that are deterministic in their control, or they can be self-driving vehicles with a totally autonomous operation. For full autonomous operation, self-learning, or cognitive features, is key to ensure the proper and safe behavior and thereby adapting to the context in real-time.

Data and information can be produced and consumed within an organization to solve a particular isolated problem, operation or optimization. But in an increasing number of IoT scenarios and applications, data can be used from external sources to augment the application, for instance by incorporating weather forecast data in smart agriculture. IoT applications that also involve multiple stakeholders or actors, within a specific organization or between organizations and companies, also require data to be shared and exchanged. The logistics example above is typically such a scenario. IoT data and information thus become a shared and sometimes tradeable asset. As a result, emerging are data marketplaces that can be industry-specific, but also in integrated environments like Smart Cities; see Chapter 3 and

Part 3. Open Data is also driven in the public sector with the Public Sector Information (PSI) legislation [37] in the European Union's strive towards a Digital Single Market as one example, which is very applicable to IoT. Considerations on sharing data are described further below.

5.3.2 MACHINE INTELLIGENCE OVERVIEW

We use the term *Machine Intelligence* as primarily the combination of the methods, tools, and techniques from *Machine Learning* and *Artificial Intelligence* (AI). The purpose is to create data-driven, intelligent, and nonfragile systems for automation of tasks and processes and to enhance human capabilities. We also add *Control Theory* to our definition of MI as this is a key field for a set of IoT applications. The fields of MI, AI, and Control Theory are wide fields with some prominent techniques that are applicable to IoT; see Figure 5.6.

ML is a discipline in computer science aiming at giving computers the ability to learn without explicit programming. AI on the other hand is about software and algorithms that intend to model and behave as humans behave and think, i.e., mimicking how the human brain solves problems and executes tasks. MI is used in numerous different disciplines with commercial deployments ranging from financial systems, healthcare, medical science like cancer research, and computer vision. Other

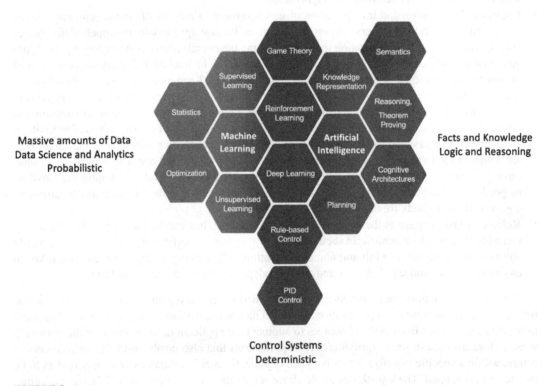

FIGURE 5.6

Machine Intelligence techniques.

examples include personal assistants like Apple Siri and Amazon Alexa, as well as computers playing games like IBM Watson playing chess and Google's DeepMind AlphaGo. IoT is just one more of the applied areas of MI. In IoT, the ambition is to have a system that behaves as the physical world, according to the laws of physics, and to make the *things* behave naturally as if they were intelligent entities but with a different reference to intelligence than the human one.

ML is based on different algorithms that learn from data and build models from the data that can be used to make predictions or decisions. ML can learn complex relationships in vast amounts of data. Different ML techniques exist, like Neural Networks and Bayesian Networks. ML can be based on different learning types like supervised learning where input and the corresponding output are provided as a training set, unsupervised learning where the task is to find structure from a given dataset, and reinforcement learning which is based on receiving feedback in the form of rewards.

AI on the other hand has a focus on providing abilities like knowledge, inferencing, reasoning, and planning. These are typical abilities attributed to the cognitive behavior of human beings. Finally, control theory is about the continuous operation of a dynamical system (mechanical, thermodynamic, etc.) with the objective of always operating at an optimal point with an appropriate stability. The field of MI is very wide, and the interested reader is referred to the comprehensive book by Russell and Norvig [38]. Different standards organizations are also addressing this field, including NIST with the work in the Big Data Working Group[31] and ISO on AI in ISO/IEC JTC 1/SC 42[32]. The work by NIST includes a reference architecture covering requirements, a stakeholder perspective, and a Conceptual Model covering the main functional domains.

Some examples of the type of tasks and problems that are commonly addressed using MI are as follows:

- **Descriptive Analytics**: This is about insights into the past. Descriptive analytics enable a better understanding of the performance or behavior of complex systems and what has actually happened, typically by creating KPIs.
  - By recording data about the behavior of an industrial machine, it can be possible to analyze the reasons for an unexpected failure of the machine.
- **Predictive Analytics**: Use current and historical facts to predict what will happen next. Examples include:
  - Forecast demand and supply in a power grid and train a model to predict how price affects electric usage to optimize the performance and minimize peaks in the electricity consumption.
  - Predictive Maintenance on electromechanical equipment in a power plant by modeling the relationship between equipment health characteristics measured by sensors and historic failures.
  - Understand how electricity and water consumption relates to regional demographics.
  - Model the effects of traffic lights on a city's road network based on data from cars and sensors in the city to minimize congestion.
- **Prescriptive Analytics**: Provide recommendations in operations to optimize a process or avoid undesired situations like failures. Prescriptive analytics builds on descriptive and predictive analytics.
  - Recommend when to take a process industry machine out of service to perform maintenance resulting in a minimal impact on the overall industrial operation.

---

[31] https://bigdatawg.nist.gov/
[32] https://www.iso.org/committee/6794475.html

- **Clustering**: Identification of groups with similar characteristics.
  - Perform segmentation of how people use and interact with machines, e.g., a consumer appliance, to find any behavioral patterns that could be used to make further enhancement of the machine. Data is in this example collected from a large number of deployed machines.
  - Mine time series data for recurring patterns that can be used in predictive analytics to detect, for example, fraud, machine failures, or traffic accidents.
- **Anomaly Detection**: From a large set of behavior data, identify outliers that represent an abnormal situation.
  - Detect fraud for smart meters by checking for anomalous electricity consumption compared to similar customers, or historic consumption for the electricity consumers.
- **Action Programming**: Defining a set of rules or parameters that automate tasks depending on different types of information.
  - Home automation. Turning on and off home lighting depending on time of day or people's presence, or capturing still images from a camera based on motion detection.
  - Central heating in a building. Depending on the desired indoor temperature regulate the water boiler power feed and the water circulation.
- **Task Planning**. Task planning takes a system start state and a desired end state, and together with a possible set of actions generate a plan how to reach the end state given a set of objectives or constraints.
  - Planning of how goods shall be delivered from a set of suppliers to a set of customers through a network of distributors consisting of trucks and logistics centers optimizing timely delivery and reducing fuel costs.
- **Knowledge representation/Digital Twin**: Modeling and representing the physical state and behavior of an asset.
  - Maintaining a real-time information model of an office building capturing different physical and contextual properties of the different floors, offices, conference rooms, facilities, appliances, and people whereabouts.

5.3.3 CONSIDERATIONS WHEN USING MI FOR IOT DATA

There are considerations when working with IoT data. IoT solutions can use data from any kind of source, e.g., dynamic data such as individual sensors, mobile device sensors, and social networks, as well as more static data that are relevant for analysis, e.g., Geographic Information System (GIS) data, public city data, or national statistics. How IoT data needs to be captured, processed, stored, managed, and consumed varies, and the data need the proper characterization that, again, is dependent on the type of IoT application at hand. Today there is a plethora of open source technologies, that enable effective data processing and are suitable for IoT applications [39]. The topic of IoT Data Management specifically is discussed in Section 5.5.

A traditional way to generally characterize "Big Data" is to talk about the four "Vs": Volume, Velocity, Variety, and Veracity. IoT data can to a certain extent be qualified by this characterization.

- **Volume**: To be able to create good analytical models it is no longer enough to analyze the data once and then discard it. Creating a valid model often requires a longer period of historic data. This is required in order to perform, e.g., descriptive or predictive analytics. Also, the number of IoT devices that generate data in a deployed solution matters.

- **Velocity**: Some IoT devices simply report a reading or an event notification of interest, but other IoT devices can generate continuous streams of data. Velocity is about analyzing streaming data, for instance real-time operational data from a farm of wind turbines. In some instances, the value of IoT data is strongly related to how fresh it is to be able to provide the best actionable intelligence. Example techniques include event stream processing and Complex Event Processing (CEP).
- **Variety**: Given the multitude of data sources of relevance for IoT, it is apparent that the variety will be very high. Data can also be unstructured, like a social network feed discussing perceived air pollution, or be structured and follow a defined data model. Data can be of different data formats or syntax (e.g., XML or JSON) and can also be semantically annotated, i.e., defining the meaning of the data, and enhanced with metadata (e.g., timestamp, location, source). Data can also be modeled using ontologies and semantic web technologies[33], e.g., using RDF and OWL [40]. Variety requires the appropriate considerations on how to process and integrate data to ensure it can be used in the same model or context.
- **Veracity**: It is imperative that we can trust the data that is used. There are many pitfalls along the way, such as erroneous timestamps, untimely delivery of data, nonadherence to standards, formats with ambiguous or missing semantics, and wrongly calibrated sensors, as well as missing data. This requires rules that can handle these cases as well as fault-tolerant algorithms that, for example, can detect outliers (anomalies). Also, the quality of the data needs to be considered – the accuracy of the data might be sufficient for one application example but not for another. Furthermore, as data can be coming from external sources, data provenance becomes important to ensure that the data is trustworthy and that liability can be enforced if needed.

But IoT also transcends the "Big Data" characterization. For some usages, IoT data is consumed as it is produced, for instance when a sensor threshold is reached and an alarm is triggered. In this case, there is no need to store the IoT event for future purposes, and the event itself can be business-critical and is hence not a subject for a statistical process generating a higher valued insight – the event itself is the single data point of value. As can also be seen, one could add one more "V" to the list above – Volatility:

- **Volatility**: IoT data itself can be nonpersistent. Volatility can be viewed as another perspective on the volume and velocity characteristics. IoT events can be consumed upon generation like in the automation scenario. Data relevance can also be degrading over time (temperature reading) thus removing the need for storing all data. In a number of cases, the data itself can be discarded, but the aggregate insights from the data kept.

In light of the above considerations, it also becomes obvious that it is important to understand and have a strategy for *where* the data is being processed – "does one bring data to the processing or the processing to the data?", i.e., whether data should be processed close to where it is generated (at the *thing*) or close to where overall enterprise objectives matter (at the "cloud"). This is a combined consideration of the volume, velocity, variety, veracity, and volatility aspects. The role and importance of distributed computing at the Edge, Fog, and Cloud is further discussed in Section 5.4.

Last but not least are the consequences of the need for user privacy. This means that data can be subject to anonymization both in terms of removing the user identity as well as the uniqueness of

---

[33] https://www.w3.org/standards/semanticweb/

user data. This could limit the possibility of cross-referencing different data sources, but might not be enough. Privacy is not only a concern for persons, but also for enterprises as insights to enterprise IoT data can reveal strategic information about the corporation and its business processes.

5.3.4 A FRAMEWORK OF MI FOR IOT

As was discussed and illustrated by the previous IoT application examples, different types of use cases require the employment of different MI technologies. By studying the variety of use case types and their different needs to process IoT data, extract insights and knowledge, perform reasoning, and automate control processes, one can deduce an MI reference framework applied to IoT [41] similar to the reasoning on a reference architecture in Section 4.1. It builds on the discussion above and provides a uniform functional perspective. Such a framework is conceptually illustrated in Figure 5.7 and is briefly described in the following. The presented framework should not be seen as a reference architecture but a conceptual or functional reference framework. Its focus is on the end-user stakeholder perspective; see Chapter 8.

Apart from using MI to create business value, a secondary consideration for such a framework is its Lifecycle Management. This includes operational aspects such as determining what logic to deploy where in a distributed execution infrastructure and how to ensure robustness. Furthermore, an MI

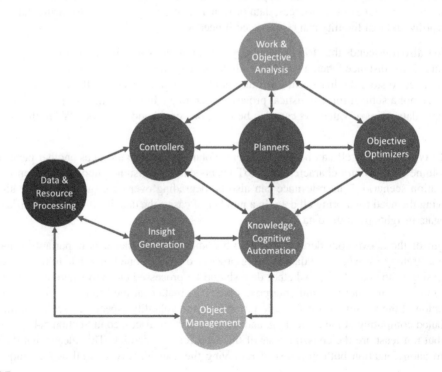

FIGURE 5.7

A framework of MI for IoT applications.

framework can also be adaptive and cognitive to handle changing operational conditions, like adjusting controller parameters or continuously doing model training.

The MI framework provides the previously discussed functionality needed to realize the variety of use case applications. An objective of the framework is to partition functionality into application-independent building blocks that can be implemented as service-oriented software components, e.g., realized as microservices [42]. The main functional domains include Data and Resource Processing, Insight Generation, Knowledge and Cognitive Automation, Controllers, Planners, Objective Optimizers, Work and Objective Analysis, and lastly Object Management. These functional domains are described in the following paragraphs.

Data and Resource Processing. By data and resources we mean sensor data, actuator services, and their representations. Sensor data includes individual data items and events as well as streaming data with varying speed, volume, and dynamicity. Resources are abstractions of sensors and actuators in the system, also allowing a dynamic mapping between the resource abstraction and the real resource; see Chapter 8. Data collection, management, filtering, curation, and preparation are crucial in IoT and are further discussed in Section 5.5.

Insight Generation. Insight generation is a key capability in a number of IoT use cases, like Predictive Maintenance. This includes clustering, descriptive, and predictive analytics, as discussed above. Generally, depending on the IoT use case, different foresight timeframes apply. For example, trajectory forecasting of moving objects can be real-time, whereas machine degradation is more long-term. Forecasting can be data-driven or model-driven depending on the problem requirements. Model training is a necessity and can be based on training sets or via reinforcement learning. Typical forecasting models can for instance be statistical or neural networks-based, Bayesian or non-Bayesian [38]. Sensor fusion is another technique. In general, fusion concerns combining data and information from diverse sources so that the resulting information is more accurate than if one had relied on a single source. An example is the localization of an object that can rely on a combination of Ultra-Wide Band (UWB) transponders and contextual information sensed by the object itself and when fused provide a higher degree of location accuracy.

Knowledge and Cognitive Automation. Knowledge management involves representing, modeling, structuring, and sharing knowledge about a physical asset or infrastructure. Knowledge is the collected set of data, inferred knowledge, and insights. Knowledge is generally of two types: declarative knowledge (also referred to as propositional knowledge) and procedural knowledge (also referred to as imperative knowledge). Declarative knowledge describes what an entity is and how it is structured and formally expressed using ontologies. Procedural knowledge describes how an entity behaves, for example, in response to stimuli; and the formal description format is typically via state machines. Knowledge is described by a set of ontologies. Ontologies in IoT are not only for the actual real-world model of the asset and expert knowledge but also knowledge about system and application objectives, such as KPIs, a work order, task plans, and constraints of the IoT system itself. The real-world model is typically captured by a hierarchical or graph structure. Techniques are also explored based on how cognition is performed in the mind of humans and other animals, leading to concepts like cognitive automation or cognitive architectures. As IoT solutions many times can go across domains and involve different actors, semantic interoperability of data, information, and knowledge are essential for achieving many business applications. Semantic interoperability enables information to be understood by systems and across domains without needing manual information interpretations [43].

Controllers. Control is a core automation point in any IoT system involving actuators. Control software commands the desired asset operation. A common characteristic to all controllers is the deterministic behavior of controlling operations based on input from an a priori desired operational behavior. The use of different controller types is based on functional and nonfunctional characteristics meeting application needs. Whereas many control systems in robotics and other real-world continuous and industrial systems use proportional, integral, and derivative (PID) controls, other IoT use cases, such as home automation, often use rule-based systems for event-driven control. A PID controller is a control loop feedback mechanism using a mathematical function that takes the deviation between the desired state and the measure state as input for control. Proportional control means that proportional feedback of the deviation is provided to determine the control value. A derivative part of the deviation dampens the error. An integral part of the deviation provides errors to be removed over time. Examples include inverse kinematics for robot control and temperature control of a fluid system. This requires knowledge about the physical behavior and properties of the physical asset of interest. Rule-based controllers are based on a set of predefined rules that are trigger–action pairs, where a trigger is a condition and an action is a predefined workflow typically containing commands to the devices or related services – for example, following the simple logic of "if this, then that". In sufficiently complex, dynamic, and non-deterministic situations one can enhance the usability and maintainability of both PID and rule-based control systems by making them use task planning technologies to help infer the actions to be taken.

Task Planning. Task planning can be defined as the process of generating a sequence of actions with certain objectives. Planning can be applied to a variety of problems such as route planning of autonomous vehicles, optimization of logistic flows, and automation of field personnel. The planning problem is normally represented by three key elements – states, actions, and goals. The state identifies the model of the system, actions represent different operations that affect the system's state, and goals are states to achieve or maintain. Deriving the task plan is to take the current state, the desired state, and the possible actions and from these generate a plan as a sequence of possible or proposed actions. A plan can also be a partially ordered list of tasks. One example to perform task planning is using AI planners where the world and the problem are modeled using a Planning Domain Definition Language (PDDL).

Objective Optimizers. Automating complex system operations by leveraging data-driven strategies designed to analyze alternatives under multiple conflicting views or KPIs is challenging. First, KPI evaluations are not always reliable and might be subject to changes over time; second, the costs incurred in adapting solutions under operation must be accounted for. In such cases, it is difficult to track how the underlying trade-offs (such as return vs. risk or throughput vs. cost) will evolve over time, and decision making preferences are hard to elicit and represent computationally. In the absence of clear preferences and priorities over the KPIs, general problem solving strategies and architectures must be designed for automating general data-driven Multiobjective Optimization (MOO) [44] systems under uncertainty. MOO can play a key role in applications where conflict resolution is expected. For instance, in supply chain control applications, the proposed system can monitor the profitability for the whole chain as well as the overall product shortage risk. Those two KPIs are clearly in conflict as optimization at an extreme for one results in a risk for the other. A key difference between task planning and optimization is that in the latter does not assume that desired goal states will be input by the system stakeholders. This stems from the fact that it can be impossible for humans to cope with the underlying complexity of explicitly specifying goal states while simultaneously fulfilling all service-level objectives. In such cases, it is possible to leverage simulation-based MOO to automatically explore the space of all candidate goal states that not only fulfill all service-level objectives but actually surpass them and deliver outstanding performance.

Work and Objective Analysis. The end-user's interests in the system can be specified as a set of high-level, quantifiable performance metrics by Service Level Objectives (SLOs) and work orders. SLOs are translated into KPIs, which are deemed critical for verifying service execution and detecting deviations from SLOs. KPIs can further be broken down to needed insights and, together with workflow orders, the intentions or actions of the system. The insights and actions can then be used to define the needed sensor data and actuator controls. For instance, in a logistics use case, a workflow order can request that a number of products be delivered to a certain subset of retailers within a specified deadline to keep the shortage risk under the agreed levels. The KPIs and workflow orders encapsulate information that allows the extraction of inputs to task planning, controllers, and MOO, which also includes the necessary information from the insight generation functional domain. For task planning, the extracted inputs should correspond to goal states that can be used to compute an appropriate plan. For controllers, workflow orders might specify new set levels of parameters or rules. For multiobjective optimizers, workflow orders should specify a set of KPIs to be balanced by automated trade-off analysis to comply to overall service objectives as well as mitigating conflicts.

Object Management. Object management involves identifying, localizing and cataloging physical assets that are handled by the IoT system. Identification of objects can be based on tags, e.g. QR codes or RFID. The purpose is to provide unique identification and naming of objects. A resolution infrastructure can then be used to find further information about the object, and an example is Electronic Product Code Information Services (EPCIS)[34]. Object localization needs to be tailored to the IoT needs, e.g. indoor vs. outdoor environments, as well as the accuracy of localization. Example techniques include Bluetooth beacons, UWB ranging using transponders, WiFi or cellular triangulation, and GPS.

5.3.5 INDUSTRIAL INTERNET ANALYTICS FRAMEWORK

The Industrial Internet Consortium[35] (IIC) has at the time of writing recently published an Industrial Internet of Things Analytics Framework (IIAF) [45], as part of their Industrial Internet Reference Architecture (IIRA) series of publications; see Section 8.9.1 for more details on the IIRA. Industrial IoT (IIoT) targets the integration of industrial assets and machines to enterprise information systems, business processes, and people who operate and use them. The IIAF thus puts a focus on applying analytics to data and control related to Operational Technology (OT). The IIAF provides assistance in how to develop and deploy analytics solutions in an industrial IoT context and provides guidance into business, usage, and technology perspectives according to the IIRA viewpoints. As such, the IIAF is the first work of its kind by an industrial alliance to start the modeling of analytics, AI, and other MI technologies applied to the context of the IoT.

In the work by IIC, analytics is mapped onto the IIRA consisting of the following main functional domains [45] (see Figure 5.8):

---

[34]https://www.gs1.org/standards/epcis/

[35]https://iiconsortium.org

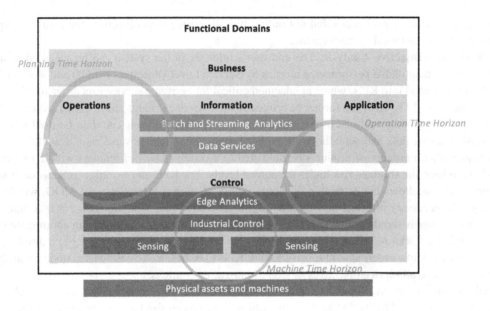

FIGURE 5.8

Analytics mapped to the IIC reference architecture (adapted from IIC).

- Control: This domain interfaces the physical assets via sensing and actuation and provides the necessary underlying communications and execution means. It is the collection of functions performed by the industrial assets and associated control systems.
- Information: The information domain performs the collection, transformation, and analysis of data to reach a higher level of intelligence of the system.
- Application: The application domain provides use case-specific logic, rules, and models to deliver system-wide optimization of operations and relies on intelligence from the information domain.
- Business: This domain integrates information across applications and business systems in order to reach the desired business objectives.
- Operations: This last domain ensures the continued operations of assets and the associated control systems.

As can be seen, applying industrial analytics is considered in three main time horizons. The control domain at the edge provides real-time operational insights in the "machine time horizon" which involves sensing–actuation control loops in milliseconds or less which can be typical in factory robot control. Machine fault detection or diagnostics is based on deriving insights from machine data to discover anomalies or understand any changing behavior of the operations. This takes place in the "operation time horizon" which requires responses in seconds or more. Finally, the "planning time horizon" focuses more on business planning, operations planning, and scheduling and also other more long-term engineering processes. This time horizon requires responses in days or more.

The IIAF provides a high level overview of industrial analytics design considerations, deployment models, and different types of analytics including the relation to Big Data and AI. As such, the IIAF touches on what is outlined in this section as well as in Sections 5.4 and 5.5. As can be understood, the use of the term "analytics" in the work by IIC has a wider definition, similar to the wide definition of MI we use.

5.3.6 CONCLUSIONS

Employing the data- and information- and knowledge-centric processing and control capabilities from the combination of ML, AI, and Control Theory to IoT sensor data and actuators is key to achieve the true benefits of any IoT solution. To generate insights, do forecasting, and automate processes involving physical assets require the right considerations in terms of needed functional capabilities combined with associated required characteristics of the solution, all depending on the particular application in focus. The underlying infrastructure support in processing data is provided by appropriate techniques and considerations pertaining to distributed computing and data management as covered in Sections 5.4 and 5.5, respectively.

5.4 DISTRIBUTED CLOUD AND EDGE COMPUTING
5.4.1 A NEW SOFTWARE DELIVERY MODEL

As discussed in Chapter 2, there is a general trend away from enterprises investing in, hosting, and operating dedicated compute facilities to run their enterprise applications. Instead, applications are increasingly hosted and executed in remotely located large Data Centers, and two examples include Microsoft Office 365 for office applications and Salesforce for Customer Relations Management (CRM) solutions. This trend is clearly also seen on the consumer side where web-based applications are abundant, such as those offered by Google and Facebook. Connected consumer electronics like wearables, web cameras, home appliances, and personal assistants are also web-enabled and rely on applications that execute remotely, e.g., Garmin collecting, analyzing, and visualizing health statistics from wearables, Apple Home providing programmable home automation, and Amazon Alexa being able to analyze spoken questions and provide answers. All this is enabled by the *cloud* paradigm.

Cloud builds on the notion of sharing computer resources over a network and allows applications to be distributed, both in terms of execution and in terms of reaching users; see Figure 5.9. The attractiveness of a cloud is the ability to enable economy of scale and cost reduction: resource utilization can be higher, and a large common pool of resources allows for elastic scaling and *pay per use* rather than upfront investments. Further, an enterprise can focus on the business without investing in and operating an in-house software infrastructure. The underlying technology evolution that enables cloud computing is virtualization of both hardware and software resources thanks to commoditization of processors and storage and new software engineering practices.

The traditional cloud model relies on applications that essentially execute in Data Centers and are accessed by the end users through client devices. However, there are increasing needs and requirements to distribute applications, both from a computing and from a storage perspective. Clouds can be distributed, i.e., involving multiple Data Centers that are networked and federated, for instance across country borders. Application execution can also be distributed *in between* the Data Center and the client devices, including partly executing on the client device itself. This is commonly referred to as Fog and Edge computing.

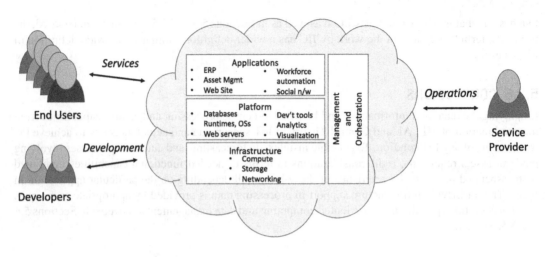

FIGURE 5.9

Conceptual overview of cloud computing.

Cloud also represents a new way to deliver software outcomes – instead of shipping software bundles for local installation, what is sold is *Software as a service* (SaaS). This is a new means to deliver value to end-users. Cloud can hence be said to enable new business models. Cloud is a prominent example of how a technology paradigm can drive business model innovation, enable cost efficiency, and support new applications in an effective way.

5.4.2 CLOUD FUNDAMENTALS

Cloud computing is a model for enabling ubiquitous, on-demand network access to a shared pool of configurable resources such as networks, servers, storage, applications, and associated services. The resources can be provisioned, configured, and made rapidly available to users with minimal management effort or cloud service provider interaction. As mentioned, cloud relies on virtualization, which simply put means that the resources of a server, or physical machine, can be divided up by a layer of virtualization software to create several separate Virtual Machines. These Virtual Machines can then be used by applications as if they were different physical machines. Other related examples of how applications can benefit from virtualization can be found in the networking domain, namely with Software Defined Networking (SDN) and Network Function Virtualization (NFV).

In essence, SDN [46] is about separating network control functions (control plane) from packet or payload forwarding (user plane) using virtualization. The objective of SDN is to place the control functions centrally whereas packet forward remains distributed in the network infrastructure. The benefit is the ability to provide network control programmability that is decoupled from the underlying infrastructure.

NFV [47] on the other hand makes use of virtualization to virtualize various network node functions and turn them into building blocks, so called Virtual Network Functions (VNFs), that can be combined in various ways to provide different communication services. A VNF can consist of one or more Virtual

Machines running different software on top of a cloud infrastructure. The benefit is that one can avoid to have specialized hardware nodes for the different network functions which will lead to reduced operating and capital expenses as well as less time necessary to market for new services. Both SDN and NFV can be seen as applications of the cloud computing paradigm.

NIST [48] provides a definition of cloud computing that covers the essential characteristics, service models, and deployment models of cloud computing. This definition serves as a baseline for describing the fundamentals but also benefits for updating following evolving concepts since the definition was done. NIST identifies five characteristics that capture the essence of cloud computing which are adapted in the following:

- On-Demand Self-Service. A cloud user can unilaterally provision computing capabilities, such as server time and network storage, as needed, or automatically, without requiring human interaction with each cloud service provider.
- Ubiquitous Network Access. Capabilities are available over the network and accessed through standard mechanisms that promote use by heterogeneous thin or thick client platforms. These may be mobile phones, tablets, laptops, workstations and IoT devices.
- Resource Pooling. The cloud service provider's computing resources are pooled to serve multiple users using a multitenant model, with different physical and virtual resources dynamically assigned and reassigned according to user demand. There is a sense of location independence in that the user generally has no control or knowledge over the exact location of the provided resources, but may be able to specify location at a higher level of abstraction (e.g., country, state, or datacenter). Examples of resources include storage, processing, memory, and network bandwidth.
- Rapid Elasticity. Capabilities can be elastically provisioned and released, in some cases automatically, to scale rapidly outward and inward commensurate with demand. To the user, the capabilities available for provisioning often appear to be unlimited and can be appropriated in any quantity at any time.
- Measured Service. Cloud systems automatically control and optimize resource use by leveraging a metering capability, at some level of abstraction, appropriate to the type of service (e.g., storage, processing, bandwidth, and active user accounts). Resource usage can be monitored, controlled, and reported, providing transparency for both the provider and user of the utilized service.

Cloud computing can be offered by service providers in different service models following an increasing level of abstraction. This gives users a choice of different degrees of flexibility, control, or ease in the development and launch of applications, something that is important when considering a strategy for selecting the right models. The service models follow the service-oriented paradigm of "Everything as a Service" or XaaS. Four main service models in order of increasing levels of abstraction can be identified for cloud computing, and they are Infrastructure as a Service (IaaS), Platform as a Service (PaaS), Software as a Service (SaaS), and finally Function as a Service, which is closely related to the concept of serverless computing:

- IaaS: This model basically provides virtualization of hardware. The provider is responsible for raw compute an storage capabilities and offers Virtual Machines, containers, and other resources such as hypervisors (e.g., Xen, KVM) and storage as services to customers. Pools of hypervisors support the Virtual Machines and allow users to scale resource usage up and down in accordance with their computational requirements. Users are responsible for installing and maintaining an OS image and application software on the cloud infrastructure. The provider manages the underlying cloud infrastructure, while the customer has control as well as responsibility over OS, storage, deployed applications, and possibly some networking capabilities. In this model, the user is provided with flexibility and control at the expense of ease in developing and launching applications. Users are typically charged by the provider for the allocated and used resources.

- PaaS: This refers to cloud solutions that provide both a computing platform including OS and a solution stack as a service via the network. The customers themselves develop the necessary software using development tools, services, and APIs provided by the provider, who also provides the networks, the storage, and the other distribution services required. Again, the provider manages the underlying cloud infrastructure, while the customer has control over the deployed applications and possible settings for the application hosting environment. This model eases application development at the expense of flexibility as the user and application developer is constrained by the services provided by the platform provider.

- SaaS: This refers to complete application software that is provided to consumers on demand and as a service, typically via different thin clients and many times through a web browser or application-specific program interface. The end-user has limited ability to change anything beyond user-specific application configuration settings. The users do not manage the cloud infrastructure in any way. This is handled by the service provider. Examples include office and messaging applications, social networking applications, email, or CRM and ERP tools housed in the cloud.

- Function as a Service (FaaS): FaaS [49] is a recent development that is based on the concept of running code in the cloud without the need for provisioning or managing servers. The term serverless computing is also used for this concept, though it is a slight misnomer as server hardware and server processes are still required. In FaaS, there is no continuous process on a server that waits for an API call, but instead a specific event triggers the execution of a particular piece of code that implements the desired function. FaaS offers further cost efficiency, shorter start-up times, and the increased ability for an application to scale up or down. There are different platforms emerging offering FaaS; see [50].

These different service models can also be mapped to the different layers of Figure 5.9, where IaaS, PaaS, and SaaS correspond to the *Infrastructure*, *Platform*, and *Applications* layers, respectively, and where FaaS is also in the *Applications* layer.

As mentioned, cloud is commonly associated with publicly available cloud-based services provided by companies like Microsoft Azure and Amazon Web Services. However, there are several different deployment models for cloud, each having their own applicability considerations to cater for:

- Private Cloud. The cloud infrastructure is provisioned for exclusive use by a single organization comprising multiple consumers (e.g., business units). It may be owned, managed, and operated by the organization, a third-party, or some combination of them, and it may exist on or off premises.

- Community Cloud. The cloud infrastructure is provisioned for exclusive use by a specific community of consumers from organizations that have shared concerns (e.g., mission, security requirements, policy, and compliance considerations). It may be owned, managed, and operated by one or more of the organizations in the community, a third party, or some combination of them, and it may exist on or off premises.

- Public Cloud. The cloud infrastructure is provisioned for open use by the general public. It may be owned, managed, and operated by a business, academic, or government organization, or some combination thereof. It exists on the premises of the cloud provider.
- Hybrid Cloud. The cloud infrastructure is a composition of two or more distinct cloud infrastructures (private, community, or public) that remain unique entities, but are bound together by standardized or proprietary technology that enables data and application portability (e.g., cloud bursting for load balancing between clouds).

The cloud execution environment can be realized in several different ways. As mentioned, IaaS relies on VMs running a deployed OS of choice. A choice for PaaS is to provide OS-level virtualization which provides isolated user space instances referred to as containers, and a popular openly available example is to use Docker containers[36]. An application running in a container only has access to a subset of the underlying available resources, devices, data, etc. that are needed to run the containerized application. Typically, container applications provide different *microservices* [42]. Essentially, a microservice is an application-independent building block following the SOA paradigm where a specific end-user application is realized by the composition of diverse microservices. These services can be programmed in different programming languages and be executing on different platforms and in different cloud environments. They are typically linked together using REST interfaces in a network environment including even the Internet.

Orchestration of resources and software is needed on different levels of the cloud platform. Orchestration includes services for Lifecycle Management of the software, i.e., distribution services, deploying and managing software for instance based on available and needed resources, services ensuring reliable operation, fault recovery, and achieving scalability, all with the necessary security measures. Popular tools for OS and VM orchestration in an IaaS include OpenStack[37] and, for container orchestration in a PaaS, Kubernetes[38], which has support for Docker containers.

5.4.3 COMPUTING AT THE EDGE

As previously mentioned, many common office and consumer applications are well suited for implementation in large centralized Data Centers. But there is an increasing need to both build distributed cloud solutions as well as taking computing to the edge. Typical reasons can be legal, business-strategic, cost, or technical such as performance. One aspect is where data and information are stored, something of relevance, e.g., from a data protection perspective. Another aspect is where data is actually processed. It is for the latter popular to talk about "taking data to computing, or the computing to the data". In IoT, there are a number of reasons why it makes sense to distribute data processing and application logic closer to where IoT data is generated or where IoT control is actuated.

The intention of enabling distributed processing is not to move away from the Data Center approach, but rather to extend the cloud computing paradigm to the edge with the assumption that computation and storage can take place anywhere and not only at dedicated Data Center servers. From this point of view, a datacenter can rather be seen as a high-density node facility part of a totally

---

[36]https://www.docker.com

[37]https://www.openstack.org

[38]https://kubernetes.io

distributed system. It is important to point out that such a distributed system does not automatically imply that it is fully decentralized. Control can still be centralized but processing and storage can be distributed. What defines the *edge* can be discussed, but NIST provides a useful reasoning toward a terminology [51], including placing cloud, fog, and edge in that relative order from the Data Center outwards to the end devices. Our adaptation of the terminology has the following tiers, but it shall be noted that the definition of fog and edge is still under discussion at the time of writing. NIST states that fog works with the cloud, whereas edge is defined by the exclusion of cloud and fog, and that fog is hierarchical where edge tends to be limited to a small number of peripheral layers.

- Cloud. This is typically a large Data Center with global applicability and reach.
- Distributed cloud. This is typically distributed but still large Data Centers that cover regional or national needs. The Data Centers in a distributed cloud solution can be federated. The need for distributed cloud can be purely technical, as load balancing, but also a national strive to ensure privacy of citizen data.
- Fog. Fog computing places the cloud computing paradigm of virtual resources *between* the smart end-device and traditional cloud or Data Centers. Fog is intended to support vertically isolated latency-sensitive applications by providing ubiquitous, scalable, layered, federated, and distributed computing, storage, and network connectivity.
- Edge. Edge is the network layer encompassing the smart end-devices and their users to provide, for example, local computing capability on a sensor, metering or other device that is network-accessible.

As shown in Figure 5.10 this is a tiered approach to cloud and distributed computing.

The main motivations for why distributed processing is of particular relevance for IoT can be viewed from four main perspectives, which in summary are the following.

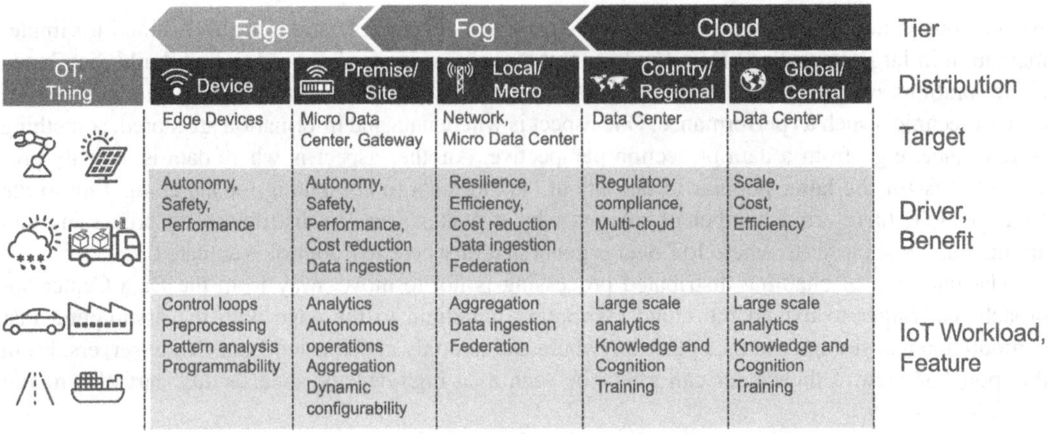

FIGURE 5.10

Tiered distributed computing for IoT.

- Physical assets and infrastructures, Things and Operational Technology (OT). IoT data is generated at the assets of relevance, and typical needs to process data at its origin include fully autonomous and local operation, for instance in manufacturing plants or for self-driving vehicles. Another reason to have data processing close to the source is the process of curation and preparation of data (see also Section 5.5 for a further discussion on these topics), including data normalization and annotation of semantics and metadata, e.g., origin and timestamps to ensure provenance. IoT resources may also require virtualization and resource management close to the assets.
- Performance. Processes can be time-sensitive and also mission-critical. Some applications can require a very low "time-to-insight" or "time-to-action". Resilience and uptime can be increased by removing round-trips to Data Centers. Much faster control loops can be achieved.
- Cost reduction. By aggregating data and using local processing, data volumes can be reduced and the level of detail can be kept at an appropriate level. The use of network, processing, and storage resources can be optimized and the cost hence reduced.
- Legal, privacy, security, safety, data. Compliance and regulations can provide constraints to how data is transferred and stored with respect to country borders. Transfer of data and storage of data in Data Centers by definition implies an exposed security threat. Data may also need to be ingested to the system in a hierarchy and involve data federation across domains, i.e., data is not only processed "in the endpoints". Safety can require autonomous operation to be legally compliant, for instance industrial robots related to safety of humans entering manufacturing cells.

Fog is hence by definition the tier between Data Centers and smarter end-devices, thus at the level of the WAN infrastructure and the Internet itself. As already mentioned, networks are under transformation to become virtualized through NFV and SDN that separates network control from the infrastructure. This virtualization can then enable the transformed network to become a fog fabric also catering for application workloads, such as distributed IoT applications. An example of the synergy between the same underlying technologies applied to Data Centers and to network infrastructure virtualization is ETSI Multi-access Edge Computing (MEC)[39] [52]. ETSI has released a set of specifications on MEC that are publicly available. The Open Fog Consortium[40] is another prominent consortium working on defining an open and interoperable fog computing architecture [53].

The practices of cloud computing, e.g., virtualization techniques and orchestration, also form the basis for computing toward the edge including into the device domain. This is based on the assumption that the OS used is a variant of Linux. However, as explained in Section 5.1, very basic devices using microcontrollers are constrained and cannot run an OS like Linux. For these devices to be part of an extended "compute fabric" for distributed applications, one has to resort to other methods. One example is to use an actor model programming, which is similar to serverless computing or FaaS. The actor model is also interesting for IoT in terms of its data centric processing philosophy. An example realization of serverless computing fitting constrained devices is provided by the Calvin platform [54].

---

[39]http://www.etsi.org/technologies-clusters/technologies/multi-access-edge-computing
[40]https://www.openfogconsortium.org

5.4.4 CONSIDERATIONS AND CONCLUSIONS

The cloud paradigm is equally applicable to IoT solutions and to many other applied areas. The benefits of sharing resources from a pool that allow elasticity in scaling and pay per use are appealing. Of the different service models, the IoT PaaS is of particular interest. The IaaS service model requires that the IoT solution is more or less built from scratch, including installing and maintaining the OS as well as application software. The SaaS service model on the other hand comes with predefined applications more or less off the shelf, thus allowing very little flexibility beyond application configuration settings.

In the IoT PaaS setup, one does not have to care about deploying and maintaining the compute platform, but is rather offered a solution stack as a service. This approach offers an attractive balance of prepackaged functionality and flexibility in realizing applications. The solution stack typically comes with a number of different functional capabilities. It is customary that an IoT PaaS solution stack includes different tools for device integration, MI, security, and exposure:

- Device protocol adapters. IoT devices generally use different protocols. It is common that IoT PaaS offerings contain different protocol adapters to connect to devices.
- Device Management. Example functions include the provisioning and configuration of devices and firmware upgrades.
- Data Management. This includes everything from data ingest, data curation and preparation, and data pipelining to data storage.
- Connectivity management. This manages the underlying networks that can integrate different connectivity technologies and from different network service providers even on a global scale.
- Security and identity management. Security features include authentication of both users and devices as well as authorization of access to different resources such as IoT data, DM, and other system features. Identity management is required to secure different and independent parts of the system, e.g., network access and enterprise data.
- Analytics and control. Analytics to understand data is key and different tools for processing and extracting insights can be part of a PaaS solution. This could also include means for programmability of actions based on the insights, e.g., setting triggers for alarms or sensing–actuation control loops.
- Visualization. Powerful visualization of complex data is usually important for domain experts to draw conclusions and generate human-based insights from business processes.
- Exposure and integration. This aims at providing the right APIs for integration of IoT capabilities and information into business systems like ERP and CRM and for providing the right level of abstraction to build different IoT applications.

The choice of public, private or hybrid deployment models is largely a trade-off between cost, trust, confidentiality of operations, and the level of desired control and independency. Where to place processing and storage in terms of Data Center vs. the edge is in essence a combined consideration of performance, cost, and autonomy of applications. The choice of global Data Center vs. regional or national Data Centers is mainly a consideration of legal aspects around treatment of customer and citizen data.

Finally, as mentioned, cloud computing is also a good example of an enabler of new business models including how cloud can help lower the barrier for different organizations to collaborate and integrate across value chains. Cloud can also enable rapid creation and rollout of new applications including a continuous integration and continuous deployment workflow, thus meeting time to market expectations.

5.5 DATA MANAGEMENT
5.5.1 INTRODUCTION

Modern enterprises need to be agile and dynamically support multiple decision making processes taken at several levels. In order to achieve this, critical information needs to be available at the right point, in a timely manner, and in the right form [55]. All this info is the result of data being acquired increasingly by IoT interactions, which in conjunction with the processes involved, assist in better decision making.

Some of the key characteristics of IoT data include:

- **Big Data:** Huge amounts of data are generated, capturing detailed aspects of the physical processes as sensed/actuated by IoT devices.
- **Heterogeneous Data:** The data is produced by a huge variety of sensors and is itself highly heterogeneous, differing on sampling rate, quality of captured values, etc.
- **Real-World Data:** The overwhelming majority of the IoT data relates to real-world processes and may be dependent on the environment they interact with.
- **Real-Time Data:** IoT data can be generated in real-time and overwhelmingly edge/fog architectures exist to process them locally or communicate them to enterprise systems in a near real-time manner. The latter is of pivotal importance since many times their business value depends on the timely processing of the info they convey.
- **Temporal Data:** The overwhelming majority of IoT data is of temporal nature, measuring the environment over time.
- **Spatial Data:** Increasingly, the data generated by IoT interactions are not only captured by mobile devices, but also coupled to interactions in specific locations, and their assessment may dynamically vary depending on the location.
- **Polymorphic Data:** The data acquired and used by IoT processes may be complex and involve dimensions that can have different meanings depending on the semantics applied and the process they participate in, i.e., contextual.
- **Volatile Data:** IoT data may not be in batch loads, consistent, etc., but streamed in, short-lived, and with deviations that need to be investigated (e.g., due to a malfunctioning sensor or false device configuration, etc.)
- **Variable Value:** The value of the data may be derived from a single short-lived message, e.g., to convey an alarm, or in correlation with context info, etc. Often new value can be acquired by putting existing data into context and correlating them with other actions.
- **Proprietary Data:** Up to now, due to monolithic application development, a significant amount of IoT data is stored and captured in proprietary formats. However, increasingly due to the interactions with heterogeneous devices and stakeholders, open approaches for data storage and exchange are used.
- **Security and Privacy Data Aspects:** Due to the detailed capturing of interactions by IoT, analysis of the obtained data has a high risk of leaking private information and usage patterns, as well as compromising security. Ownership of data is an important aspect as well.

Billions of devices are expected to interact and generate data at exponential growth rates in the IoT era. Hence, Data Management is of critical importance as it sets the basis upon which any other processes can rely upon and operate. Several aspects of Data Management need to be addressed in order to fully take advantage of the IoT data and the insights they can unlock.

5.5.2 MANAGING IOT DATA FLOW

The data from the moment it is sensed (e.g., by a wireless sensor) up to the moment it reaches a backend system has been processed manifold (and often redundantly), either to adjust its representation (e.g., transformation) in order to be easily integrated by the diverse applications, or to perform computations on it (e.g., aggregation) in order to extract business value and associate it with respective business needs (e.g., business process affected, etc.).

A number of data processing points between the machine and, e.g., the enterprise that act on the datastream (or simply forward it) based on their end-application needs and existing context may exist [55]. IoT data processing is hence possible to take place anywhere between the Edge and the Cloud, which is illustrated in Figure 5.14 and discussed in Sections 5.4 and 5.6.

Dealing with IoT data (from generation to utilization in business) may be generally decomposed into several stages:

- Data generation.
- Data acquisition.
- Data validation.
- Data storage.
- Data processing.
- Data remanence.
- Data analysis.

Not all of the stages are necessary in every enterprise solution, and they may be used in a different order than the one followed below. Additionally, the degree of focus in each stage heavily depends on the actual usage requirements put upon the data as well as the available infrastructure.

5.5.2.1 Data generation

Data generation is the first stage during which data is generated actively or passively by a device or system or as a result of their interactions. The sampling of data generation depends on the device capabilities and constraints as well as potentially the application needs. Typically there exist default data generation parameters, which are further configurable in order to allow the system operation to be optimal with respect to operational costs, e.g., frequency of data collection vs. energy used in the case of WSNs. Not all data acquired may actually be communicated, as some of it may be assessed locally and subsequently discarded, while only the assessment result may be communicated upstream.

5.5.2.2 Data acquisition

Data acquisition deals with the collection of data (actively or passively) from the device or system or as a result of their interactions [56]. The data acquisition systems typically communicate with distributed devices over wired or wireless links to acquire the needed data and need to respect security, protocol, and application requirements. The nature of acquisition varies, e.g., it could be continuous monitoring, interval-poll, event-based, etc. The frequency of data acquisition overwhelmingly depends on, or is customized by, the application requirements (or their common denominator).

Data acquired at this stage (for nonclosed local control loops) may also differ from the data actually generated. In simple scenarios, due to customized filters deployed at the device, a fraction of the generated data (e.g., adhering to the time of interest or over a threshold) may be communicated [55].

Additionally, in more sophisticated scenarios, data aggregation and even on-device computation of the data may result in communication of KPIs of interest to the application, which are calculated based on a device's own intelligence and capabilities [56].

5.5.2.3 Data validation

Acquired data must be checked for correctness and meaningfulness within the specific operating context. The latter is typically performed based on rules, semantic annotations, or other logic. Data validation in the era of IoT, where the acquired data may not conform to expectations, is a must, as data may be intentionally or unintentionally corrupted during transmission or altered or may not make sense in the business context. As real-world processes depend on valid data to make business-relevant decisions, this is a key stage, which sometimes does not receive as much attention as it should.

Several known methods are deployed for consistency and data type checking; for example, imposed range limits on the acquired values, logic checks, uniqueness checks, and correct timestamping. In addition, semantics may play an increasingly important role here, as the same data may have different meanings in various operating contexts, and via semantics one can benefit while attempting to validate them. Another part of the validation may deal with fallback actions such as requesting the data again if checks fail, or attempts to "repair" partially failed data.

Failure to validate may result in security breaches. Tampered data fed to an application is a well-known security risk as its effects may lead to attacks on other services, privilege escalation, Denial of Service, database corruption, etc., as we have witnessed on the Internet over the last decades, but increasingly also in industrial settings [57]. As full utilization of this step may require significant computational resources, it may be adequately tackled at the network level (e.g., in the cloud), but may be challenging in direct IoT interactions, e.g., between two resource-constrained devices communicating directly with each other.

5.5.2.4 Data storage

Massive data generated by IoT interactions falls also in the "Big Data" domain. Machines generate an incredible amount of information that is captured and needs to be stored for further processing. As this is proving challenging due to the size of information, a balance between its business usage vs. storage needs to be considered; that is, only the fraction of the data relevant to a business need may be stored for future reference. This means, for instance, that in a specific scenario (typically for streaming data used to make a decision), once a decision is made, only the processed result could be stored while the original data could be discarded.

Although storage is getting cheaper with time, storing massive amounts of data may not always be meaningful. On the other side, if at a later point in time some of that data is needed because new research has revealed some business value, their lack may diminish the competitive advantage of an enterprise. Hence, one has to carefully consider not only the business value of such data in current processes, but also potentially other directions that may be followed in the future by the company as different assessments of the same data may provide other, hidden, competitive advantages in the future. Due to the massive amounts of IoT data, as well as their envisioned processing (e.g., searching), specialized technologies such as Massively Parallel Processing databases, distributed file systems, and cloud computing platforms are needed.

5.5.2.5 Data processing

Data processing enables working with the data that is either at rest (already stored) or is in motion (e.g., stream data). The scope of this processing is to operate on the data at a low level and "enhance" them for future needs. Typical examples include data adjustment during which it might be necessary to normalize data, introducing an estimate for a value that is missing, and reordering incoming data by adjusting timestamps. Similarly, aggregation of data or general calculation functions may be operated on two or more data streams and mathematical functions applied on their composition.

Another example is the transformation of incoming data; for example, a stream can be converted on the fly (e.g., temperature values are converted from 30°C to 86°F) or repackaged in another data model. Missing or invalid data needed for a specific time-slot may be predicted and used until, in a future interaction, the actual data comes into the system. This stage deals mostly with generic operations that can be applied with the aim to transform data to better match the requirements set (e.g., for ingestion, export to applications, legal requirements). This step takes advantage of low-level (such as database-stored procedures) functions that can operate on the data at massive levels with very low overhead, little network traffic, and few other limitations.

5.5.2.6 Data remanence

IoT data may reveal sensitive business aspects, and hence their Lifecycle Management should include not only the acquisition and usage, but also the end-of-life of data. Sometimes systems or services are decommissioned but no proper attention is paid to the data they are associated with. However, even if the data is erased or removed, residues may still remain in electronic media and may be easily recovered by third parties – often referred to as data remanence. Several techniques have been developed to deal with this, such as overwriting, degaussing, encryption, and physical destruction.

For IoT, points of interest are not only the databases where the IoT data is collected, but also the points of action, which generate the data, or the individual nodes in between, which may cache the data. At the current technology pace, those buffers (e.g., on-device) are expected to be less at risk since their limited size means that after a specific time has elapsed, new data will occupy that space; hence, the window of opportunity for third-party exploitation may be rather small. In addition, for large-scale infrastructures the cost of potentially acquiring "deleted" data may be large; hence, their hubs or collection endpoints, such as potentially low-cost databases, may be more at risk. Moreover, in light of the lack of cross-industry IoT policy-driven Data Management, it might be difficult not only to control how the IoT data is used, but also to revoke access to it and "delete" them from the Internet once shared. The latter poses an issue since today sharing of data is a double-edged sword: on the one hand sharing may create value, but on the other hand sharing implies loss of control with respect to future usage.

5.5.2.7 Data analysis

Data available in the repositories can be subjected to analysis with the aim of obtaining any hidden information and potentially using it for acquiring insights and supporting decision making processes. The analysis of data at this stage heavily depends on the domain and the context of the data. For instance, Business Intelligence tools process the data with a focus on the aggregation and KPI assessment. Data Mining focuses on discovering knowledge, typically in conjunction with predictive goals. Statistics can also be used on the data to assess them quantitatively (descriptive statistics), find their main characteristics (exploratory data analysis), confirm a specific hypothesis (confirmatory data analysis), and discover knowledge (Data Mining) and for ML. This stage is the basis for any sophisticated applications that take advantage of the information hidden directly or indirectly in the data and can be used, for example, for business insights. IoT has the potential to revolutionize modern businesses, and we analyzed some of these aspects more in Section 5.3, where we focus on data science and knowledge management.

5.5.3 IOT DATA CONSIDERATIONS

IoT Data Management is still at an early stage, especially when looking at its whole lifecycle from cradle to grave. The real paradigm shift the IoT data brings depends heavily on a single aspect, i.e., data sharing. Although there are benefits acquired by processing the IoT data at local loops, their real benefit is brought into the foreground when these are shared at large scale. The latter can act as an enabler to better understand complex systems of systems and better manage them. The Cooperating Objects vision [58], which assumes cooperation among devices and systems as the key driving force for interaction, sheds some light on the benefits and challenges [59] that will emerge in all layers of such an IoT-empowered infrastructure. As an indicative example, a smart city can be used where huge amounts of data from a smart city infrastructure, citizens, businesses, and individual assets need to be considered, analyzed, and, after decisions are taken, enforced. IoT data hold the key to do so and enable an efficient and sustainable future. Some such aspects are addressed in the use cases (see Part 3), e.g., in Chapter 12 in which the impact of smart grid data on a smart city is discussed. There are several other examples of applications that could in multiple ways benefit diverse domains and their market needs [59]. Open source software that enables big data collection and processing in a distributed manner has flourished in the last years [39].

The IoT infrastructure in place heavily depends on real-world processes, implying also that a big percentage of data will be generated by machines that interact with the real-world environment, while the rest will be purely virtual data. For the first part, where machines are involved, there is a real cost for the infrastructure that has to be met. Hence, it is expected that stakeholders in the future will further diversify, and we will see the emergence of infrastructure providers who will operate and manage many of the machines generating this data, which can then be communicated to others (e.g., analytics specialists to take advantage of the insights offered). The end-beneficiaries might acquire information, but do not necessarily need to have access or to process the data by themselves. Hence, as we see, there is a rise of specialists in the various stages of IoT Data Management that will cooperate with application providers, users, etc. for the common benefit. Such ecosystems are expected to be of key importance in the future IoT era. This transition is already at an early stage and boldly contradicts the existing initial IoT efforts, where the application developer, the data collector, and the infrastructure operator roles are largely performed by the same stakeholder (or a very small number of them).

Because of expected wide sharing of data and usage in multiple applications, security and trust are of key importance. Security is mandatory for enabling confidentiality, integrity, availability, authenticity, and nonrepudiation of data from the moment of generation to consumption. Due to the large-scale IoT infrastructure, heterogeneous devices, and stakeholders involved, this will be challenging. In addition, trust will be another major issue, as even if data is securely communicated or verified, the level of trust based on them will impact the decision making process and risk analysis. The recent replaying of data in the Stuxnet worm [57] has demonstrated that although data may appear legitimate, they still need to be independently verified to make sure that the whole chain from generation to consumption is not tampered with. Managing security and trust [60] in the highly federated IoT-envisioned infrastructures poses a significant challenge, especially for mission-critical applications that also exercise control.

Privacy is also expected to be a significant issue in IoT infrastructures due to its multidomain impact (see Chapter 6). Currently, a lot of emphasis is put on acquiring the data, and no real solutions exist for large-scale systems to share data in a controlled way. Once data is shared, the originator has no more control over its usage. This calls for policy-driven Data Management for their whole lifecycle so that, for instance, data can be invalidated, or even removed from the IoT global ecosystem once wished so, as we have discussed earlier in the data remanence stage. A typical example here constitutes the usage of private citizen data, which could be controllably shared as wished; it should also be possible to (partially) revoke that right at will. Similarly, shared data infrastructures raise practical domain-specific concerns, e.g., in self-driving cars [61]. Understandably, this is an issue that will not be solved only by technology, but will need to be accompanied by the appropriate legislation frameworks.

Finally, we are still at the dawn of an era that has to deal with huge data and unveil the hidden information patterns behind them. Being able to search and apply intelligent algorithms that may unveil those hidden patterns is expected to be a significant business advantage. Data Science in the IoT era is a cross-discipline approach building on mathematics, statistics, high-performance computing, modeling, ML, engineering, etc. that will play a key role in understanding the data, assessing their information at large scale and hopefully enabling the better studying of complex systems of systems and their emergent characteristics. The latter brings hope that features of several real-world processes in the context of IoT, such as cascading failures, dynamic behaviors, nonlinear relationships, feedback loops, and nested systems, will be better studied, understood, and applied in real-world domains (e.g., smart city, markets, enterprises, planet ecosystem).

5.5.4 CONCLUSIONS

Data and its management hold the key to unveiling the true power of IoT. To do so, however, we have to think and develop approaches that go beyond simple data collection and enable the management of their whole lifecycle at very large scale, while in parallel considering the special needs and the usage requirements posed by specific domains or applications. Mastering the challenges of Data Management will enable data analysis to flourish, and this in turn will empower new innovative approaches to be realized for the benefit of citizens, business, and society.

5.6 BUSINESS PROCESSES IN IOT
5.6.1 INTRODUCTION

A business process refers to a series of activities, often a collection of interrelated processes in a logical sequence, within an enterprise, leading to a specific result. There are several types of business processes, such as management, operational, and supporting, all of which aim at achieving a specific mission objective. As business processes usually span several systems and may get very complex, several methods and techniques have been developed for their modeling, such as the Business Process Model and Notation (BPMN) [62], which graphically represents business processes in a business pro-

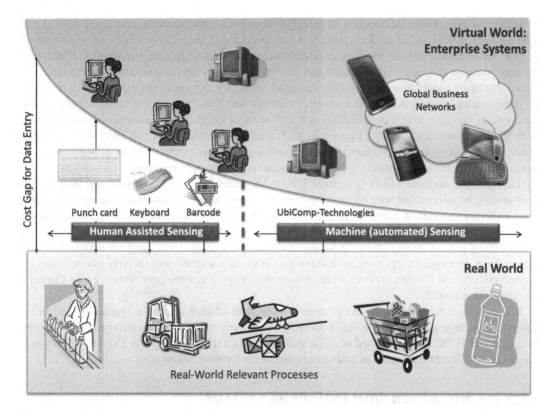

FIGURE 5.11

The decreasing cost of information exchange.

cess model. Managers and business analysts model an enterprise's processes, in an effort to depict the real way an enterprise operates, with the goal to improve transparency and subsequently efficiency.

Several key business processes in modern enterprises heavily rely directly or indirectly on interaction with real-world processes. These relate mostly to monitoring, but also to some extent to control (management), in order to take business-critical decisions and optimize actions across the enterprise. The introduction of modern ICT has significantly changed the way enterprises (and therefore business processes) interact with the real world.

Over the last decades, as depicted in Figure 5.11, we have witnessed a paradigm change with the dramatic reduction of effort required for real-world data acquisition. This is attributed mostly to the automation offered by machines embedded in the real-world processes. Initially all these interactions were human-based (e.g., via a keyboard) or human-assisted (e.g., via a human-controlled barcode scanner); however, with the prevalence of RFID, WSNs, and advanced networked embedded devices, all information exchange between the real-world and enterprise systems can now be done automatically without any human intervention at blazing speeds.

In the IoT era, remote devices can be clearly identified and continuously connected. In addition, with the help of services (at device, cloud, and enterprise level), they can be easily integrated, which leads to active participation of the devices to the business processes. This easy integration is changing the way business processes are modeled and executed today as new requirements come into play. Existing modeling tools though are hardly designed to specify aspects of the real world in modeling environments and capture their full characteristics. In this direction, the existence of SOA-ready devices (i.e., devices that offer their functionalities as a web service) simplifies the integration and interaction as they can be considered as a collection of traditional web services that run on a specific device. Nevertheless, there is ongoing research on inclusion of semantics in order to include, in an easier and more accurate way, the IoT aspects in business process modeling and execution.

The industrial adoption of IoT (e.g., of WSNs) is hampered by the heterogeneity and lack of common integration approaches with business process modeling languages and backend systems. There are, however, promising approaches such as the one provided by makeSense [63,64], which tackles this problem space with a unified programming framework and a compilation chain that, from high-level business process specifications, generates code ready for deployment on WSN nodes. A layered approach for developing, deploying, and managing WSN applications that natively interact with enterprise information systems such as a business process engine and the processes running therein is proposed and assessed.

IoT empowers business processes to acquire very detailed data about the operations and be informed about the conditions in the real world in a very timely manner. Subsequently, better Business Intelligence [65] and more informed decision making can be realized. The latter enables businesses to operate more efficiently, which translates to a business-competitive advantage.

5.6.2 IOT INTEGRATION WITH ENTERPRISE SYSTEMS

M2M communication and the larger vision of the IoT pose a new era where billions of devices will need to interact with each other and exchange information in order to fulfill their purpose. Much of this communication is expected to happen over Internet technologies [66] and tap into the extensive experience acquired with architectures in and experiences with the Internet/Web over the last several decades. More sophisticated, though still overwhelmingly experimental, approaches go beyond simple integration and target more complex interactions where collaboration of devices and systems is taking place.

As shown in Figure 5.12, cross-layer interactions and cooperation can be realized:

- At the M2M level, where the machines cooperate with each other (machine-focused interactions).
- At the Machine to Business (M2B) layer, where machines cooperate also with network-based services, business systems (business service focus), and applications.

Figure 5.12 shows several devices at the bottom layer, that can communicate with each other over short-range protocols (e.g., over ZigBee, Bluetooth), or even longer distances (e.g., over Wi-Fi, LTE, LoRa). Some of them may host services (e.g., RESTful services) and even have dynamic discovery capabilities based on the communication protocol or other capabilities (e.g., Web Service Eventing or WS-Eventing in the Device Profile for Web Services (DPWS)). Some of them may be very resource-constrained, which means that auxiliary gateways (as shown in Figure 11.1) could provide additional support such as mediation of communication and protocol translation. Independent of whether the de-

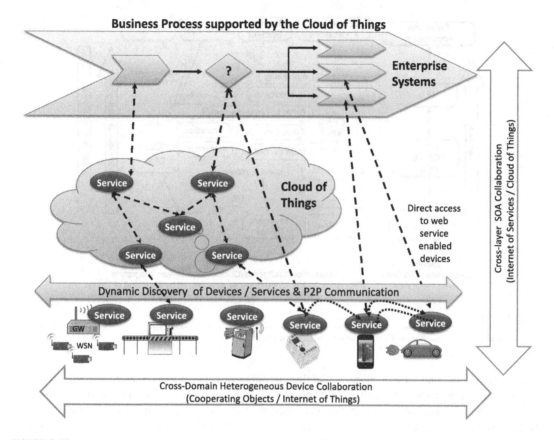

FIGURE 5.12

A collaborative IoT infrastructure.

vices are able to discover and interact with other devices and systems directly or via the support of the infrastructure, the IoT interactions enable them to empower several applications and interact with each other in order to fulfill their goals.

Promising real-world integration is done using a SOA approach by interacting directly with the respective physical elements, for example, via web services running on devices (if supported) or via more lightweight approaches such as REST. In the case of legacy systems, gateways and service mediators are in place to enable such integration challenges. However, in the era of sophisticated embedded networked devices, open interactions, and virtualized resources, a dilemma is emerging which manifests to which functionalities can be abstracted and migrated to the cloud (with all the benefits and constraints it poses) and which of these should still remain on the device itself. This is not an easy decision to take, as benefits and constraints exist in both directions, so practically, we expect an amalgamation of them to occur [67].

Many of the services that will interact with the devices are expected to be IP-based services, for example cloud services. The main motivation for enterprise services is to take advantage of the cloud

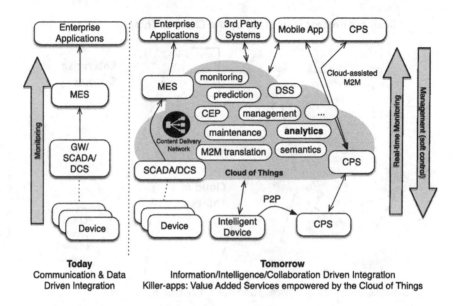

FIGURE 5.13

The Cloud of Things as an enabler for new value-added services.

characteristics such as virtualization, scalability, multitenancy, performance, and Lifecycle Management. Similarly, we expect to see that a large number of devices, and generally Cyber-Physical Systems, will make their functionality available on the cloud [67]. As such, we are moving towards an infrastructure where the cloud and its services (as depicted in Figure 5.13) take a prominent position towards empowering modern enterprises and their business processes.

A key motivator for hybrid on-device and cloud services is the minimization of communication overhead with multiple endpoints. For example, transmission of data to a single or limited number of points in the network (device to cloud communication), and subsequently letting the cloud do the load balancing and further mediation of communication. For instance, as depicted in Figure 5.13, a Content Delivery Network (CDN) can be used in order to get access to the generated data from locations that are far away from the IoT infrastructure (geographically, network-wise, etc.).

To this end, the data acquired by the device can be offered without overconsumption of the device's resources, while in parallel, better control and management can be applied. Typical examples include enabling access to the full historical data, preprocessing of information, transparently upgrading the cloud services, or even not providing access to internal systems for security reasons. This clear decoupling of "things" and the usage of their data is expected to further empower information-driven business processes and applications that can operate over federated infrastructures.

5.6.3 DISTRIBUTED BUSINESS PROCESSES IN IOT

Today, as shown in a simplified manner on the left part of Figure 5.14, the integration of devices in business processes merely implies the acquisition of data from the device layer, its transportation to the

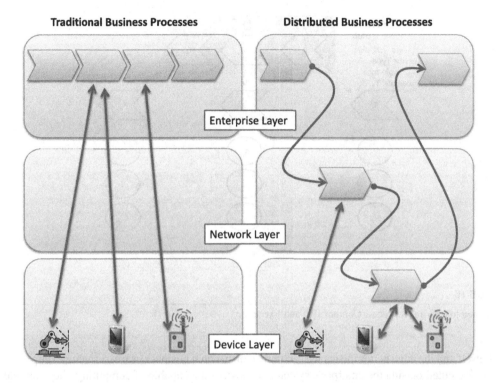

FIGURE 5.14

Traditional (centralized) vs. Distributed Business Logic.

backend systems, its assessment, and, once a decision is made, potentially the control (management) of the device, which adjusts its behavior. However, in the future, due to the large scale of IoT infrastructures, as well as the huge data that it will generate, such approaches are too costly and may not make always business sense.

Transportation of data from the "point of action" where the device collects or generates them, all the way to the backend system to then evaluate their usefulness, will not be practical for communication reasons, as well as due to the processing load that it will incur at the enterprise side; this is something that the current systems were not designed for. Enterprise systems trying to process such a high rate of non- or minor-relevancy data will be overloaded. Such approaches were necessary up to now due to the lack of advanced processing and storage capabilities at the device level, but with the latest advances and miniaturization this is no longer a significant hurdle.

A strategic step is to minimize communication with enterprise systems to only what is business-relevant. With the increase in resources (e.g., computational capabilities) in the network, and especially on the devices themselves (more memory, multicore CPUs, etc.), it makes sense not to host the intelligence and the computation required for it only on the enterprise side, but actually distribute it on the network, and even on the edge nodes (i.e., the devices themselves), as depicted on the right side of Figure 5.14. Partially outsourcing functionality traditionally residing in backend systems to the net-

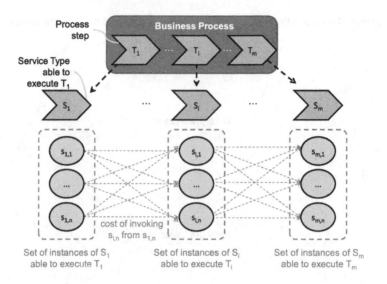

FIGURE 5.15

Distributed IoT Business Process Composition and Execution.

work itself and the edge nodes means we can realize distributed business processes whose subprocesses may be executed outside the enterprise system. As devices are capable of computing, they can realize the task of processing and evaluating business-relevant information they generate either by themselves (on-device) or in clusters.

Distributing the computational load in the layers between enterprises and the real-world infrastructure is not the only reason; distributing Business Intelligence is also a significant motivation. Business processes can bind during execution of dynamic resources that they discover locally and integrate them to better achieve their goals. Being in the world of service mash-ups, we will witness a paradigm shift not only in the way individual devices, but also how clusters of them, interact with each other and with enterprise systems in increasingly seamless ways [68].

Modeling of business processes [69] can now be done by focusing on the functionality provided and that can be discovered dynamically during run-time, and not on the concrete implementation of it; we care about what is provided but not how, as depicted in Figure 5.15. As such, we can now model distributed business processes that are executed on enterprise systems, in-network, and on-device. The vision [69] is to additionally consider during run-time the requirements and costs associated with the execution in order to select the best of available instances and optimize the business process in total according to the enterprise needs, e.g., for low impact on a device's energy source or for high-speed communication.

5.6.4 CONSIDERATIONS

Existing tools and approaches need to be extended to make the business processes IoT-aware. The current terminology in modeling tools is focused on the enterprise context and does not sufficiently include notation of physical entities such as IoT devices and their capabilities. Although distributed execution of processes exists (e.g., in BPMN) and experimental research provides some interesting directions [64,63], additional work is needed to be able to select the devices in which such processes execute and consider their characteristics or dynamic resources, etc. [69]. The dynamic aspect is of key importance in the IoT, as much of it is expected to be mobile and availability is not guaranteed, which means that availability in modeling time does not guarantee availability at run-time and vice versa. Even if the latter is true, this might again change during the execution of a business process; hence, fault tolerance needs to be considered.

IoT infrastructures are expected to be of large scale. Hence, scalability is an aspect that needs to be considered in the business process modeling and execution. In addition, event-based interactions among the processes play a key role in IoT, as a business process flow may be influenced by an event, or as its result, trigger a new event. Such considerations need to be seen also under the light of real-time interactions, which may be mandatory in several domains (e.g., industrial automation). The same holds for the Quality of Information acquired at each step, as resource-constrained devices may have a higher probability of delivering nonerror-free information (e.g., due to malfunctions). The latter would enable a different modeling of processes that depend on this information and not consider it always as correct or trustworthy. To solve this today, additional Data Management steps are considered, as discussed in Section 5.5.

5.6.5 CONCLUSIONS

Modern enterprises operate on a global scale and depend on complex business processes. Business continuity needs to be guaranteed, and therefore efficient information acquisition, evaluation, and interaction with the real world are of key importance. The infrastructure envisioned is a heterogeneous one, where millions of devices are interconnected and ready to receive instructions and create event notifications and where the most advanced ones depict self-behavior (e.g., self-management, self-healing, and self-optimization) and collaborate with each other. This can lead to a paradigm change as business logic can now be intelligently distributed to several layers such as the network, or even the device layer, creating new opportunities, but also challenges that need to be assessed. Future Enterprise systems will be in a position to better integrate state and events of the physical world in a timely manner, and this will hence lead to more diverse, highly dynamic, and efficient business applications.

5.7 DISTRIBUTED LEDGERS AND APPLICATIONS

A relatively new technology – blockchain – has started to be discussed in relation to IoT. Originally used for the digital cryptocurrency Bitcoin, the use of blockchain has spread far beyond this original application. Within the scope of IoT, however, it is useful to take a broader perspective of this technology and understand the wider landscape of Distributed Ledger Technology (DLT).

As illustrated in Figure 5.16, there are three main types of DLT – Permissionless, Public Systems; Private, Permissioned Systems; and hybrid Systems. Each version is useful to achieve different objectives and meet different requirements.

Distributed Ledger Technology

"Blockchain"

PERMISSIONLESS, PUBLIC, SHARED SYSTEMS

PERMISSIONED, PUBLIC, SHARED SYSTEMS

PERMISSIONED, PRIVATE, SHARED SYSTEMS

DATABASES

Cross Stakeholder Decentralization

FIGURE 5.16

Landscape of Distributed Ledger Technology (Imperial College London).

As illustrated above, each has its unique properties, and each has different forms of access control for reading and editing the information on the blockchain. As we move from right to left across the types of the ledger, the level of decentralization increases, while transaction speed decreases.

Permissionless, Public, Shared Systems are those that allow anyone to join the network, anyone to write to the network, and anyone to read the transactions from those networks. These systems have no single owner – everyone on the network has an identical copy of the "ledger". In the media, the prototypical example of this is Bitcoin, but there are many others. Due to the unique design goals of operating in a completely open environment without any points of centralized trust, and in which potentially malicious actors are not only allowed to submit transactions but also to participate in transaction validation, these systems add an extra component that prevents these activities (e.g., proof of work). This approach is computationally expensive, uses a significant amount of electricity, does not scale well, and requires large numbers of network participants in order to be able to generate "trust" (see Chapter 6). However, this approach does allow large numbers of participants to collaborate based on the codes only in a decentralized manner. Bitcoin and Ethereum are the two best-known examples, but there are many others.

Permissioned, Private, Shared Systems are those that have whitelisted access. Only those people with permission can read or write to such systems. They may have one or many owners – often consortia are formed to manage the ownership. Examples include Hyperledger.

Permissioned, Public, Shared Systems are a form of hybrid systems that provide for situations where whitelisted access is required, but all the transactions should be publicly viewable. Examples of this are government applications where only certain people should be able to write to the network, but all transactions can be publicly verified.

5.7.1 DLT, IOT, AND DATA OWNERSHIP

By the end of 2017, amongst the world's top eight largest tech companies by market capitalization, three of them directly generate revenue from their users' data: Alphabet, Facebook, and Tencent; three of them produce hardware and software that generate enormous amounts of data: Apple, Microsoft, and Samsung; two of them apply user-generated data to accurately target what exactly to sell to you: Amazon and Alibaba. On 8 January 2018, these eight companies have a combined market capitalization of 4.9 trillion US dollar, or almost half of China's GDP.

This is the power of data in our modern economy. As AI becomes a utility of the 21st century, the importance of data is only going to be more paramount. But, who should own the data? Do Facebook and Google have the right to own their users' data and profit from it? Will their users never question about the ownership of something that is proven so valuable? The debate over data ownership is not new, and the tech titans' self-claimed data ownership is not unchallenged. So far, a large part of the debate was through certain regulators' rulings with limited impact compared to the total size of profit these companies are enjoying. The key reason why the debate of ownership has not been as meaningful as it should be is because there is one crucial piece that has been missing: tradeability. Regulator rulings are important for data privacy. But without the ability to trade the data each individual generates, the value of data ultimately can only be realized by these tech companies.

When tradeability is achieved, data privacy will no longer be a privacy issue, but an economic issue. Even without it, we are already subconsciously assigning an economic value to the data we are generating. Millions of small enterprise owners in China and India are happy to use their data to have their microloans approved within seconds by ML-empowered SME lenders, while the western middle class is less comfortable to give away their data as borrowing a 1000-US dollar loan is not the most urgent problem in their lives.

Every data generator has a price for the data they generate, perhaps low enough to be free as long as they can get small loans, perhaps very high such as when they encounter fatal disease and their genetic data could be vital to save their lives. Differentiated economic value of data has not been discussed because of the absence of a micropayment and microincentive system. Blockchain is an effective tool to assign tokens of economic value of each dataset, document immutable records of each data transaction, and eventually build a global, open data marketplace.

Try to reimagine eBay; in addition to the physical goods, it is also a place where people could auction off their sets of data and set a price for them. eBay does not have to be a place where data producers and purchasers transact the actual data, but can instead be just a place for the parties to make a financial agreement and financial transaction. eBay then issue a data transaction certificate on eBay's enterprise blockchain to prove such commitment. The certificate becomes a global standard to prove a mutual commitment for data transactions. Academics, pharmaceutical companies, insurance companies, etc. would be able to conduct much larger-scale research with richer data while giving the data ownership back to the data producers. If a pharmaceutical company's research has insufficient responders in an assumingly sufficient data marketplace, then the company would know the market does not respond to such data collecting kindly. With such global standard, eBay becomes one of the two most valuable companies, with Alibaba dominating China's data market, which could be as large as the rest of the world added up together. There might well be a secondary data market. Obviously, there would be a lot of ethics and governance considerations before we get there, but technically, with blockchain, we are more or less there.

But such exchange might not require a marketplace. Individual companies with the right tools and infrastructure could directly reach out to data producers to conduct continuous transactions. Take the Austin-based medical startup Nano Vision as an example. It uses a chip that can mint a dedicated cryptotoken to assign real value to molecular data and create an economic system that could fundamentally change data-driven medical researches. Instead of isolating data in laboratories, Nano Vision uses their physical chips to understand molecular data in real-time from homes, hospitals, classrooms, etc. They use blockchain to secure and authenticate data and attribute each piece of data to its correct source in order to compensate all participants for their contributions. Mass molecular data generating infrastructures with a microincentive system empowered by blockchain, with an overlay with ML, could provide transformative advancements in medical research and epidemics predictive research.

Few individuals would argue that data ownership should squarely belong to the data producers. But to have the tool and a sufficient marketplace to give the pricing and tradeability back to the data producers has not been possible until most recently. If the largest companies in the world created nation-size wealth by harvesting user data, who is here to say it would not be disrupted once the right tools and mindsets are in place? The economic incentive is one of the most dynamic forces on this planet. It will be a matter of time that some brilliant and ambitious minds march into this space and redistribute the wealth created by data.

The regulator's role then would be setting limits to protect children and the vulnerable demographic part of our societies from predatory behaviors of data collectors.

Geopolitically speaking, national AI capacities will play a central role in the world power race in the coming decades. Without data, AI is just empty algorithms like a skyscraper with nothing but the foundation. In the coming years, an astronomic amount of data will be produced by IoT. Blockchain has the potential to rewrite how big data is generated. How we embed the micropayment and microincentives into the IoT data generation process could significantly transform the current paradigm of the most valuable companies in the world. At the moment, the top eight largest companies in the world indeed have generated an enormous amount of data. A majority of some of the most valuable data is unstructured and untagged. Such data without tagging is challenging for AI to process. When data collectors compensate data generators using blockchain-empowered microincentive system, it would also give the buyers (data collectors) the right to demand a certain quality of the data and to require the sellers (data generators) to tag their data, for example, and make the dataset more AI friendly.

The leader of the Fourth Industrial Revolution will be the nation that leads AI. China might not have a data collection problem because people assume their data is supervised or owned by the government. China's online payment system has already been way ahead compared to the rest of world. In 2016, China's online payment volume was 8.6 trillion US dollar and grew four-fold in one year. The US volume was 112 billion US dollar, growing at 39% per year. For the rest of the world, in order to accumulate such mass and high-quality data, they must create an effective marketplace for data through tradeability and ownership back to the data generators and stay competitive in the race of AI. It is strategically important and ethically necessary. In the coming 10 years, we might witness a wealth redistribution driven by a redistribution of data ownership empowered by blockchain.

SECURITY

6

CHAPTER OUTLINE

6.1 INTRODUCTION

Security must be considered at design time for IoT systems regardless of the intended application. Early security research for embedded systems, particularly those concerning Wireless Sensor Network (WSN) applications, considered security as critical to trustworthiness and acceptability given that the key applications driving development were military and healthcare applications. We have now come to accept that these systems will pervade all aspects of our personal and professional lives, with many becoming autonomous.

It is thus obvious that we must do our utmost to ensure that these systems are, at the very least, as secure as we can possibly make them against prospective threats from malicious actors. This requires a mentality shift from common practices that tend to implement security as an afterthought rather than by design. This means accepting the overheads involved, which are typically quantifiable in terms of

Internet of Things. https://doi.org/10.1016/B978-0-12-814435-0.00018-3

the additional complexity of the system, computational and storage requirements, cost of development, etc.

The reality is that many IoT systems will be deployed in the public domain, with devices significantly outnumbering those charged with their monitoring and maintenance (Bell's law). They must thus be able to operate *autonomously*; one of the security requirements discussed in the following subsections concerning basic principles of secure operation.

Autonomous operation with regard to security is a very interesting concept. It is hard to think of any electronic device in the public domain that has not been *hacked*. As such, and bearing in mind the need for physical access to hardware to effectively attack many contemporary information security systems, physical security of devices is an increasingly important topic to ensure that malicious actors cannot physically manipulate the electronic systems of IoT devices. This must also be taken together with the very physical nature of the tasks these devices are designed to undertake: sensing and actuating. In the case of the former, new attacks based on overloading physical transducers to stimuli can influence a system's behavior in regard to the latter (assuming that the system trusts the data from its sensors), allowing adversaries new ways to manipulate systems and launch attacks designed to starve system resources.

This chapter describes the basic security principles that must be understood and addressed in the design of any IoT system. It begins by exploring the potential threats to devices, networks, and systems in the IoT and explains at an accessible level the various methods available to protect against threats. Many of these methods are either suggested or specified by the standards that govern the operation of the IoT system. The chapter concludes by looking to future developments in security for IoT systems.

6.2 BASIC PRINCIPLES

The primary security principles that IoT system designers must be aware of and ultimately guarantee include message confidentiality, data integrity, freshness, efficiency, autonomy, and authentication.

Message confidentiality is one of the most important aspects of security for the IoT and is typically the first problem addressed when attempting to secure a network. Confidentiality accounts for ensuring that no acquired data can leak to adversaries or unauthorized networked neighboring nodes. Many applications require in-network distribution of keying information, which necessitates the construction of a secure channel over which this information can be sent, and ensure that malicious parties cannot gain access to this information. Encryption is the standard method of ensuring message confidentiality.

Data integrity is required to ensure that an adversary cannot change or alter any information sent across a network. Although the information may be encrypted, that does not prevent an attacker from attempting to alter transmitted messages. Manipulation of packets can be prevented through the provision of data integrity, usually enforced through the presence of a strong authentication mechanism.

Freshness of data is important in IoT systems. This ensures that data received is recent and that no old messages have been replayed, hence thwarting the classic replay attack. The problem of freshness is usually countered through the employment of a timestamp or counter, which can be easily inserted into a transmitted packet.

Efficiency concerns the relation to the resource consumption associated with the implementation of security architectures in IoT systems. The chosen protocol should fit within acceptable performance parameters. The originally intended functionality of the system should not be infringed upon through the presence of any security protocol. If a proposed scheme (e.g., in a standard or specification) is too resource hungry, in terms of processor requirements, memory space needed, and power consumption, such that it hampers the performance of the network or significantly decreases the longevity of the deployment, steps must be taken to ensure equivalent security strength or to reassess the proposed system architecture (i.e., with regard to device design, communication medium, energy provision, standard adherence, etc.).

Autonomy applies to ensuring that every device in an IoT deployment may operate independently and be flexible enough to accommodate fluctuations in the network architecture. This can be as a result of devices entering and exiting the network and the ever-changing topology that entails. For larger deployments, the preinstallation of a shared key between all the networked devices is infeasible and insecure, and as such key management is a significant problem.

Authentication is an extremely important requirement for a secure IoT system. In addition to an adversary attempting to manipulate a transmitted message, it may also attempt to insert packets. As such, it is necessary to authenticate every transmitting device in the system before accepting that a message is valid.

Availability of IoT devices must be ensured such that the application continues to function as intended. In many cases, it will be required for a central application server or network coordinating device to reach all other devices in the network. A number of attacks, including hole, capture, and jamming attacks, described later, can render networked devices unavailable.

6.2.1 ENCRYPTION

Encryption has been used for many centuries as a method of ensuring the confidentiality of secret communication. Technically, it is the transformation of information, or plaintext, to make it unreadable, or indecipherable, to any person who does not have the key (a shared secret). The transformed text is known as ciphertext. This is achieved using a certain (shared) algorithm, referred to as a cipher. There are many different algorithms available for performing encryption, using many combinations of keys, ciphers, and other tools. The ones relevant to contemporary IoT systems are discussed in this work.

6.2.1.1 Ciphers

In cryptographic terms, a cipher is defined as an algorithm for performing encryption or decryption operations. This is a predefined set of functions that must be followed precisely on each occasion. Conventionally, it is a mathematical function, or set of functions (generally, two related functions; one for encrypting and one for decrypting). When using a cipher, the original information is the plaintext, and the outputted encrypted information is the ciphertext. The ciphertext should be indecipherable to anyone who does not possess knowledge of the cipher/algorithm or the key.

The key, or shared secret between trusted parties, is a preagreed "cryptovariable", without which it should be impossible to decipher the plaintext. It is possible to classify ciphers in terms of the state of the information upon which they operate, or whether or not the same key is used for the encryption and decryption parts of a particular algorithm [70].

If a cipher performs operations on a block of information, usually of a fixed size, it is known as a block cipher. If the cipher operates on a continuous stream of information, this is known as a stream cipher. In other words, a stream cipher operates on a single bit at a time, whereas a block cipher operates on a group of bits.

The classification of ciphers based upon keying type generates two kinds of algorithms. If the same key is used for encrypting and decrypting the plaintext, this is known as a Symmetric Key Algorithm (SKA). If a different key is used for encryption than that for decryption, it is known as an Asymmetric Key Algorithm. If one key cannot be deduced from the other, it is known as a public key infrastructure (PKI).

6.2.1.2 Symmetric ciphers

There are a wide variety of symmetric ciphers available for implementation with IoT systems. The most popular and studied of these include Skipjack, RC5, RC6, and AES (also known as Rijndael). These ciphers have been implemented in many IoT subnetwork architectures and are thought to be the most secure and lightweight options available. Other newer options are available, some of which have found their way into recent standards, e.g., MISTY1 and its successor KASUMI, specified in the 3GPP standards.

In addition to the specification of a cipher, one must consider the optimal mode of operation for this cipher. Modes of operation for symmetric ciphers include Electronic Codebook (ECB) mode, Cipher Block Chaining (CBC) mode, Output Feedback (OFB) mode, Cipher Feedback (CFB) mode, Counter (CTR) mode, Propagating Cipherblock Chaining (PCBC) mode, and Counter with CBC-MAC (CCM) mode, to name a few. The implementation of some modes of operation for symmetric ciphers can be used to provide authentication in addition to encryption, and also to avoid having to use padding to ensure a multiple of a block size to perform accurate encryption and decryption.

In 2006, Law et al. presented a survey and benchmarking of the various symmetric ciphers available for WSNs. This work also included analysis of the ciphers using the various modes of encryption (CBC, CFB, OFB, and CTR). The work concluded that the use of the CTR mode is recommended, with the Rijndael cipher (AES) [71]. This exercise considered only software implementations of the security algorithms, which can be significantly improved in regard to their efficient performance through the use of dedicated hardware accelerators [72].

6.2.1.3 Asymmetric ciphers

Mostly synonymous with typical Internet security, asymmetric ciphers that may be useful in IoT contexts including RSA and ECC. Asymmetric algorithms' security can be evaluated in terms of the best-known solving speeds of the difficult mathematical problem upon which they are based. The solving time for ECC is fully exponential, whereas that of RSA is subexponential. This means that the time required to solve the problem used in ECC increases exponentially as the size of the problem increases linearly. Solving the underlying integer factorization-based problem of RSA has a less than exponential solving time. ECC can provide the equivalent level of security for a key size of 512 bits as RSA can provide with 15360 bit keys (a 30:1 ratio), delivering 256 bits of security which would require 10^{66} MIPS years to break using brute force methods [73].

There are a number of algorithms to be considered in terms of asymmetric functionality. These include Diffie–Hellman (DH), Digital Signature Algorithm (DSA), and their elliptic curve counterparts, ECDH and ECDSA, respectively. Elliptic Curve Integrated Encryption Scheme (ECIES) is often specified for encryption in the ECC domain. It is worth remembering that these asymmetric ciphers also require the implementation of a symmetric cipher.

6.2.2 AUTHENTICATION

Authentication is complimentary to encryption and can prevent many of the attacks described in the following section that could be launched by an IoT system adversary. Authentication is a mechanism whereby the identity of a device or agent in the network can be identified as a valid member, and as such data authenticity can be achieved. Again, there are a number of methods of ensuring authentication based upon cryptographic primitives.

6.2.2.1 Symmetric authentication

A Message Authentication Code (MAC) can be used to authenticate a message. This is similar to a one-way hash function, but it includes the use of a key. Only a holder of the key can verify the hash. The hash function is turned into a MAC by encrypting the hash value with a symmetric encryption algorithm. MACs can also be generated from block cipher algorithms. An example of this is CBC-MAC. In this case, a message is encrypted with some block cipher algorithm in CBC mode, implying a chain of blocks is created such that each block depends on the correct encryption of the previous block. In order to calculate the CBC-MAC for a message, the message is encrypted with a zero-initialization vector. This method has been a primary method of authentication for symmetric schemes in low-power wireless networking.

A Message Integrity Code (MIC) is used interchangeably in terminology with MAC, but differs slightly in that a secret key is not used in its operation. A MIC should be encrypted during transmission if it is to be used to ensure integrity. A given message should always produce the same MIC when the same algorithm has been used to derive it.

Hash functions. A cryptographic hash function is a procedure that returns a fixed-size bit string (hash value) for a block of data (usually the message to be sent). Any change to the message, accidental or intentional, should change the hash value; also known as the "Message Digest". A hash function should maintain ease of computation for any given data. It should be almost impossible to reconstruct plaintext from a given hash. It should be impossible to alter the plaintext without changing the hash value and it should be highly unlikely for two different messages to have the same hash value. The most commonly used hash functions are MD5 and SHA-1.

6.2.2.2 Asymmetric authentication

A digital signature is an example of an asymmetric authentication mechanism. It differs from the previously described MAC in that a digital signature is generated using the private key of a public/private key pair. This demonstrates asymmetric functionality, and seeing the private key should only be known to the sender, the digital signature proves that the message could only have emanated from the holder of that private key and is therefore authentic.

There are numerous Digital Signature Algorithms. These include algorithms such as DSA, SHA with RSA, ECDSA, and ElGamal. During the initial development of low-power IoT security schemes based around WSN-type technologies, it was thought that digital signatures were too computationally expensive. Notwithstanding, there are a number of implementations that have been proven feasible for the resource-constrained IoT technology.

6.2.2.3 Application of authentication

There are a number of methods to achieve authentication within IoT environments. These range from device-to-device protocols, where each node authenticates its neighbor's identity, to broadcast protocols (CBC-MAC, etc.), which enable a sender to broadcast critical data and/or commands (in-network reprogramming, for example) to sensor nodes in an authenticated way ensuring that an attacker cannot forge any message from the sender. As traditional broadcast techniques (like public key-based digital signatures) were originally considered too expensive for IoT-type devices, a number of lightweight protocols were developed in the academic community.

6.3 THREATS TO IOT SYSTEMS

IoT will be susceptible to a plethora of threats, old and new. Accepting that all of the potential attack vectors cannot be known in advance, it is worth considering the types of attack that traditional computing systems have encountered and the set of feasible threats considered for underlying IoT technologies. These broadly include Denial of Service, Sybil, privacy, "hole", and physical attacks.

IoT systems may be comprised of complex heterogeneous subsystems that combine wired, wireless, and Internet computing technologies and need not subscribe to a single standard or specification. Therefore, it is difficult to holistically assess the various threats at the system level. On the other hand, many attacks can be grouped and assessed by to which *layer* of the protocol stack they apply.

In the following, we discuss the types of attack that may affect IoT systems, relate them to the layer(s) of the stack to which they apply, and subsequently propose countermeasures. This type of approach should allow a system designer to attempt a realistic risk and security assessment of a proposed or existing IoT system.

6.3.1 DENIAL OF SERVICE (DOS) ATTACKS

DoS attacks are the most common and easiest to implement attacks on IoT systems. They can be seen in many forms and are defined as any attack that can undermine the network or systems' capacity to perform expected functions. For any wireless network, for example, "jamming" the channel with an interrupting signal is an effective DoS attack, as are flooding (multiple packet transmission) or collision (timed flooding) attacks, for example.

At the physical layer, jamming is a popular DoS attack. It can be either intermittent or constant jamming, both of which will have a detrimental impact upon the network. The number of devices required to effectively perform such an attack may be as small as a fraction (one malicious node may be enough to disrupt the intended operation) of the total number of devices and/or agents in the system, where the adversary interferes on the same communication frequency as other devices in the network or system.

At the data link layer, collision attacks are a common DoS attack, whereby an adversary will intentionally violate the communications protocol (this may only mean changing a small part of the packet which would lead to an error in the checksum, for example), in an attempt to generate collisions. This would require the retransmission of any packet affected by the collision. This contributes to unnecessary consumption of device and network resources. Misdirection is another attack at this layer, whereby a malicious device may refuse to route messages at all, potentially disconnecting parts of networks.

The transport layer is susceptible to DoS in the form of flooding. This attack is as simple as sending multiple connection requests to a device, agent, or server. Because resources must be allocated to handle requests, overloading with malicious requests will quickly deplete resources, particularly for highly constrained IoT devices, which can render devices useless and thus kill the overall utility of the system. De-synchronization is another attack at the transport layer. A pair of devices, for example, may be forced into a synchronization recovery protocol by disrupting some of the packets being transmitted between these two devices and maintaining correct timing. Again, this leads to excess resource consumption and possible exhaustion.

The application layer is susceptible to a path-based DoS attack, whereby the attacker may insert spurious or replayed packets into the network. As the packet is forwarded to the destination, energy and bandwidth are consumed by forwarding nodes. This attack may starve the network of authentic data transmission, as resources along the path to a base station or application server are consumed.

6.3.2 SYBIL ATTACKS

The Sybil attack can be described as that of a malicious node taking on multiple identities. This node can then launch a number of attacks, which could include negative reinforcement, or stuffing of the ballot box of a voting scheme, for example. This attack is most effective at the higher layers of the communication protocol stack.

At the physical layer, the Sybil attack is performed by compromising a legitimate device or fabricating a new one. Identities are therefore acquired either through theft or the fabrication of new ones. The malicious device can then behave as if it was a number of devices, participating at different points in the network or system.

At the data link layer, the Sybil attack has two variations; negative reinforcement and stuffing of the ballot box. Data aggregation is employed in many low-power IoT networks to reduce power consumption. If a device is continually providing large amounts of inaccurate data, then an aggregated data packet returned by the network would be corrupted. Similarly, a voting scheme may be corrupted via a Sybil attack, whereby the malicious node's various identities could mean that it could stuff the ballot box in its own favor, for example.

At the network or routing layer, a Sybil attack could be used to compromise most multipath routing schemes. If the information provided by the malicious node, and its multiple identities, is taken into account in the routing tables of nodes, then the decisions taken by the nodes to route messages to other network members would be at risk of failure.

6.3.3 PRIVACY ATTACKS

There are many reasons why the privacy of IoT messaging and data should remain intact, regardless of the nature of the application. The reality of IoT is that there are increasingly large volumes of information easily available via remote access, and as such an adversary may obtain information in an anonymous manner, with minimal risk. Attacks against device, message, and data privacy take the form of monitoring and eavesdropping, traffic analysis, and camouflage.

The most obvious attack to privacy is by monitoring and eavesdropping. An adversary can easily learn the contents of a communicated message solely through listening on a wired or wireless channel, the latter of which is much more difficult to detect. Should an adversary be eavesdropping on network traffic, and decipher the content of a message containing important control information about the configuration of the network or IoT system, it could lead to a breach of the integrity of the entire system. This can have significant consequences, particularly where third-party applications are built upon the compromised system.

Traffic analysis is a combination of monitoring and eavesdropping. Through monitoring the number of packets sent or received by specific devices, it could lead to inferred information as to the role of that device in the network or system. The adversarial device could then behave like a legitimate one, attracting packets and rerouting them incorrectly, for example. These packets could be sent to nodes performing analysis of the network data, etc.

6.3.4 "HOLE" ATTACKS

There are many types of so-called *hole* attacks. These include worm-hole attacks, black-hole attacks, and sink-hole attacks. A worm-hole attack can disrupt routing by convincing nodes that are usually many hops away from a base station that they are only one or two hops away. This assumes, of course, that a networked device has already been compromised by an adversary. There are no definitive ways to ensure security against a worm-hole; notwithstanding, some attempts have been made to better understand the problem [74,75]. In a worm-hole attack, malicious nodes may reroute messages received in one particular part of a network over a low-latency link and replay them to a different part of the network. Resulting from the nature of wireless transmission, it is possible for an adversary to create a worm-hole for packets that are addressed to other nodes (since it can overhear other wireless transmissions) and tunnel them to a colluding adversary in another location. This tunnel can be created using a variety of means, including: an out-of-band hidden channel (possibly a wired link), packet encapsulation, or high-powered transmission. The tunnel creates the illusion that the endpoints are very close to one another by ensuring that packets appear to arrive sooner and via less hops, compared with packets sent over regular routes. This permits an adversary to undermine the correct operation of the routing protocol by controlling various routes in the network [76].

Black-hole and sink-hole attacks are similar in that they advertise zero routes to base stations or gateway devices through a specific malicious node. Some routing protocols will then elect to route a number of packets via this node (as it appears to be the lowest cost, and this is usually sought), and a large amount of network data could be lost. The neighbors of the malicious node select this route and compete for bandwidth, and through this process, energy is wasted and resources are consumed. When a hole or partition is created in the network, it is considered to be a black hole.

6.3.5 PHYSICAL ATTACKS

The majority of IoT applications require devices to be deployed in uncontrollable environments with regard to third-party accessibility. Whether in the home, a business, or harsh, hostile environments, it is expected that devices will function autonomously. As a result of their small size, unattended operation, distributed deployment, and vast numbers, they are uniquely susceptible to a range of attacks that compromise physical integrity.

If a device is destroyed, in some cases this may be detrimental to the operation of a system, whereas in other cases the loss of a small fraction of devices may not hinder the operation of the application. If, however, a device can be physically tampered with, it is probable that an adversary can extract sensitive information from that device. Cryptographic secrets may be compromised, circuits altered, or the device reprogrammed. It has been shown that devices can easily be compromised within minutes, and warnings are provided by most manufacturers. It is also worth bearing in mind that scarcely a single example of an electronic device exists that has not been "hacked". Examples range from smartphones and to ATMs to industrial controllers deployed in industrial settings behind "air gaps", e.g., Stuxnet [57].

Entirely new and unforeseen attack vectors are likely to emerge as new devices become ubiquitous. Given that complicated social engineering has been shown effective to compromise industrial control systems, devices deployed in uncontrolled environments are infinitely more accessible and so it is fathomable that much simpler methods of manipulation could be used to disrupt the normal performance of an IoT system. For example, overloading a device with spurious sensor stimuli could render the sensor useless, or a sufficient amount of fake data could be generated to influence the behavior of a linked controller without necessarily making it obvious that the system is being attacked, and thus the deployed application will ultimately fail.

Tamper-resistant hardware is a plausible solution, coupled with advances in encrypted controller cores, memory, etc., but it is still likely that a sufficiently motivated, skilled, and resourced adversary will be able to compromise any device and thus potentially influence the operation of any IoT system. Improving the physical security of a device will incur additional costs which may be prohibitive for potential application designers attempting to develop cost-effective solutions for certain industries. As the criticality of the application increases, so does the need for security at every level of the system.

It is worth noting several recent examples of threats that extend beyond the capabilities of the general public. State-sponsored efforts to undermine the effectiveness of industrial technologies, ultimately resulting in catastrophic effects, must be considered once we apply IoT technologies to our critical infrastructures; e.g., Stuxnet [57].

6.4 MITIGATING THREATS TO IOT APPLICATIONS

There are a number of steps that system designers can take to ensure that their IoT systems and applications are as robust as possible to attack. Considering the heterogeneity of the application space, certain attacks and mitigation strategies may not be relevant, and therefore the system designer should carefully consider the suite of techniques adopted to ensure that their application or system is as secure as possible. This section takes a layered approach to examine the various threats and suggest potential defenses. Where wireless connectivity and end-to-end systems spanning heterogeneous communication infrastructures are elements of the overall system architecture, the system designer should always implement the security mechanisms specified in the associated standards. These are discussed in more detail in the next section.

6.4.1 APPLICATION LAYER AND PHYSICAL ATTACKS

Application layer attacks are designed to compromise the regular operation of the system. There are several methods by which an adversary can achieve this objective, including launching attacks at various other layers if a device is physically captured. Attacks include overwhelming sensors with spurious stimuli (a physical attack), path-based Denial of Service attacks, eavesdropping and physical capture, and reprogramming.

Detecting and mitigating a physical attack is very difficult to predict and understand at design time. The normal operation of the system will likely depend on the interpretation of sensor data. Therefore, having a realistic model of the overall system's expected behavior may assist in determining if a signal in sensor data that might otherwise require a response is in fact a real event, or if it has been fabricated by an adversary. Developing an understanding of the potential failure modes of the system (e.g., random or systemic failures) in addition to understanding the probable ranges of sensor readings and their calibration status will help system administrators to understand if and when their application may be under attack.

Physical capture attacks, reprogramming, and/or malware that is otherwise capable of manipulating the system's performance are essential to prevent insofar as possible. Encrypted microcontroller cores and memory may assist in the prevention of such attacks, as may tamper-resistant casings or enclosures for devices that can, for example, alert a system administrator if a device has been interfered with. If the code running on a device can be manipulated or altered, it is extremely difficult to detect in real-time if the device has come under the control of an adversary and/or what level of damage to the system is being caused.

Concerning eavesdropping on network traffic or path-based Denial of Service attacks, cryptographic mechanisms including encryption, authentication, and replay rejection are likely to mitigate. The most appropriate ciphers and security architectures for the subsystem will be specified in the appropriate standard. It is certainly worth remembering that adversaries rarely attack the security primitives; rather they tend to exploit weaknesses in the system's implementation. Therefore, the system designer should take care to ensure that the system's overall implementation is as robust as possible to attack.

Devices that are physically addressable and accessible over public networking infrastructure, i.e. Internet-connected, should always adopt Internet-standard security and should never be online without disabling factory default security settings. "Internet of Things" (IoT) devices have already been shown to represent a significant threat to the Internet when insecurely connected, such as the Mirai and Dyn attacks in 2016 [77,78].

6.4.2 TRANSPORT LAYER

At the transport layer, the most probable attacks are Denial of Service attacks caused by flooding and/or desynchronization. While resources are always wasted dealing with DoS attacks, e.g., filtering and ignoring rogue packets, they must still be thwarted, and effective authentication methods are required to achieve this.

6.4.3 NETWORK LAYER

Several DoS attacks are possible at the network layer. These include neglect and greed, homing, misdirection (aka spoofing), black-hole, and flooding attacks. Again, these require a combination of encryption, authentication, authorization, redundancy, and traffic monitoring to protect against and mitigate. Sybil attacks and worm-hole attacks may also be preventable by using effective authentication and authorization policies.

6.4.4 DATA LINK/MEDIUM ACCESS CONTROL LAYER

Denial of Service at the MAC layer may be attempted using collision, exhaustion and unfairness (with respect to resource utilization). Several techniques may be used to prevent such DoS attacks, including error correcting codes, rate limitation, and frame size reduction. Interrogation and Sybil (aggregation and voting) attacks are also possible at the MAC layer and may be defended against by implementing appropriate encryption and authentication mechanisms.

6.4.5 PHYSICAL LAYER

At the physical layer, in addition to the aforementioned tampering attacks, DoS in the form of jamming and Sybil attacks are possible. There is no way to prevent jamming attacks in wireless networks (i.e., where an adversary overloads a portion of the wireless spectrum with spurious power), but it may be possible to mitigate them in terms of using spread spectrum or mode changing techniques. With regard to Sybil attacks, authentication is again the primary defense.

6.5 SECURITY IN ARCHITECTURES AND STANDARDS

The most up-to-date, comprehensive, and yet concise description of the state-of-the-art and challenges associated with security for IoT is arguable the similarly named Internet Draft (work in progress) developed by the IETF Thing-to-Thing Research Group (T2TRG) (draft-irtf-t2trg-iot-seccons-15[1]). The authors summarize several important security aspects for IoT covering the lifecycle of a *thing*, threats and mitigation techniques for secure IoT system design, IP-based security protocols applicable to IoT, and residual challenges and potential solutions to ensuring a secure IoT.

From an architectural perspective, Chapter 7 describes those fragments maintained and developed by organizations, alliances, and technical communities globally, all of which are sensitive to the need for securing IoT devices, systems, and applications. Some are more prescriptive than others in terms of specifying precisely which security mechanisms should be applied. Considering the heterogeneity of the multitude of underlying technologies governed by different standards and specifications, and taking into account the unique characteristics, constraints, and performance requirements of each use case, this is not surprising. This section focuses on what we believe are the most popular and important security primitives, primarily driven by standardization efforts of the IETF and 3GPP, noting that there are several industry alliances and bodies leveraging constrained IP stacks based on IETF initiatives. These include Thread[2], the Industrial Internet Consortium (Chapters 5 and 8), OMA SpecWorks (Chapter 5), the Open Connectivity Foundation (Chapter 7), and the Fairhair Alliance[3].

6.5.1 IETF

The IETF IoT-related specifications are discussed in detail in Section 7.3.3 and summarily illustrated in Figures 7.2 to 7.5. Depending on the stack an implementer chooses, there are several layers at which relevant security mechanisms should be applied. In almost all cases, it is wise to consider both the end-to-end security of the information to be transmitted, which can be implemented from the transport layer upwards, and the security of the network itself, which may be implemented across the networking, data link, and physical layers. It is worth remembering that protocols designed for traditional IP-enabled

---

[1] https://tools.ietf.org/id/draft-irtf-t2trg-iot-seccons-15.txt
[2] https://www.threadgroup.org
[3] https://www.fairhair-alliance.org

computer networking are often too resource-intensive for implementation within the constraints of IoT devices.

It is worth considering an example scenario to illustrate the choices at hand and the complexity involved. Suppose that an implementer wishes to connect a network of devices that communicate using low-power Wireless Personal Area Network technology (e.g., IEEE 802.15.4 at the physical and link layers) to the Internet via a gateway (or border router) in order to communicate with a remote application server. The implementer may choose to use the 6tisch Working Group recommended stack (Figure 7.3), thus implementing CoAP over RPL and using 6LoWPAN compression with 6top adaptation to enable IEEE 802.15.4e TSCH at the link layer. To simplify matters, assume that the compression and adaptation are done securely and are compatible. Therefore, security decisions must be taken with regard to CoAP, RPL, and TSCH.

With regard to CoAP, there are several options that can be considered; these include DTLS (1.2 or 1.3), TLS (1.2 or 1.3), and Object Security for Constrained RESTful Environments (OSCORE[4]). It is described in draft-selander-ace-object-security[5]. There are several dependencies and levels of overhead to be taken into account.

RPL offers three security modes, namely *unsecured*, *preinstalled*, and *authenticated*, uses AES-128-CCM as the underlying cryptographic algorithms, and insists that RPL ICMPv6 (i.e., IPsec) MAC and signature calculations are performed before lower-layer compression [80]. The unsecured mode does not preclude the use of link layer security, but the preinstalled and authenticated modes require a preinstalled key to participate in the network. In the case of authenticated mode, two keys are needed for the joining device to act as a router.

Fortunately, in this case, the 6TiSCH stack also relies on OSCORE for its end-to-end security (draft-ietf-6tisch-minimal-security-05[6]), which simplifies matters. One of the drawbacks of following this prescriptively is that the framework requires a shared symmetric key to enable joining between a new device and an existing central entity, the provisioning for which is not fully defined at the time of writing.

6.5.2 3GPP AND LOW-POWER WIDE AREA NETWORKS

3GPP (Chapters 3 to 5) is responsible for three variants of LPWAN technology that leverage cellular connectivity. These are LTE-M, NB-IoT, and EC-GSM-IoT (Chapter 5), which are directly comparable with LoRa and Sigfox, among others. These technologies are all similarly constrained to those most considered by the IETF and may or may not leverage IP networking capabilities (e.g., it is optional for LTE-M and NB-IoT).

In regard to security, they are somewhat different, for example in that they take their unique identities from International Mobile Subscriber Identities (IMSIs), and use their SIM/UICC/eUICC for device authentication which enables network level authentication and session key distribution based on the LTE Authentication and Key Agreement protocol. This is a challenge–response protocol that uses symmetric cryptography wherein LTE-M, NB-IoT, and EC-GSM-IoT support algo-

---

[4]OSCORE is a method for application-layer protection of CoAP using CBOR Object Signing and Encryption (COSE), defined in RFC 8152 [79].
[5]https://tools.ietf.org/id/draft-ietf-core-object-security-12.txt
[6]https://www.ietf.org/id/draft-ietf-6tisch-minimal-security-05.txt

rithm negotiation to enhance overall security. Data confidentiality is provided by the EEAx algorithms (based on SNOW, a stream cipher, and AES, respectively), as described in Release 13 (Chapter 5).

Comparing the 3GPP-specified LPWAN options to those such as LoRa (WAN) and Sigfox is not favorable in terms of the latter. For example, network authentication, identity protection, and data confidentiality do not exist for Sigfox and are only partially supported in LoRaWAN [81]. On the other hand, there are IETF Working Groups looking at these technologies (Chapter 7), which may result in better security in the near future.

6.6 SECURITY FOR A SAFE IOT

As objects, systems, and systems-of-systems become increasingly interconnected, the safe operation of these systems (and those systems and/or actors with which they interact) must be taken into account by design. Factoring safety into the design of an IoT system is an extraordinarily difficult endeavor for many application scenarios and use cases. This section briefly examines three application contexts where the safety of the system is of utmost importance. Specifically, these relate to industrial robotics, automotive systems, and Smart Cities. The safety concerns in each case are highlighted.

Before proceeding, a satisfactory definition of safety is required. The Industrial Internet Consortium defines *safety* as the "condition of the system operating without causing unacceptable risk of physical injury or damage to the health of people, either directly, or indirectly as a result of damage to property or to the environment."[7] This is within a broader discussion of elements of trustworthiness of industrial applications of IoT technologies, which includes security, reliability, resilience, and privacy in addition to safety; where trustworthiness is defined as the "degree of confidence one has that the system performs as expected with characteristics including safety, security, privacy, reliability and resilience in the face of environmental disruptions, human errors, system faults and attacks"[8]. A comprehensive treatise of these elements is beyond the scope of this book, but certain useful lessons can be extracted considering the following application areas.

6.6.1 SAFETY IN INDUSTRIAL AUTOMATION AND ROBOTICS

There are numerous example scenarios where humans and robots may be required to interact, such as a hypothetical manufacturing scenario where a robot arm is required to use a sharp tool to perform a cutting task close to a human [82], or where multiple robot arms are used to pick and place items on a production line where humans are also present. In either case some interesting safety factors come into play. The authors of [82] have developed a comprehensive survey of safety methods and standards for safe Human–Robot Interaction (HRI). These are informed by several sources, not in the least the ISO, who have begun to release documents to assist. The ISO 10218 describes methods to enable safe collaboration, including speed and separation monitoring and power and force limiting, in addition

---

[7]http://www.iiconsortium.org/pdf/IIC-Security-WP.pdf
[8]http://www.iiconsortium.org/vocab/

to other safety requirements [83]. The authors of [82] describe safety in HRI in terms of *physical* and *psychological*. The former is self-explanatory, where the latter is discussed in terms of discomfort caused by "robotic violation of social conventions and norms" during interaction, but in contrast with physical safety, psychological harm can be sustained via "distal interaction via a remote interface". The authors break related works into several subtopics, including safety through control, safety through motion planning, safety through prediction, and safety through consideration of psychological factors, in addition to discussing how these can be improved and integrated.

From a security point of view, safety cannot be guaranteed if the robots in question may be influenced by malicious actors. If they happen to be connected to the Internet, then there are significant information security requirements that must be satisfied before these systems could be considered safe for close interaction with human operators. Any change or compromise to the safety mechanisms implemented could be very harmful to a human operator and/or the plant/process involved. Therefore, the system should be as secure as possible in terms of preventing any malicious actor from accessing and/or influencing the system.

6.6.2 SAFETY IN AUTOMOTIVE SYSTEMS

The automotive industry has very well-established safety guidelines. As technology advances, cars (trucks, trains, and even unmanned aerial systems) are increasingly autonomous and Internet-connected. Accordingly, there are numerous new safety concerns. These relate to both the autonomous driving capabilities permissible and the conditions and threats under which to the remote hijacking of these "systems" can occur. SAE International specifies six levels of autonomy ranging from no automation to full automation under all road and environmental conditions [84]. If vehicles can be remotely hijacked by any malicious actor, the safety of the operator, passenger, and/or environment, including third parties, cannot be guaranteed. Therefore, it is critical to ensure the highest possible levels of security are implemented for automotive systems.

6.6.3 SAFETY IN SMART CITIES

Smart Cities (Chapter 14) are increasingly complex and interconnected systems mostly comprise a city's infrastructure. This includes transport, water, energy, and civil and communication infrastructure – all of which can be considered critical. The idea that cities can leverage the interconnectedness of these infrastructures to improve services and provide enhanced efficiency in managing increasingly populous urban environments is being adopted globally [85]. Eventually, cities will become incredibly complex Cyber-Physical Systems. Data stemming from sensors connected to all manner of urban artifacts will be used to monitor and manage the complementary infrastructure(s), often using Internet-connected electromechanical actuators. There are several safety considerations that must be considered in such scenarios. It is important to ensure that these systems cannot be compromised by malicious actors. Any compromise to the integrity of these systems could lead to catastrophic consequences considering the criticality of these systems and the numbers of people who could be affected by attacks to these infrastructures. Without water, energy, transport or communication infrastructure, a city would grind to a halt and chaos could reign. Securing remote access to connected city infrastructure is paramount.

6.7 PRIVACY IN IOT

Devices connected to the IoT generate, process, and share sensitive data that are critical for security and thus must be kept private. As such, they are a very attractive target for a wide variety of attacks [86,87]. One of the main reasons behind the amount of vulnerabilities are the characteristics of the devices forming the IoT. In particular the heterogeneity of devices, the fact that many of them are very constrained in terms of resources, and the decentralization of the designs bring up many security and privacy challenges that have been widely studied in the literature [88–90,61].

The ubiquity of the IoT makes large volumes of information easily available to eavesdroppers, making privacy protection in IoT increasingly challenging. Most research regarding privacy in IoT-related applications is oriented to protecting the content of the messages flowing in the system through means of cryptography. This line of research mainly dedicates efforts to develop lightweight authentication and encryption mechanisms that can be used when typical alternatives are too demanding for the devices involved in the system [91].

Another security problem stemming from the decentralization of IoT systems is the establishment of trust among actors. In this respect, one line of solutions explores the development of context-aware architectures [92] that enable devices to make local decisions regarding security policies. An alternative is the deployment of reputation schemes that enable devices to decide about the level of trust that can be placed upon other actors which with they interact [93].

6.8 FUTURE DEVELOPMENTS IN SECURITY

Future developments to bear in mind for a secure IoT are likely to center around the challenges already known to the community. draft-irtf-t2trg-iot-seccons-15[9] lists many of these, where resource constraints and heterogeneous communication, already discussed in Chapter 5, pose many challenges for systems implementers. In particular, fragmentation of packets leading to increased overheads, coupled with lossy links, etc., make practical security implementation very difficult. Many implementations to date simply ignore the correct use of security primitives. Resistance to DoS attacks, particularly those designed to exhaust resources, will be particularly difficult to mitigate. As the number of link layers built upon emerging heterogeneous physical media continues to grow, the challenges for delivering security remains. As many are designed for ultra-low-power operation, make use of lesser bandwidth portions of the spectrum, and use smaller message sizes, securing the messages becomes increasingly difficult.

With regard to end-to-end security, a number of so-called middleboxes will be required to bridge heterogeneous physical media and translate between protocols. The IETF considers that the meaning of end-to-end security in terms of integrity and confidentiality (where these middleboxes would be required to either decrypt and/or partially modify messages) could be affected. One particularly interesting recent development, homomorphic encryption techniques, allows certain operations to be performed on secure messages, but these remain limited to arithmetic operations and are too limited at present for practical consideration. Ultimately, there is no known *perfect* solution to ensure the confidentiality and integrity of two IP-connected devices in this problem space.

---

[9]https://tools.ietf.org/id/draft-irtf-t2trg-iot-seccons-15.html

There are several operational issues that must be taken into consideration assuming that an appropriate bootstrapping of a security domain phase has been completed. These relate to key management, group management, and so forth. Some solutions have been developed by the IETF, but a significant amount of work remains to be done.

The ability to securely update software running on connected devices remains a difficult problem to solve. If it is assumed that the devices may be required to operate for periods that extend towards a decade or more, then it is probable that the device will need to be updated securely. This can come down to evolving understanding of embedded flaws in the system, or that the underlying security protocols have been obsoleted due to advances in cryptanalysis. This may be exacerbated by advances in quantum computing that could lead to easily finding keys by analyzing passively captured packets.

Several other challenges that need additional attention include advances in understanding the risks posed to individuals' privacy given the number of devices with which they will interact and their capacity to build detailed pictures of peoples' private lives. This feeds into the aforementioned trustworthiness discussion, but is also closely related to recent initiatives such as the General Data Protection Regulation (GDPR)[10].

Reverse engineering of devices, as mentioned early in this chapter, will be a major cause for concern, as it is relatively straightforward for skilled attackers to launch side channel attacks on devices, e.g., extracting keys, finding weaknesses in the source code, modifying source code, possibly extracting proprietary information, and/or configuration of personal data (depending on the application, of course).

While there is much discussion of security and trustworthiness more broadly with regard to IoT, it is clear that there remains a substantial amount of research and development to be carried out in this space and that those tasked with implementing security for these systems face an extremely difficult challenge. While no system may be perfect, it is crucial to take securing applications into account at design time – devices, networks, and information – wherever possible and with the best available security mechanisms.

---

[10]https://www.eugdpr.org

ARCHITECTURE AND STATE-OF-THE-ART

CHAPTER OUTLINE

7.1 INTRODUCTION

In Chapter 5, we outlined the technology building blocks that form the basis of Internet of Things (IoT) solutions. Chapters 7 and 8 provide a detailed conceptual overview of an IoT reference model and architecture introduced in Chapter 4. The term "Architecture Reference Model" (ARM) is borrowed from the IoT Architecture (IoT-A) European research project because the objective of the following chapter is quite similar to that of the IoT-A ARM [19]. A reference model is a model that describes the main conceptual entities and how they are related to each other, while the reference architecture aims at describing the main Functional Components of a system as well as how the system works, how the system is deployed, what information the system processes, etc.

Internet of Things. https://doi.org/10.1016/B978-0-12-814435-0.00019-5

An ARM is useful as a tool that establishes a common language across all the possible stakeholders of an IoT system. It can also serve as a starting point for creating concrete architectures of real systems when the relevant boundary conditions have been applied, for example, stakeholder requirements, design constraints, and design principles.

Before delving into a description of an ARM in Chapter 8, this chapter presents the state-of-the-art of fragments of IoT reference models/high-level architectures/ARMs as well as the corresponding state-of-the-art in IoT frameworks. The fragments typically originate from different Standards Development Organizations (SDOs), alliances, and community activities which address partially the concerns addressed by an ARM and provide information on some parts of a whole system. The order of presentation is not important. The choice of groups or activities to focus on in this chapter is based on several factors such as maturity, influence to other SDOs, alliances or activities, impact, and popularity.

7.2 ITU-T

The Telecommunication sector of the International Telecommunication Union (ITU-T) has been active on IoT standardization since 2005 with the Joint Coordination Activity on Network Aspects of Identification Systems (JCA-NID), which was later renamed to Joint Coordination Activity on IoT (JCA-IoT) in 2011. During the same year apart from this coordination activity on IoT, ITU-T formed the specific IoT Global Standards Initiative (IoT-GSI) activity in order to address specific IoT-related issues.

The IoT-GSI concluded its activities in July 2015 and a new Study Group 20 (SG20)[1] on "IoT and its applications including smart cities and communities" was formed. The SG20 formed two working parties (similar to Working Groups in other SDOs) to address seven questions on topics such as requirements, use cases, architecture, security, privacy, end-to-end connectivity, applications and supporting platforms, research and emerging technologies, terminology and evaluation, and assessment of smart cities and communities. The SG20 deliverables still reference the main architecture deliverable (ITU-T Recommendation Y.2060[2] [94,215]) as a base for all the rest of the recommendations produced by SG20 and therefore it will be described briefly in this section.

The ITU-T Recommendation Y.2060 provides an overview of the IoT space with respect to ITU-T. This recommendation describes a high-level overview of the IoT domain model and the IoT functional model as a set of Service Capabilities similar to ETSI M2M (see Appendix A).

The ITU-T IoT domain model includes a set of physical devices that connect directly or through gateway devices to a communication network that allows them to exchange information with other devices, services, and applications. The physical world of things is reflected in an information world of virtual things that are digital representations of the physical things (not necessarily a one-to-one mapping because multiple virtual things can represent one physical thing). The devices in this model include mandatory communication capabilities and optional sensing, actuation, and processing capabilities in order to capture and transport information about the things.

---

[1]https://www.itu.int/en/ITU-T/about/groups/Pages/sg20.aspx
[2]https://www.itu.int/rec/T-REC-Y.2060-201206-I

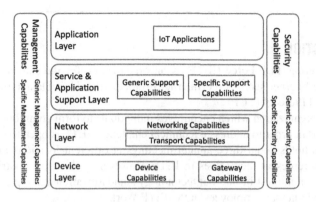

FIGURE 7.1

ITU-T IoT Reference Model (redrawn from Figure 4, ITU-T Y.2060 [215]).

Regarding the Service Capabilities (Figure 7.1), starting from the Application Layer the ITU-T IoT model considers this layer as the host of specific IoT applications (e.g., remote patient monitoring). The Service and Application Support Layer (otherwise known as Service Support and Application Support Layer) consists of generic service capabilities used by all IoT applications, such as data processing and data storage, and specific service capabilities tailored to specific application domains, such as e-health or telematics. The Network Layer provides networking capabilities such as Mobility Management, Authentication, Authorization, and Accounting (AAA), and Transport Capabilities such as connectivity for IoT service data. The Device Layer includes Device Capabilities and Gateway Capabilities. The Device Capabilities include, among others, the direct device interaction with the communication network and therefore the Network Layer Capabilities, the indirect interaction with the Network Layer Capabilities through Gateway Devices, any ad hoc networking capabilities, as well as low-power operation capabilities (e.g., capability to sleep and wakeup) that affect communications. The Gateway Device Capabilities include multiple protocol support and protocol conversion in order to bridge the Network Layer capabilities and the device communication capabilities. In terms of Management Capabilities, these include the typical FCAPS (Fault, Configuration, Accounting, Performance, Security) model of capabilities as well as Device Management (e.g., device provisioning, software updates, activation/deactivation), network topology management (e.g., for local and short-range networks), and traffic management. Specific management functionality related to a specific application domain is also included among the Management Capabilities. With respect to the Security Capabilities, this layer represents a grouping of different Security Capabilities required by other layers. The capabilities are grouped generically, such as AAA and message integrity/confidentiality support, and specifically, such as ones that are tailored to the specific application, e.g., mobile payment.

Comparing the ITU-T and ETSI M2M/oneM2M model/architecture, the two approaches are very similar in terms of Service Capabilities for M2M. However, the ITU-T IoT domain model with physical and virtual things and the physical and virtual world model shows the influences of more modern IoT architectural models and references such as the IoT-A [19] while the ETSI M2M/oneM2M focused more on the Telecommunication aspects of M2M.

7.3 IETF

7.3.1 INTRODUCTION

The Internet Engineering Task Force (IETF) is an organization for the standardization of the technologies used in the Internet such as the Internet Protocol (IP), User Datagram Protocol (UDP), or Transmission Control Protocol (TCP).

IETF is an open organization which encourages individuals to form common interest and common goal Working Groups in order to address a technical challenge. IETF is open to everyone without any participation fee and holds about three annual meetings at different locations in the world. Individuals are supposed to represent themselves in IETF. However, each active contributor belongs to an organization that has an interest in Internet technology standardization and typically the interest of an organization is voiced via their employees in the IETF Working Groups.

Individuals with a common interest or a common problem hold Bird of a Feather (BoF) meetings in order to submit an application to the IETF government bodies to form a Working Group and work on the common problem. When a Working Group has reached its goals (the ones set upon creation) it concludes and dissolved but there is a possibility that it may resume in the future if, e.g., a protocol needs updates after its specification is done. An IETF Working Group produces documents with three possible levels which reflect the maturity of a proposed solution: (a) individual drafts, known as Internet Drafts, (b) Working Group drafts, also known as Active Internet Drafts, and (c) IETF standards, also known as Request For Comments (RFC). The naming convention of an IETF document matches the level of the document. Individual drafts are drafts of standards written by one or more individuals that share a common view of the problem and the solution. Individual drafts have the name of the first author in the document name, reflecting that the solution to the specified problem is probably an opinion of one or more individuals. If the solution becomes more mature and is accepted by the majority of the Working Group, then the individual draft becomes a Working Group draft and the set of contributing individuals may change in size. The name of a Working Group draft document does not include any names of individuals. When a Working Group draft addresses all the objectives of the problem at hand it goes into a round of reviews within IETF and it becomes an RFC. An RFC is given a name "RFCXXXX"[3]; for example the RFC describing IP is RFC791.

7.3.2 IETF IOT-RELATED WORKING GROUPS

At the time of writing of this book, IETF has defined at least the following Working Groups for addressing IoT-related issues:

1. **6lowpan**[4]: The 6lowpan Working Group (see Figure 7.2A) has concluded its tasks and produced a number of RFCs (RFCs 4919, 4944 (6LoWPAN), 6282, 6568, 6606, 6775). The main RFC, RFC4944, is 6LoWPAN, an IPv6 adaptation layer over Low power Wireless Personal Networks (WPANs) and especially IEEE 802.15.4. RFC 4944 includes the specification of Header Compression for IPv6 UDP and TCP headers, fragmentation/reassembly of packets that do not fit in single IEEE 802.15.4 packets and packet formats for mesh routing (while the actual mesh rout-

---

[3]The general reference for an IETF RFC with a serial number "XXXX" is https://doi.org/10.17487/rfcXXXX.
[4]https://datatracker.ietf.org/wg/6lowpan/about/

ing protocol specification is left open). Among other notable RFCs is RFC6282 which specifies more details on the IPv6 and UDP Header Compression (6LoWPAN HC) and an efficient IPv6 Neighborhood Discovery (6LoWPAN ND) protocol over constrained networks. When the 6lowpan Working Group was active, its members attempted to define an adaptation layer specification on top of Bluetooth Low Energy (BTLE, BLE) and DECT Ultra Low Energy (DECT ULE). However, these specifications were moved before the conclusion of the 6lowpan Working Group to another currently active Working Group, the 6lo Working Group.

2. **6lo**[5]: The 6lo (IPv6 over Networks of Resource-constrained Nodes) Working Group (see Figure 7.2A) focuses on the work that facilitates IPv6 connectivity over networks of constrained nodes with limited power, memory, and processing resources. The limited node resources have an impact on the protocol state, code space, energy consumption, and networking availability. Examples of 6lo work items that the 6lo Working Group focuses on are IPv6-over-"foo" adaptation layer specifications using 6lowpan, information and data models (e.g., MIB modules), and protocol optimizations (e.g., Header Compression). The 6lo closely coordinates with the 6man (IP-over-foo for nonconstrained devices), the lwig and intarea Working Groups. The 6lo Working Group has produced a number of RFCs (RFCs 7388, 7400, 7428, 7668, 7973, 8025, 8065, 8066, 8105, 8163) and there are currently several Working Group drafts being developed. The most notable RFCs and Working Group drafts focus on the following link layers: BLE (RFC7668), Bluetooth Mesh (BTMesh, expired Working Group draft[6]), DECT ULE (RFC8105), Master-Slave/Token-Passing (MS/TP, RFC8163), G.9959 ITU-T(RFC7428) and Near Field Communication (NFC, Working Group draft).

3. **6tisch**[7]: The 6tisch (IPv6 over the Timeslotted Channel Hopping (TSCH) mode of IEEE 802.15.4e) Working Group focuses on providing the necessary adaptation layers for IPv6 over the TSCH mode of IEEE 802.15.4e. The IEEE 802.15.4e is an amendment of the IEEE 802.15.4 PHY/MAC layer specification that introduces the time-slotted access for the IEEE 802.15.4 MAC layer in order to enable industrial automation and process control. It is related to WirelessHART and ISA100.11a enables the application of IPv6 (an Information Technology (IT) protocol) to the Operational Technology (OT) space. The Working Group has produced two RFCs at the moment, RFC 7554 and RFC 8180, which describe the problem and the minimal configuration. There are several Working Group drafts developed at the moment that outline the terminology for this problem and describe the architecture and the 6TiSCH Operational sublayer (6top), as well as specify some security details of the adaptation layers. The nodes communicate with Time Division Multiple Access (TDMA) and use the 6top to coordinate the use of the different time-slots.

4. **CoRE**[8]: The CoRE (Constrained RESTful Environments) Working Group focuses on providing specifications of RESTful application level protocols for accessing resources on constrained devices. The Working Group has produced a number of RFCs (RFCs 6690, 7252, 7390, 7641, 7959, 8075, 8132) and there are currently several Working Group drafts being developed. The most notable RFC, RFC7252, specifies an efficient, client-server messaging protocol over UDP called Constrained Application Protocol (CoAP). CoAP is used by applications to access simple

---

[5]https://datatracker.ietf.org/wg/6lo/about/

[6]https://www.ietf.org/archive/id/draft-ietf-6lo-blemesh-02.txt

[7]https://datatracker.ietf.org/wg/6tisch/about/

[8]https://datatracker.ietf.org/wg/core/about/

resources such as sensors or actuators on constrained devices using RESTful methods such as GET/PUT/POST/DELETE. Other RFCs and Working Group drafts focus on issues such as the discovery of resources (RFC6690 on CoRE link formats and a Working Group draft on Resource Directory), subscribe-notify functionality (OBSERVE method, RFC 7641), publish-subscribe functionality (Working Group draft), group and block communication (RFC7390, RFC7959), interfaces (Working Group draft), guidelines for HTTP-CoAP adaptation on gateway type of devices (RFC8075), object security (OSCORE, Working Group draft), resource representation data models (SenML, YANG, Working Group drafts), management interface (CoAP Management Interface (CoMI), Working Group draft), CoAP over TCP, TLS, WebSockets (Working Group draft), etc. Although the CoRE Working Group defines a RESTful approach to resource access it does not define more details such as standard resource names (e.g., /temp) or types (e.g., "temperature") for specific products (e.g., a temperature sensor node from Honeywell) as this is in the scope of industry and market alliances such as the IP for Smart Objects Alliance (IPSO) or Open Mobile Alliance (OMA). Just a few months before the writing of this book the IPSO Alliance has merged with OMA and the resulting organization is now called OMA Specworks.

5. **roll**[9]: The roll (Routing Over Low power and Lossy networks) Working Group focuses on providing routing protocols for lossy networks of constrained devices. Examples of low-power lossy networks are ones based on IEEE 802.15.4, Bluetooth, Low Power Wi-Fi, wired or other low-power Powerline Communication (PLC) links. The Working Group focused on providing routing solutions for connected home, building, and urban sensor networks and has produced several RFCs (RFCs 5548, 5673, 5826, 5867, 6206, 6550, 6551, 6552, 6719, 6997, 6998, 7102, 7416, 7731, 7732, 7733, 7774, 8036, 8138) on the topic. The main one is RFC 6650 (RPL: IPv6 Routing Protocol for Low-Power and Lossy Networks) focusing on routing using ICMPv6 where each device can be part of multiple routing topologies such as a star or tree or full-mesh. Another important RFC is RFC 7732, which defines a Multicast Protocol for Low-Power and Lossy Networks (MPL). Also worth mentioning is RFC 8138, which describes a 6LoWPAN Routing Header (6LoRH) compression for RPL used by other Working Groups such as 6tisch.

6. **ace**[10]: The ace (Authentication and Authorization for Constrained Environments) Working Group focuses on providing an authentication and authorization framework inspired by the Web Authorization Protocol (OAuth) 2.0 and customized for constrained environments. The ACE specification (currently a Working Group draft) provides an extensible framework for which more detailed specifications are expected to be described in different profile specifications. The framework assumes the existence of efficient transfer protocols such as CoAP and aims at providing authorization for accessing resources with RESTful methods (GET, PUT, POST, DELETE) identified by a URI and hosted on Resource Servers on constrained nodes. So far the Working Group has produced one RFC, RFC7744, that describes the use cases motivating the ACE framework and several Working Group drafts describing the framework, a Datagram Transport Layer Security (DTLS) profile, and a specification of an efficient format for the authorization tokens.

7. **lwig**[11]: The IETF lwig (Light-Weight Implementation Guidance) Working Group focuses on providing guidelines for the implementers of the IETF IP protocol suite on the smallest devices. The

---

[9]https://datatracker.ietf.org/wg/roll/about/
[10]https://datatracker.ietf.org/wg/ace/about/
[11]https://datatracker.ietf.org/wg/lwig/about/

goal of the Working Group is to be able to build minimal yet interoperable IP-capable devices. The purpose of the Working Group is to collect experiences from implementers of IP stacks on constrained devices. The output of lwig is a set of documents that describe implementation techniques for reducing complexity, memory footprint, or power usage while ensuring conformance to the relevant specifications and interoperability with other devices. The topics for this Working Group are chosen from these protocols: IPv4, IPv6, UDP, TCP, ICMPv4/v6, MLD/IGMP, ND, DNS, DHCPv4/v6, IPSec, 6LoWPAN, CoAP, RPL, SNMP, and NETCONF. Currently the Working Group has produced two RFCs: (a) RFC7228, which includes a terminology for constrained devices and a classification of the different device resources, and (b) RFC7815, which provides guidelines for the implementation of the Internet Key Exchange version 2 (IKEv2) protocol on constrained devices. The Working Group includes several Working Group drafts that currently focus on guidelines for the implementation of IETF CoAP, energy-efficient implementations, TCP usage, etc.

8. **cbor**[12]: The cbor (Concise Binary Object Representation Maintenance and Extensions) Working Group aims at updating the Concise Binary Object Representation (CBOR) format specification originally developed and released in 2013 as RFC7049. CBOR extends the JavaScript Object Notation (JSON) data interchange format to include binary data and an extensibility model that features extremely small parser code size and small message size. These goals make CBOR suitable for inclusion in other IoT-related protocol specifications.

9. **ipwave**[13]: The ipwave (IP Wireless Access in Vehicular Environments) Working Group focuses on Vehicle-to-Vehicle (V2V) and Vehicle-to-Infrastructure (V2I) use cases and the adaption of the IEEE 802.11 Outside the Context of a Basic Service Set (802.11-OCB, formerly known as "IEEE 802.11p") for the transmission of IPv6 packets. The Working Group was formed in September 2016 and did not have time to produce any RFC. At the time of writing, there are two Working Group drafts, one describing the use cases and problem statement and another specifying an IPv6 adaptation layer on top of IEEE 802.11 Outside the Context of a Basic Service Set (IEEE 802.11-OCB). The use cases and problem statement draft includes topics such as V2V and V2I use cases, vehicular network architectures, standardization activities, IP address autoconfiguration, routing, mobility management, DNS naming service, service discovery, security, and privacy.

10. **lpwan**[14]: The lpwan (IPv6 over Low Power Wide-Area Networks) Working Group focuses on enabling IPv6 connectivity over the following Low-Power Wide-Area technologies: Sigfox, Long Range (LoRa), Wireless Smart Ubiquitous Networks (Wi-SUN) and Narrowband IoT (NB-IoT). According to the Working Group description, IETF 6lo techniques cannot be easily adapted to fit the characteristics of these technologies which are constrained with respect to networking features (low bandwidth, high latency, etc). The Working Group aims at producing descriptions of the compression and segmentation/reassembly of UDP/IP packets and compression of CoAP messages. The Working Group was formed in September 2016 and did not have time to produce any RFC. At the time of writing, there are few Working Group drafts that contain the problem description and compression specifications of UDP/IPv6 and CoAP.

---

[12]https://datatracker.ietf.org/wg/cbor/about/
[13]https://datatracker.ietf.org/wg/ipwave/about/
[14]https://datatracker.ietf.org/wg/lpwan/about/

11. **homenet**[15]: The homenet (Home Networking) Working Group focuses on the use of IPv6 in the evolved landscape of the modern residential network, which may include multiple different networks with respect to link layer technologies (such as low-powered sensor networking technologies) or security. More specifically the Working Group addresses the issues of prefix configuration for routers, managing routing, name resolution, service discovery and network security for multisegment residential networks. The Working Group had an initial goal of producing an architecture document and recommendations for changes in existing standards for addressing the aforementioned issues. The architecture document was published as an informational RFC7368, while the Working Group has produced three more RFCs, 7695, 7787, and 7788, that focus on distributed prefix assignment, the Distributed Node Consensus Protocol (DNCP) for ensuring that all the nodes in the home network have the same view of some specific state (e.g., the specific state of a home automation device), and the Home Networking Control Protocol (HNCP). The DNCP is a generic state synchronization protocol, agnostic with respect to the structure of the state transported among nodes (e.g., Type-Length-Value (TLV) tuples) and an abstract protocol, which needs specific profiles (state structures, configuration options) in order to be implementable. The HNCP defines the specific structure of the state to be shared among home router nodes and advanced hosts in order to facilitate network discovery, prefix assignment, name resolution, and service discovery. The Working Group has also been working on several Working Group drafts on routing, naming, and service discovery.

12. **dice**[16]: The dice (DTLS In Constrained Environments) Working Group has concluded its activities. It focused on providing specific profiles for the Datagram Transport Layer Security (DTLS) on constrained environments. Constrained environments include constrained devices, e.g., with limited memory resources or constrained networks, e.g., small-sized packets. It produced one RFC, RFC7925, which provided guidelines for the use of Transport Layer Security (TLS) and DTLS in IoT deployments for the common cases of sensor data collection of actuator control command dispatching.

13. **cose**[17]: The cose (CBOR Object Signing and Encryption) Working Group has concluded its activities. It focused on specifying CBOR-based object signing and encryption formats. The motivation of COSE was similar to the JSON Object Signing and Encryption (JOSE) Working Group. That is to provide an object signing and encryption specification using the efficient CBOR data representation for constrained environments. This specification is described in the only RFC produced by this Working Group, RFC8152. More specifically the specification describes the creation and processing of signatures, Message Authentication Codes, and encrypted content using CBOR for serialization as well as the representation of cryptographic keys using CBOR.

7.3.3 IETF FRAGMENTS OF AN ARCHITECTURE

As seen from the list of the IoT-related Working Groups, the number of Working Groups is large, and their scope is quite diverse. In this section, an attempt is made to relate the different Working

---

[15] https://datatracker.ietf.org/wg/homenet/about/

[16] https://datatracker.ietf.org/wg/dice/about/

[17] https://datatracker.ietf.org/wg/cose/about/

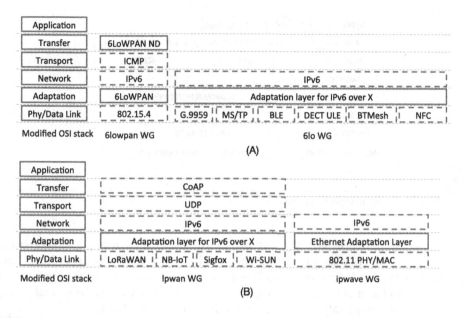

FIGURE 7.2

IETF 6lowpan, 6lo, lpwan, and ipwave Working Groups and specification scope.

Group results in a few common frames. Since IETF produces documents that often describe protocols, architectures, or data models, these serve as the selected common frames.

Some of the existing specifications, or specifications under development, define protocols that could be mapped on a communication stack similar to the International Standards Organization (ISO) Open System Interconnection (OSI) model (Figure 7.2A). However, the figure describes these protocols in the context of a modified OSI stack with two more layers: (a) an Adaptation layer between the Network layer and each underlying Phy/Link layer and (b) a Transfer layer, which includes protocols whose functionality lies in between the Transport and Application layers. An example of an Adaptation layer protocol is 6LoWPAN (Figure 7.2A) and examples of Transfer layer protocols are HTTP and CoAP (Figure 7.2B).

It should be noted however that these layers are not strictly defined by IETF. They are a convention used in this book for a better presentation of the different protocols. In some cases, a copy of a layer, e.g., "Layer", is also used in a figure in order to clarify cases such as the use of an Application layer protocol for a lower-layer encapsulation. Encapsulation is a pattern often used in layered communication protocols where the frames or messages of a protocol on a certain layer are encapsulated in frames or messages of a lower layer. In general, in the context of communication protocols based on message exchanges, a frame is defined as a series of bits or bytes with three main parts, a header, describing what this message is about (metadata about the message), a payload or the main content of the message, and an optional trailer typically used as a checksum for the first two parts of the message. Encapsulation is the implementation of the fact that in a layered communication stack the functionality of a lower layer is used for implementing the functionality of the layer above it. As a result in certain

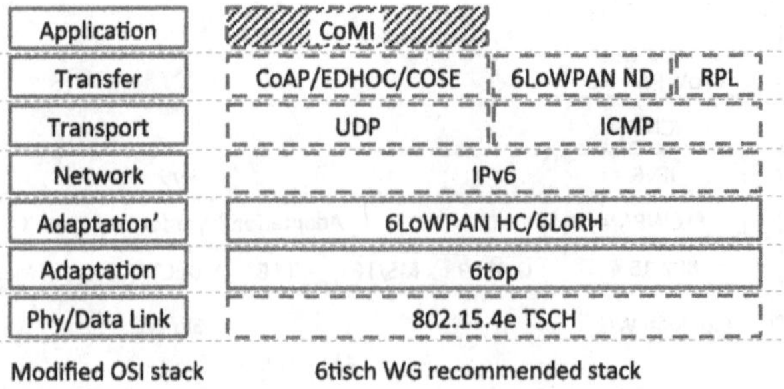

FIGURE 7.3

IETF 6tisch Working Group and specification scope.

cases, a Transfer layer protocol such as CoAP may be encapsulated in an Application layer protocol (Figure 7.4) which is in turn encapsulated in a Transport layer protocol. In this case, CoAP appears to the applications on the Application layer as a Transfer layer protocol, i.e., below the Application layer; however, the encapsulation of CoAP in an Application layer protocol violates the general principle of a higher-layer protocol using only services/functionality from a layer below it.

The text below summarizes the important IoT-related IETF specifications with respect to the protocol layer or the architectural element that they describe. To the extent possible the descriptions start from protocols defined in lower layers and move upwards in the stack. Solid white-filled rectangles are used to denote that the respective Working Group has defined the specific protocol layer stated in the rectangle, dashed rectangles show the protocol layers that are not defined by the respective Working Group (but defined by other Working Groups) but they are assumed as existing and recommended by the Working Group, and solid rectangles with a fill pattern denote a data model or profile defined by the respective Working Group as opposed to a full protocol layer definition.

Figure 7.2A and B show the contributions of the 6lowpan, 6lo, lpwan, and ipwave Working Groups. The main theme is that these Working Groups have been defining adaptation layers for IPv6 for different PHY/MAC technologies. The 6lowpan Working Group defined the adaptation layer between IEEE 802.15.4 to IPv6 and also 6LoWPAN Neighborhood Discovery (6LoWPAN ND) on top of ICMP as described earlier. The 6lo Working Group has defined adaptation layers for IPv6 over G.9959, MS/TP BLE, DECT ULE, BTMesh, and NFC. The lpwan Working Group aims to define adaptation layers of IPv6 over LoRaWAN, NB-IoT, Sigfox, and Wi-SUN and the recommended layers on top of IPv6 are UDP and CoAP. The ipwave Working Group defines an adaptation layer of IPv6 over IEEE 802.11-OCB.

Figure 7.3 shows the recommended stack for the 6tisch Working Group. The Working Group defines one adaptation layer, the 6top, and reuses the ideas of the 6lowpan Working Group 6LoWPAN HC and 6LoRH to provide an adaptation solution for IPv6 for IEEE 802.15.4e TSCH. The recommended stack on top of IPv6 includes UDP/CoAP/EDHOC/COSE/CoMI, ICMP/6LoWPAN ND, and RPL.

Figures 7.4 and 7.5 show the contribution of the CoRE Working Group, which is the single Working Group with the most RFCs and Working Group drafts. Figure 7.4 shows the main specifications of the CoRE Working Group, which include the Constrained Application Protocol (CoAP) initially defined over UDP with DTLS supporting the secure version of the protocol. The IETF CoAP RFC7252

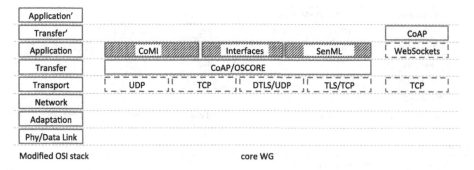

FIGURE 7.4

IETF CoRE Working Group and specification scope.

describes the Transport and Transfer Layers, which essentially define the transport packet formats, reliability support on top of UDP, a RESTful application protocol with GET/PUT/POST/DELETE methods similar to HTTP with CoAP clients operating on CoAP server resources, and finally the secure version of the protocol. A CoAP server is just a logical protocol entity, and the name "server" does not necessarily imply that its functionality is deployed on a very powerful machine; a CoAP server can be hosted on a constrained device. Recently the WG has defined CoAP over TCP and the use of TLS to secure the underlying TCP transport as well as transporting CoAP within WebSockets which are transported over TCP. This is the reason that Figure 7.4 introduces additional Transfer and Application layers (Transfer', Application' respectively) in order to show this encapsulation of CoAP in WebSockets which in turn is transported in TCP frames. The specifications for CoMI, Interfaces, and SenML do not strictly define protocols or protocol behavior but define interfaces (CoMI, Interfaces), some architecture fragments (CoMI), and a data model (SenML) for the CoAP endpoints and the information generated by them or required by them. Moreover, at the time of the writing of this book, there is a Working Group draft for OSCORE (Object Security for Constrained RESTful Environments) which provides an object security solution based on CBOR and COSE as a complement to transport security based on DTLS or TLS. OSCORE provides authentication, encryption, integrity, and replay protection for CoAP, is designed for message traversal over multiple different underlying protocols (such as HTTP and CoAP in case of the message traversing an HTTP/CoAP Proxy), and can secure both unicast and multicast communication requests with unicast responses.

Before describing some details about these three specifications, it is worth noting that the IETF stack or specifications for IoT do not currently include any specifications that are similar to the profile specifications of other IoT technologies such as ZigBee (please refer to Chapter 5). By a profile specification, we mean a document that describes a list of profile names and their mappings to specific protocol stack behavior, specific information model, and specific serialization of this information model over the relevant communication medium. An example of an excerpt of a profile specification, e.g. a "Temperature" profile, would mandate that: (a) the profile should support a resource called /temp, (b) the resource /temp must respond to a GET method request from a client, and (c) the response to a GET method request shall be a temperature value in degrees Celsius formatted as a text string with the format "<temperature value encoded in a decimal number > °C" (e.g., "10 °C"). It must be noted that the device profiles are used for ensuring interoperability between market products, and therefore

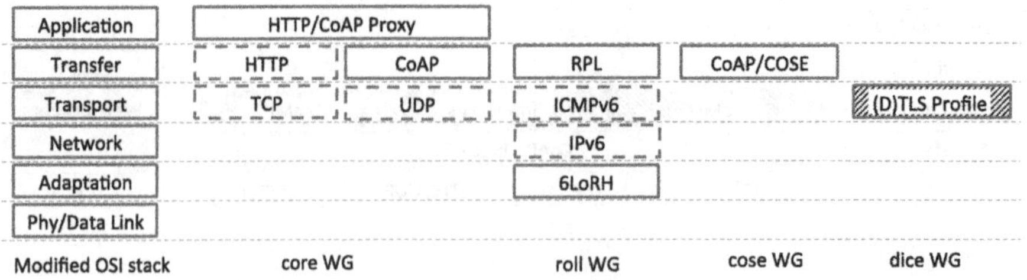

FIGURE 7.5

IETF Working Groups and Specification Scope.

it is not the responsibility of IETF to specify such details. Therefore, a step towards the specification of profiles was taken by the Internet Protocol for Smart Objects (IPSO) Alliance, which is mainly a market promoting alliance. For this purpose, IPSO has published the Smart Objects Guidelines in two forms (Starter [95] and Expansion Packs [96]), which provide an object model for commonly used sensors and actuators. The common object model is based on the Lightweight M2M (LWM2M 1.0.1) specification from the OMA [97].

The Interfaces specification [98] from the CoRE Working Group outlines in a paper specification what typically a Web Application Description Language (WADL)[18] file specifies in detail in a machine-readable form. A WADL file describes the specific interface of a RESTful web service, i.e., the types of REST methods (e.g., GET, PUT, POST, DELETE) allowed, the type of parameters expected by the specific REST endpoint, and the response format or content type. The Interfaces specification defines a few types of "standard" IoT resources such as sensors, actuators, parameters, and collections of resources. It defines the allowed methods for each type of resource and the return or requested content type of requests or responses to these resources. These interface specifications are identified with specific identifiers which are used in the CoRE Link Format (RFC6690[19], see below). The specification is one step towards market profiles for different devices and resources but does not provide anymore details.

The CoMI specification [99] describes an interface and parts of an architecture for enabling the management of CoAP endpoints in the same way as the management of networked entities. The specification assumes the use of YANG data model (RFC7950[20]) for the requests and responses between a management client and a management server on a CoAP device.

The CoRE Working Group also specifies the media types for representing simple sensor measurements and device parameters using Sensor Measurement Lists (SenML) [100]. SenML describes the data model and the content media types for the responses to CoAP requests sent to the sensor and parameter type of resources. The representations are defined in JSON, Concise Binary Object Representation (CBOR), Extensible Markup Language (XML), and Efficient XML Interchange (EXI), which share the common SenML data model.

Figure 7.5 shows the contribution of the CoRE Working Group to the specification of an HTTP/CoAP proxy for requests that originate from HTTP Clients and are directed to CoAP Servers.

---

[18]https://www.w3.org/Submission/wadl/
[19]https://doi.org/10.17487/rfc6690
[20]https://doi.org/10.17487/rfc7950

Figure 7.7 shows the architectural elements and the request traversal over the stacks for an HTTP/CoAP proxy.

Figure 7.5 also shows the stack layers defined by three other IETF Working Groups, the roll Working Group, the cose Working Group, and the dice Working Group. The roll Working Group defined among others the IPv6 Routing Protocol for Low-Power and Lossy Networks (RPL, RFC6650[21]) and the 6LoRH (RFC8138[22]). The cose Working Group defined the CBOR Object Signing and Encryption (COSE) protocol. The dice Working Group did not specify a protocol layer per se but rather TLS and DTLS profiles for IoT devices.

It can be noted that some IoT-related IETF specifications do not necessarily specify communication stack protocols but architecture elements or data models which are, along with protocols, fragments of an architecture.

Apart from the core of the specifications, the IETF CoRE Working Group includes several other interesting RFC and Working Group draft specifications that sketch parts of an architecture for IoT.

The CoRE Link Format RFC6690[23] describes a discovery method for the CoAP resources of a CoAP server. For example, a CoAP client sending a request with the GET method to a specific well-defined server resource (./well-known/core) should receive a response with a list of CoAP resources and some of their capabilities (e.g., resource type, interface type). As seen earlier the accompanying draft specification, the CoRE interface specification [98], describes interface types and corresponding expected behavior of the RESTful methods (e.g., a sensor interface should support a GET method). The response serialization (e.g., if the response is a temperature value in degrees Celsius) is specified by the SenML specification [100].

The IETF CoRE Working Group has also produced a draft specification for a Resource Directory [101]. A Resource Directory (Figure 7.6A) is a CoAP server resource (/rd) that maintains a list of resources, their corresponding server contact information (e.g., IP addresses or Fully Qualified Domain Name (FQDN)), their type, the interface, and other information similar to the information that is specified by the CoRE Link Format RFC 6690[24].

An RD plays the role of a rendezvous mechanism for CoAP Server resource descriptions, in other words, for devices to publish the descriptions of the available resources and for CoAP clients to locate resources that satisfy certain criteria such as specific resource types (e.g., temperature sensor resource type).

While the Resource Directory is a rendezvous mechanism for CoAP Server resource descriptions, IETF does not have a corresponding function for a rendezvous mechanism for CoAP Server resource presentations. An individual draft specifying a Mirror Server [102] (Figure 7.6B) was not developed into a Working Group draft and therefore expired. This architecture functional gap is expected to be filled in by the CoAP Publish-Subscribe or pub-sub specification which is a Working Group draft at the time of writing of this book (Figure 7.6C). A CoAP server resource (/ps), also called a CoAP pub-sub broker, serves as the endpoint for CoAP clients to publish their resource representations and other CoAP clients to receive these representations if they have previously subscribed to them. This functionality is especially useful when the publishing clients have intermittent connectivity or the devices

[21] https://doi.org/10.17487/rfc6650
[22] https://doi.org/10.17487/rfc8138
[23] https://doi.org/10.17487/rfc6690
[24] https://rfc-editor.org/rfc/rfc6690.txt

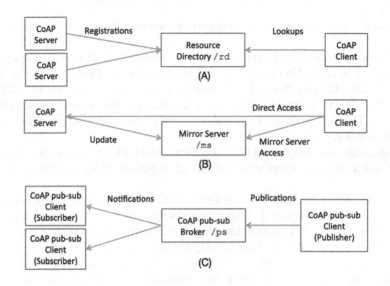

FIGURE 7.6

IETF CoRE Functional Components: (A) Resource Directory, (B) Mirror Server, (C) Pub-Sub Broker.

hosting the publishing clients have long sleep cycles for energy efficiency conservation purposes. The publisher uses a hierarchical topic name in order to identify the pub-sub topic.

Because CoAP as an application protocol is not yet widely deployed, while HTTP is ubiquitous, the IETF CoRE Working Group has included the fundamentals of a mapping process between HTTP and CoAP in the IETF CoAP specification as well as a set of guidelines for the interworking between HTTP and CoAP as RFC 8075[25] (Figure 7.7A).

The interworking issues appear when an HTTP Client accesses a CoAP Server through an HTTP-CoAP proxy (Figure 7.7B). The mapping process is not straightforward for a number of reasons. The main reason is the different transport protocols used by the HTTP and CoAP: HTTP uses TCP while CoAP uses UDP. The guidelines recommend addressing schemes (e.g., how to map a CoAP resource address to an HTTP address), the mapping between HTTP and CoAP response codes, the mapping between different media types carried in the HTTP/CoAP payloads, etc. As an example, consider the case that an HTTP Client sends an HTTP request to a CoAP server (Figure 7.7B) through a Gateway Device hosting an HTTP-CoAP Cross Proxy. The Gateway Device connects to the Internet via an Ethernet cable using a LAN, and on the CoAP side, the CoAP server resides on a Sensor/Actuator Network (SAN) based on the IEEE 802.15.4 PHY/MAC. The HTTP request needs to include two host addresses, one for reaching the HTTP-CoAP Proxy and one for reaching the specific CoAP Server in the SAN. Moreover, the request needs a resource name for the resource endpoint on the CoAP Server. The default recommended address mapping is to append the CoAP resource address (e.g., `coap://s.example.com/light`) to the HTTP-CoAP proxy address (e.g., `https://p.example.com/hc/`), resulting in `https://`

---

[25] https://rfc-editor.org/rfc/rfc8075.txt

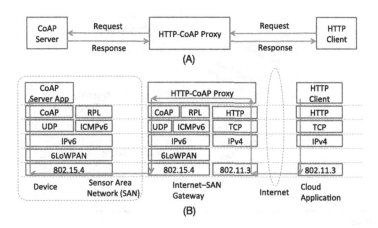

FIGURE 7.7

IETF CoRE HTTP Proxy: (A) possible configurations, (B) example layer interaction upon a request from a HTTP Client to a CoAP Server via an HTTP Proxy.

`p.example.com/hc/coap://s.example.com/light`. The request is in plaintext format and contains the method (`GET`). It traverses the IPv4 stack of the client, reaches the gateway, traverses the IPv4 stack of the gateway, and reaches the HTTP-CoAP proxy. The request is translated to a CoAP request (binary format) with a destination CoAP resource `coap://s.example.com/light`, and it is dispatched in the CoAP stack of the gateway, which sends it over the SAN to the end device. A response is sent from the end device and follows the reverse path in the SAN in order to reach the gateway. The HTTP-CoAP proxy translates the CoAP response code to the corresponding HTTP code, transforms the included media, creates the HTTP response, and dispatches it to the HTTP client. While the described example scenario seems straightforward, in practice, the HTTP-CoAP proxy needs to handle all problematic situations and peculiarities of the CoAP and HTTP protocols, e.g., asynchronous behavior of the Observe mode[26] of CoAP. An interested reader can refer to the relevant specifications for further information.

The IETF ace Working Group has specified an authorization framework for constrained environments. Authorization means that a client is granted access to a resource hosted on a device, the Resource Server (RS), and this exchange is mediated by one or multiple Authorization Servers (ASs). A fully specified authorization solution includes this framework and a set of profiles. The framework describes the architecture and interactions in generic terms while the profiles of this framework are additional specifications that define the use of the framework with concrete transport and communication security protocols (e.g., CoAP over DTLS). ACE is based on four building blocks: OAuth 2.0 (RFC6749[27]), CoAP (but not excluding other underlying protocols such as MQTT, BLE, HTTP/2, QUIC), CBOR, and COSE.

Figure 7.8 shows the architecture of ACE and the basic interactions. A client (C) intends to access a resource representation on an RS. The client contacts an AS to obtain a token. A token can be

[26]https://rfc-editor.org/rfc/rfc7641.txt
[27]https://rfc-editor.org/rfc/rfc6749.txt

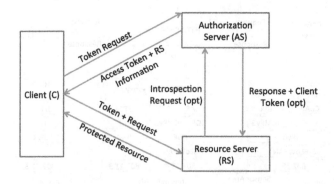

FIGURE 7.8

IETF ACE interactions.

either an access token or a proof of possession token. An access token is a data structure representing authorization permissions issued by the AS to the client. A proof of possession token is a token bound to a symmetric or asymmetric cryptographic key and is used by the RS to authenticate the client. The client receives the access token and potentially some information about the RS capabilities (RS information). The client presents the token and the specific access request to the RS. The RS may optionally validate the token by using an introspection request to the AS. If the token is self-contained and the RS can validate the token itself, then the introspection request is not necessary. If the token is valid the resource access request is granted and the RS responds with the resource representation protected by the chosen security protocol.

7.4 OMA

The OMA and the Internet Protocol for Smart Objects (IPSO) Alliance complement the IETF work with respect to semantics and device profiles. The two alliances recently merged into an SDO under the name OMA SpecWorks. However in this section we still use the old name, OMA, for brevity purposes.

OMA has proposed a simple Client-Server architecture for the realization of a simple Device and Data Management (DDM) platform. The OMA has specified version v1.0 [97] of the OMA Lightweight M2M (LwM2M)[28] client-server protocol and is currently preparing version v1.1.

The OMA LwM2M architecture follows the client-server model and the specification describes the protocol between a client and server from a client perspective. This means that the specification describes the changes occurring to the client when the server invokes the operations of the specified interfaces (Figure 7.9). The interfaces between a server in the LwM2M architecture and systems that reside outside are out of scope although practically these interfaces are important for the smooth integration of the server into a bigger IT system.

---

[28]https://www.omaspecworks.org/what-is-oma-specworks/iot/lightweight-m2m-lwm2m

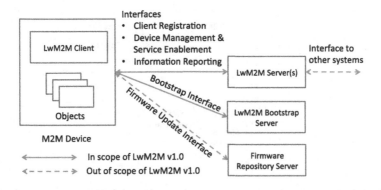

FIGURE 7.9

OMA LwM2M High Level Architecture.

There are three types of servers in the LwM2M architecture: (a) A LwM2M Server, (b) a Bootstrap Server, and (c) a Firmware Repository Server. These servers could be colocated on a single machine, but the specification leaves this choice to the developers of the system. The bulk of the specification covers the interactions between a LwM2M Server and an LwM2M Client. A LwM2M Server and a Bootstrap Server operate through the specified interfaces on LwM2M Client Objects while the Firmware Repository Server implements a firmware update protocol which is left open for OMA LwM2M. Firmware update procedures can also be implemented by the LwM2M Server through the Device Management and Service Enablement interface.

The OMA LwM2M v1.0 specification contains two main parts: (a) a generic specification of the involved entities and high-level descriptions of the interfaces and (b) a mapping between the interfaces and operations to a binding protocol, i.e., a protocol that implements the generic specification. While these two parts exist in the same document in v1.0, in v1.1 these two parts are expected to be split in two separate documents (OMA LwM2M Core and Transport specifications) in order to allow a different rate of update and evolution. It is worth noting that although the protocol implementing the OMA LwM2M Core specification is called a "Transport" specification, the current assumed protocol is CoAP which is not technically a transport layer (as in the ISO/OSI protocol stack model) protocol but a transfer layer protocol. As a result, for this book, the more suitable term "binding protocol" is used to denote the implementation protocol of OMA LwM2M.

Currently OMA LwM2M v1.0 assumes the following protocol stack, shown in Figure 7.10. The supported underlying transport layer protocols are UDP and SMS (either on the device or on a smartcard if the device actually has such hardware). If the device contains a Universal Integrated Circuit Card (UICC)/Subscriber Identity Module (SIM) that can send and receive SMS messages, then this system is considered secure (in terms of message security) and does not need to be protected. On the other hand, UDP and clear-text SMS messages are not secured by default, and therefore the Datagram Transport Layer Security (DTLS) is used. Then CoAP is used as a transfer layer on top of a secure messaging/transport layer. The LwM2M Layer describes the specifics of CoAP (e.g., which resources, which operations, options, etc. of CoAP are used) for the implementation of the generic interfaces between a LwM2M Server and a LwM2M Client. Objects are part of the data model of the LwM2M protocol, and the specification responsibility falls in both the CoAP and LwM2M layers.

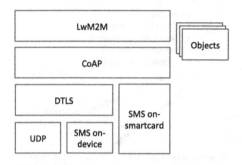

FIGURE 7.10

OMA LwM2M v1.0 Implementation protocol stack (redrawn with permission by Open Mobile Alliance from [97]).

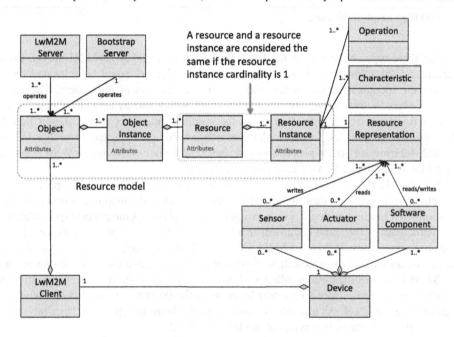

FIGURE 7.11

OMA LwM2M v1.0 Conceptual Model.

The Interface operations between a LwM2M Client and Server operate on LwM2M Client data structures, and through these data structures the Client is monitored or controlled. Figure 7.11 shows the Conceptual Model of OMA LwM2M. The notation and the figure styles follow the Universal Modeling Language (UML) notation. Chapter 8 provides a short introduction to the presentation style of classes and their relationships. In short, every device includes sensors, actuators, software components, and a LwM2M client. The LwM2M Client includes one or more Objects (data structures) on which the LwM2M Server and Bootstrap Server operate through the LwM2M Interfaces. The model

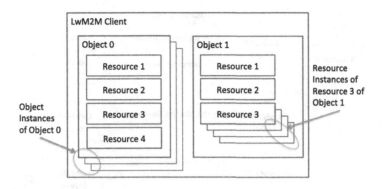

FIGURE 7.12

OMA LwM2M v1.0 Resource model (adapted and redrawn with permission by Open Mobile Alliance from [97]).

contains the resource model of LwM2M also shown in Figure 7.12. An Object contains one or more Object Instances which in turn contain one or more Resources. Each Resource may contain multiple Resource Instances but in the case that the Resource Instance count is only one, then a Resource and a Resource Instance are considered the same with respect to identification. A Resource includes a Resource Representation (similar to a resource representation in a RESTful model). A Resource contains also Operations (Read, Write, Execute) and Characteristics (e.g., Read-only, Read/Write, etc.). An Object, Object Instance, Resource, and Resource Representation may contain one or more Attributes which are metadata for the respective classes. One may note that in reality the resource model may be implemented as a set of data structures on the device memory. Then device sensors typically write the respective resource representations, device actuators read the relevant representations and affect the real world, while device software components may both read or write resource representations. Another observation is that practically a LwM2M Server and Bootstrap Server operate on device sensors/actuators and relevant software components through the shared device data structures (Resource model).

The LwM2M resource model that includes Objects, Object Instances, Resources, and Resource instances is shown in Figure 7.12. The main entities in the resource models are the Resources which represent main pieces of information that are made available by the LwM2M Client to the different LwM2M Servers. Resources are then grouped into logical groups (Objects). Each Object can be considered as a class in an object-oriented programming language. Object Instances are instances of Objects on a specific device. A Resource may have multiple Resource Instances as mentioned earlier. Objects, Object Instances, Resources, and Resource Instances have identifiers which are unique within each respective scope, i.e., Object identifiers are unique in the LwM2M Client, Object Instance identifiers are unique with respect to an Object, Resource identifiers are unique with respect to Object Instance identifiers, and Resource Instance identifiers are unique with respect to the Resource they refer to. OMA LwM2M v1.0 specifies eight Objects with identifiers "0" to "7" with their respective Resources while more are allowed to be created by LwM2M Servers. The description of these Objects and their Resources is hardcoded in the LwM2M v1.0 specification. These main Objects specify contact points and security credentials for the Bootstrap Server and LwM2M Server(s), access control information about Objects and Resources (i.e., which LwM2M Server can operate on which Objects

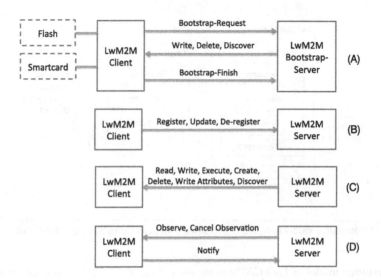

FIGURE 7.13

OMA LwM2M v1.0 Implementation protocol stack (redrawn with permission by Open Mobile Alliance from [97]). (A) Bootstrap Interface, (B) Client Registration Interface, (C) Device Management & Service Enablement Interface, (D) Information Reporting Interface.

and Resources with which operations), and several pieces of useful device configuration and information such as connectivity-related configuration and statistics, location information, firmware update configuration, and information.

A LwM2M Client needs to register with one or more LwM2M Servers so that the LwM2M Servers can perform different operations on the LwM2M Client Objects. Before being able to register the LwM2M Client needs to be provisioned with the contact information and the necessary security credentials for initiating this registration process and this is performed via bootstrapping. All these steps and several more are performed by the LwM2M Client, the LwM2M Server, or the LwM2M Bootstrap Server using the LwM2M interfaces. The main OMA LwM2M interfaces are the following (Figure 7.13):

- Bootstrap Interface: The Bootstrap Interface allows a LwM2M Client to request provisioning of bootstrap information. Bootstrap information includes contact information for LwM2M Server(s) that the client can register with, security credentials, access control rules, etc. The Bootstrap Server manages (reads/writes/deletes) the appropriate Objects containing the bootstrap information. Bootstrap Information could also be stored on a Flash memory or a Smartcard on the device that hosts the LwM2M Client.
- Client Registration Interface: Using the Bootstrap Information the Client can register/de-register with one or more LwM2M Server(s) that operate on LwM2M Objects and Resources.
- Device Management and Service Enablement Interface: This is the main interface of LwM2M. After the LwM2M Client registers, a LwM2M Server can operate on Objects, Object Instances, Resources, Resource Instances, and Attributes. The allowed operations are: read, write, execute,

create, delete, discover Objects, Object Instances, Resources, Resource Instances, and finally write Attributes.

• Information Reporting Interface: A LwM2M Server can instruct a LwM2M Client to report the indicated Resource representations periodically or when the representations match certain criteria such as exceeding certain thresholds. This is similar to the OBSERVE method of CoAP. The behavior of the LwM2M Client when instructed to Observe a certain Resource is controlled by the value of the Attributes associated with the Resource in question.

OMA hosts an Object and Resource Registry[29] which includes the eight basic Objects and their Resources. In addition to the OMA-defined Objects and Resources, OMA allows other organizations or companies to register their own Objects and Resource descriptions. IPSO Alliance, GSMA, and oneM2M are some organizations that have registered their own Object and Resource descriptions while among the enterprises that have done so we find ARM, AT&T, Cisco, Huawei, and Vodafone.

7.5 IOT-A AND IIRA

Among other contributions, the EU project Internet of Things Architecture (IoT-A)[30] developed an ARM during 2010–2013. The IoT Domain model and concepts of IoT-A are still useful as a tool for comprehending an IoT system as a whole. Since conceptually this ARM is considered as the most complete from an IT perspective, it is presented in a separate chapter, Chapter 8. Along with IoT-A, Chapter 8 also outlines the Industrial Internet Reference Architecture (IIRA) developed by the Industrial Internet Consortium (IIC)[31]. The IIRA is the state-of-the-art of an IoT reference architecture from an OT perspective. For a discussion about the differences between IT and OT the reader should refer to Chapter 8. The Reference Architectural Model Industrie 4.0 (RAMI 4.0) presented next is also a specialized OT type of reference architecture model.

7.6 RAMI 4.0

RAMI 4.0 is an architecture developed in Germany within the context of the Industrie 4.0 (I4.0) efforts.

The goal of I4.0 is to connect production with Information and Communications Technology (ICT) and enable seamless interaction driven by machine and customer data. Hence, production driven by autonomous behaviors is becoming more flexible, efficient, and resource-optimized. As I4.0 focuses on product development and production scenarios, horizontal and vertical integration is sought.

RAMI 4.0 aims to act as a unified architecture, upon which standards and use cases can be mapped and be better understood. For instance, the expectation is to be able to map requirements and standards to RAMI 4.0 and then identify commonalities and gaps, as well as overlapping standards. Subsequently

---

[29]http://www.openmobilealliance.org/wp/OMNA/LwM2M/LwM2MRegistry.html

[30]The IoT-A project was a Framework Program 7 (FP7) European Union Research project which was carried out between 2010–2013 by a partner consortium led by VDI/VDE Innovation + Technik GmbH. URL: https://cordis.europa.eu/project/rcn/95713_en.html

[31]http://www.iiconsortium.org

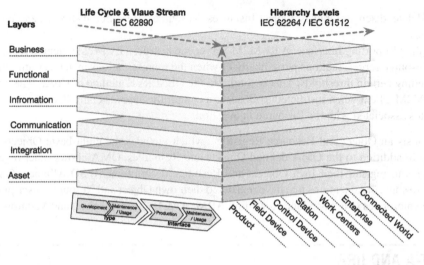

FIGURE 7.14

RAMI 4.0 Architecture.

one can select one common approach (e.g., a common standard), hence streamlining the interoperability and in parallel covering the needed requirements.

RAMI 4.0 [103], as shown in Figure 7.14, can be visualized as a 3D model, and its major aspects are shown along three axes:

- **Layers** axis: Six vertical layers, i.e., Asset, Integration, Communication, Information, Functional, and Business represent the machine properties decomposed according to their functionalities.
- **Lifecycle and Value Stream** axis: this horizontal axis refers to the IEC 62890 Lifecycle Management while also distinguishing between "types" and "instances" to differentiate between the design/prototyping and the actual manufacturing of a product.
- **Hierarchy Levels** axis: this horizontal axis categorizes according to the IEC 62264 the different functionalities within factories, i.e., product, field device, control device, station, work centers, enterprise, and connected world.

In a design typical for layered architectures, events may be exchanged among them, but only among two adjacent ones, and of course within each layer. RAMI 4.0 is designed according to Service-Oriented Architecture (SOA) principles and has the capability via its axes to provide clarity on complex processes pertaining to products and their different phases.

RAMI 4.0 defines also an I4.0 component, as shown in Figure 7.15, which can be a production system, an individual machine or station, or even an assembly inside a machine [103]. This I4.0 component and their contents follow a common semantic model and are structured so that connection with any other endpoint (i.e., other I4.0 components) is possible. To this end, an "administration shell" is used to turn any asset (i.e., the "thing") to an I4.0 component. The "administration shell" covers both the virtual representation and the technical functionality of the asset(s). From the deployment viewpoint, the asset and its administration shell can be decoupled, as the latter may be hosted in a higher-level IT system. However, if the asset, e.g., a CPS, has I4.0-compliant communication capabil-

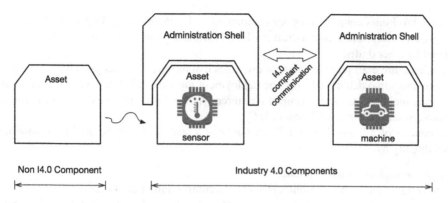

FIGURE 7.15

Industry 4.0 Component.

ity, it can also host the administration shell, e.g., in the controller of the machine, communicating via the available network interface.

I4.0 components are uniquely identified in the network, and they can interact with I4.0-compliant communication that is based on SOA principles. In that way, higher-level systems are able to access all I4.0 components in a well-defined manner, which streamlines communication and interoperability. Several other aspects pertaining to the semantics, communication, services and states, security and safety, Quality of Service, nestability, etc. are currently under active investigation.

In 2016, an effort was undertaken between the I4.0 and IIC in order to investigate common ground and align RAMI 4.0 and IIRA architectures, respectively [104]. It was identified that RAMI 4.0 focuses in-depth on next-generation manufacturing, while IIRA takes a more holistic view over several domains, e.g., energy, healthcare, manufacturing, public domain, and transportation. A merge of IIRA and RAMI 4.0 was assessed as impractical and undesirable [104]. However, what was decided is to proceed with a clear mapping between both architectures in order to enable cross-domain interoperability. Hence, such aspects could be tested in collaborative testbeds and infrastructure that would ensure technical-level interoperability. Overall, interoperability between the RAMI 4.0 and IIRA at the edge, platform, and enterprise tiers [105] is of benefit for companies acting on global markets.

7.7 W3C

The World Wide Web Consortium (W3C) has formed an Interest Group (IG)[32] on the Web of Things (WoT) out of which a Working Group was formed in early 2017. A W3C Working Group as opposed to an Interest Group has participation requirements (e.g., attending most meetings) and aims at delivering technical recommendations, software, etc., while an Interest Group is a forum of participants with common interests that does not necessarily aim at producing any kind of deliverable.

---

[32]https://www.w3.org/WoT/IG/

The WoT Working Group has produced among other documents (e.g., JSON-LD[33], WoT Scripting API) the WoT Architecture Editor's draft[34] which is work in progress and may be updated, replaced, or obsoleted by other drafts.

The Architecture draft provides a list of use cases that motivate the design, a high-level conceptual architecture, and some details showing the deployment view on different types of devices and environments. The main use cases are grouped into three types according to the deployment environment: Smart Home, Smart Factory, and Connected Car.

In the context of a Smart Home a Thing (such as a heater) should be able to be monitored and controlled directly by:

- a user device such as a mobile phone,
- another Thing in the home environment such as a control unit or controller,
- other Things that use different networking technologies, e.g., short range in the home and cellular communications if the Thing to be controlled supports different network interfaces,
- a home gateway connected to the Internet playing the role of abstracting the access (monitor and control) and management of the Thing inside the home; this is typically the case for Things that support only proprietary interfaces,
- a cloud "twin" which abstracts and mediates the monitoring and control of the Thing inside the Smart Home through its digital twin or agent in the cloud; the cloud agent could access the Thing directly or through a Smart Home gateway.

The Smart Factory use case is similar to the Home Gateway use case from the Smart Home group, in the sense that the factory offers services to the world abstracting the internal topology.

The Connected Car use case focuses on a car environment (embedded controllers in a Control Area Network (CAN) bus) being able to connect to services in a cloud through an on-board intermediate gateway.

The WoT Working Group maintains a Terminology list[35] in which essential WoT terms and concepts are explained. According to the WoT Terminology, a Thing is "an abstraction of a physical or virtual entity whose metadata and interfaces are described by a WoT Thing Description". A Thing Description (TD)[36] is a structured document describing a Thing. Currently, TDs[37] are serialized in JSON-LD and TDs can be stored and discovered using a function called a TD directory.

A TD includes metadata, a list of interactions, bindings to underlying protocols, serialization formats of the interactions, and links to related Things. At a high level a Thing is associated with Interaction Resources (Action, Property, Event), which provide services through specific APIs. The concept of an Interaction Resource is similar to the concept of Resources in the IoT-A Domain Model (please see Chapter 8): Devices host Resources which provide Services which in turn provide information about different Attributes of a Physical or Virtual Entity (Thing).

A Thing can be anything, even a location or space that does not provide any computing or communication capabilities. Therefore it is entirely optional for a TD to include metadata describing the

---

[33]https://json-ld.org

[34]https://w3c.github.io/wot-architecture/

[35]https://github.com/w3c/wot-architecture/blob/master/terminology.md

[36]https://www.w3.org/TR/wot-thing-description/

[37]https://www.w3.org/TR/wot-thing-description/

interaction capabilities of the Thing since Things do not always have interaction capabilities. When a Thing does have interaction capabilities, it typically provides a network-facing interface (WoT Interface) which includes transfer protocol bindings, media type descriptions, and security metadata in the TD.

The abstract W3C architecture is not described by using different architecture views; instead, it is described with multiple views into one architecture drawing. In a later stage, the architecture draft includes descriptions of the different deployment options. Some important concepts for the ease of understanding the abstract W3C architecture are the following:

- An important concept that binds other concepts together in the WoT is the software: the WoT architecture pictures contain a mixture of software concepts (such as firmware) and architectural concepts such as WoT Binding Templates for better understanding the architecture; the reader should clarify where a piece of software is implemented, what interactions it provides, what interfaces are used for these interactions, etc. This information is partially provided by a TD.
- Typical Web-based interactions follow the Client-Server model so in WoT the location of a client and a server plays an important role as well as the software that implements these.
- In an IoT environment there may be legacy devices that do not support Web-based technologies such as HTTP/CoAP-based transfer protocols; examples are MQTT or ZigBee devices. WoT assumes that there are devices or cloud deployments that have: (a) software to interface legacy devices through their legacy protocols as well as (b) WoT software to expose these legacy devices as Things via WoT APIs.
- There are several deployment options (devices, gateways, cloud, web browser) for a client or a server in the WoT abstract architecture.
- Different deployment options may have different communication protocol capabilities which need to be taken into account in the architecture.

In terms of software implementation, there can be two types of software stacks implementing a WoT Client and a WoT Server. Constrained-resource devices typically implement only the WoT Server while more capable devices include the implementation of both a client and a server which is called a Servient. A WoT client is said to "consume" a Thing which is "exposed" by a WoT Servient. In technical terms, a WoT client instantiates locally a software object which represents the Thing and provides the `consumedThing` API which the client uses in order to interact with the Thing locally. A Servient provides the necessary software and the counterpart `exposedThing` API which responds to requests from WoT clients. The `consumedThing` and the `exposedThing` APIs are both part of the WoT Scripting API.

Figure 7.16 shows the functional view of a WoT Servient. A Servient includes a piece of software called WoT Run-time, which implements the WoT Scripting API and uses the local platform APIs (Protocol Bindings and System APIs) to consume and expose a Thing. For example, a legacy device can have a Thing realization (i.e., a legacy device can be exposed through the WoT interfaces) through a Servient that implements the proprietary protocol for communicating to the legacy device on one side while it implements a WoT Interface on the other side. Applications running on a Servient use the WoT Scripting API to expose or consume Things along with the rest of the application logic. A Servient can be deployed on different deployment platforms such as capable devices, gateways, web browsers, and cloud infrastructure. Applications also need to maintain Security metadata such as keying material for authenticating the Things that a Servient exposes.

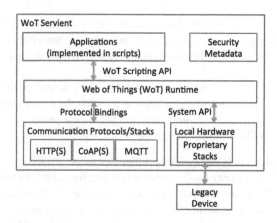

FIGURE 7.16

WoT Servient functional architecture.

For more recent and updated information on the W3C WoT Architecture the reader is referred to the W3C WoT website[38].

7.8 OGC

The Open Geospatial Consortium (OGC)[39] is an international industry consortium of a few hundred companies, government agencies, and universities that develops publicly available standards that provide geographical information support to the Web. OGC includes, among other Working Groups, the Sensor Web Enablement (SWE)[40] domain Working Group, which develops standards for sensor system models (e.g., Sensor Model Language, or SensorML), sensor information models (Observations and Measurements, or O&M), and sensor services that follow the SOA paradigm, as is the case for all OGC-standardized services. The functionality that is targeted by OGC SWE includes:

- Discovery of sensor systems and observations that meet an application's criteria.
- Discovery of a sensor's capabilities and quality of measurements.
- Retrieval of real-time or time-series observations in standard encodings.
- Tasking of sensors to acquire observations.
- Subscription to, and publishing of, alerts to be issued by sensors or sensor services based upon certain criteria.

OGC SWE includes the following standards:

---

[38]https://www.w3.org/WoT/
[39]http://www.opengeospatial.org
[40]http://www.opengeospatial.org/projects/groups/sensorwebdwg

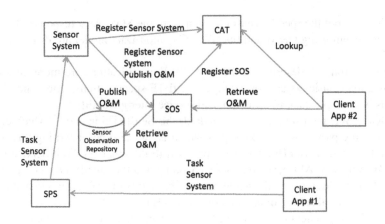

FIGURE 7.17

OGC functional architecture and interactions.

- SensorML and Transducer Model Language (TML), which include a model and an XML schema for describing sensor and actuator systems and processes; for example, a system that contains a temperature sensor measuring temperature in degrees Celsius (e.g., 30°C), which also involves a process for converting this measurement to one in degrees Fahrenheit (e.g., 86°F).
- O&M, which is a model and an XML schema for describing the observations and measurements for a sensor.
- SWE Common Data model for describing low-level data models (e.g., serialization in XML) in the messages exchanged between the OGC SWE functional entities.
- Sensor Observation Service (SOS), which is a service for requesting, filtering, and retrieving observations and sensor system information. This is the intermediary between a client and an observation repository or near real-time sensor channel.
- Sensor Planning Service (SPS), which is a service for applications requesting a user-defined sensor observations and measurements acquisition. This is the intermediary between the application and a sensor collection system.
- PUCK, which defines a protocol for retrieving sensor metadata for serial port- (RS232-) or Ethernet-enabled sensor devices.

An example of how these standards relate to each other is shown in Figure 7.17. Because OGC follows the SOA paradigm, there is a registry (CAT, Catalog) that maintains the descriptions of the existing OGC services, including the Sensor Observation and SPSs. Upon installation, the sensor system using the PUCK protocol retrieves the SensorML description of sensors and processes and registers them with the Catalog so as to enable the discovery of the sensors and processes by client applications. The Sensor System also registers to the SOS and the SOS registers to the Catalog. A client application #1 requests from the SPS that the Sensor System be tasked to sample its sensors every 10 seconds and publish the measurements using O&M and the SWE Common Data model to the SOS. Another client application #2 looks up the Catalog, aiming at locating an SOS for retrieving the measurements from the Sensor System. The application receives the contact information of the SOS and requests from the sensor observations from the specific sensor system from the SOS. As a response, the measurements from the sensor system using O&M and the SWE Common Data model are dispatched to the client application #2.

As can be seen from the description, the OGC SWE specifications are more information-centric than communication-centric, as are for example the IETF specifications. The main objective of the OGC standards is to enable data, information, and service interoperability.

In 2016 OGC SWE published another standard, the SensorThing API[41] which consists of two main parts: a sensing part[42] and a tasking part which is, at the time of writing this book, open for public comments[43]. The SensorThing API (as opposed to the other big Web Service oriented APIs) is a RESTful Web Service API based on JSON encoding and the OASIS OData[44] protocol and data encoding. The API focuses more on the semantic interoperability of IoT systems than the transportation of information, which is taken care of in protocols such as HTTP, CoAP, or MQTT. The API is expected to be used between a web client (e.g., a web application) and an observation repository for batch data access or a sensor/actuator device for near real-time data streaming.

7.9 GS1 ARCHITECTURE AND TECHNOLOGIES

The GS1 approach to addressing the reverse flow of information about the physical supply chain is based on three main principles: Identify, Capture, and Share information about products, locations, and assets in general. These principles are also used as a way to group the different GS1 standards specifications.

7.9.1 GS1 IDENTIFY

Figure 7.18 shows a simple data model of the main entities that the GS1 standards attempt to describe and model. The term "entity" is used in this context in a similar way as the word "Entity" in an Entity-Relationship model in the context of information modeling. Please note that GS1 has not provided such a model in their specification but it can be inferred by the GS1 Architecture[45] document. The reader should refer to Chapter 8 for an explanation of the modeling notation used in Figure 7.18.

Under GS1 the following types of **Entities** are important:

- **Physical**: A physical asset or item on which a data carrier is attached. A data carrier is some physical item or device which could carry a GS1 key. An example is a barcode or an RFID tag. A physical entity is also represented in the digital world with a corresponding Virtual Entity.
- **Digital**: A digital artifact such as a music file, ebook or digital coupon which only exists in the Digital world.

---

[41] http://www.opengeospatial.org/standards/sensorthings
[42] http://docs.opengeospatial.org/is/15-078r6/15-078r6.html
[43] http://www.opengeospatial.org/pressroom/pressreleases/2739
[44] https://www.oasis-open.org/committees/tc_home.php?wg_abbrev=odata
[45] https://www.gs1.org/gs1-architecture

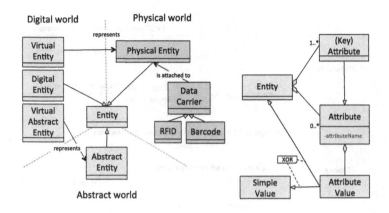

FIGURE 7.18

GS1 Identify model.

- **Abstract**: A virtual object or process, including legal abstractions (e.g., a legal party) and business abstractions (e.g., the Lot number of a specific product). More examples of an Abstract entity are a trade item class and a product category.

  Any piece of information about an Entity is captured by the use of Attributes which have a name and a value, for example, the dimensions of a product package. An Entity could also have an Attribute with a name "Name" (text string) and a value represented as another text string (e.g. "RFIDReader#1"); the value of the Attribute with a name "Name", i.e. "RFIDReader#1" indicates the actual name of the Entity and should not be confused with the name and value of an arbitrary Attribute of the Entity. Moreover, Attributes can be simple such as a text string or Entities but not both. Attributes can be static, quasistatic, or dynamic depending on the frequency of update of their value. These characteristics of an Attribute are related to the three types of data handled by GS1 systems (see the "Share" principle in Section 7.9.3).

  An important concept in an information model and its implementation (e.g., databases) is the concept of a key. A Key (Figure 7.18) is one or more attributes that are used to uniquely identify an Entity. GS1 defines standardized attributes to serve as keys for the identified Entities. An example of such a key is the Global Trade Item Number (GTIN) .

7.9.2 GS1 CAPTURE

The Capture group of GS1 standards covers the provisioning of identification information on physical data carriers such as barcodes or RFID tags as well as the reading of such identifiers from their carriers. Figure 7.19 shows a high-level view of the GS1 architecture with the three groups of specifications or the three principles of Identify, Capture, and Share are three layers. The figure is a mix of different views and should not be considered complete as it is a superposition of functional and deployment views; however, it serves as a good overview of the GS1 architecture. Please note that the figure does not show the whole coverage of the GS1 standards in each group. Interested readers should refer

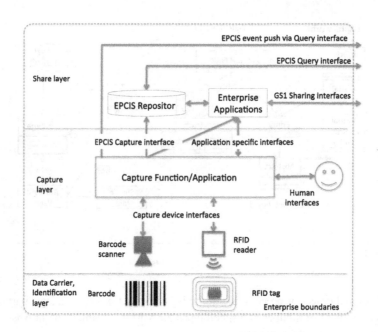

FIGURE 7.19

Simplified GS1 Architecture.

to the GS1 Architecture[46] document. At the bottom of the figure, the data carriers in the form of barcodes or RFID tags are shown. Several GS1 standard specifications cover the encoding and storage of identification information of tangible products on the data carriers. The Capture layer includes the different devices that extract the identification information from the data carriers (such as barcode scanners or RFID readers) as well as the device interfaces and protocols towards the Capture function. Examples of such protocols are the ones to disambiguate among multiple RFID tags in the vicinity of an RFID reader. The Capture function includes workflow processes to coordinate the readout of the identification information from the data carriers and the provision of services to humans through Human interfaces (e.g., screens) as well as the exposure of the captured information towards Enterprise applications through different interfaces. It is worth noting that the Capture layer abstracts the details of the data carriers towards the Enterprise applications. Moreover, the information or events about scanned or read carriers does not have any meaning if it is not put into the right business context. For example, scanning a barcode of a product at a Point of Sales (POS) generates an event that could be used either by the logic processing a purchase or by the logic processing a product return. One of the most important interfaces and functionalities in the GS1 architecture are the Electronic Product Code Information Services (EPCISs) interfaces and the EPCIS Repository (Figure 7.19). The EPCIS Capture interface is exposed by the Capture layer while the EPCIS Query interface and the EPCIS Repository belong to the Share layer. The EPCIS functionality and interfaces are presented in Section 7.9.4.

---

[46]https://www.gs1.org/gs1-architecture

7.9.3 GS1 SHARE

As stated earlier in a supply chain products flow from suppliers to manufacturers and finally to customers, while information about this flow follows the reverse direction. The information concerns business data (e.g., products, legal entities, locations) and is exchanged from one business to another. The GS1 "Share" group of standards covers this flow of information between businesses and all the important concepts and parts of this flow such as content, communication methods, discovery, and federation.

With respect to the content of information exchanged among the business entities in a supply chain the data can fall into one of these categories:

- **Master data**: These are attributes of an Entity that are static or change very infrequently throughout the lifetime of the Entity. An example is the address of a business entity or the product dimensions for a class of products (e.g., trade item class).
- **Transaction data**: This is typically business information (legal or financial) exchanged in the context of a transaction between businesses and they represent the paper documentation of corresponding transactions in old supply chains. An example of such transaction data is a document describing an order or an invoice. The transaction data are not static data and are generated by a GS1-compliant system at the rate that the businesses perform business processes.
- **Physical or Visibility event data**: These data typically capture the steps of a business process described by transaction data and they include information about physical or digital entities involved in the step. Such information includes Lifecycle Management events (creation, attaching an identification item, destruction of an item), transport events from one location to another, aggregation or disaggregation of items in bigger or smaller shipments, etc. Each Visibility event has information about the physical or digital entity involved ("What"), the time of the event ("When"), the location of the event or the expected location of the item after the event ("Where"), and finally relevant information about the involved business process, e.g., association with relevant transaction data ("Why").

It is worth noting that GS1 includes standards documents which describe the aforementioned types of information such as the Global Data Synchronization Network (GDSN)[47], GS1 XML[48], EANCOM[49], and the EPCIS[50] specification.

With respect to communication methods, GS1 defines standards to support push type of communication patterns such as unicast (one business to another), publish-subscribe and broadcast as well as pull type such as request-response for the different types of data described earlier. GS1 defines a set of standards such the Object Naming System (ONS)[51] to facilitate data discovery of data throughout the whole supply chain and not only from the originating entities (e.g., manufacturer). Finally, GS1 defines how data could be shared among different business entities in a federated manner, i.e., if multiple businesses own a piece of the whole history of the product flow, the federation allows the different entities to share relevant logistics data in a secure way.

---

[47]https://www.gs1.org/services/gdsn

[48]https://www.gs1.org/gs1-xml

[49]https://www.gs1.org/eancom

[50]https://www.gs1.org/epcis

[51]https://www.gs1.org/epcis/epcis-ons/2-0-1

7.9.4 EPCIS ARCHITECTURE AND TECHNOLOGIES

EPCIS[52] mainly belongs to the Share layer or GS1 group of standards (Figure 7.19). An EPC is an identifier of a physical object and it is typically carried on an RFID data carrier. Historically the Electronic Product Code (EPC)[53] was essential to the definition of EPCIS but not anymore. Since GS1 and EPCIS are not only concerned with physical entities but also digital entities (e.g., an audiobook) EPCIS no longer requires identification of a physical entity with an EPC or the use of RFID as a data carrier of an entity identity. Moreover, EPCIS uses identifiers to identify even classes of (physical or digital) entities as opposed to instances of entities, which was the main objective of EPC.

EPCIS aims to facilitate the smooth sharing of Visibility event data (Section 7.9.3) within one enterprise or across enterprises. Although the GS1 architecture (Figure 7.19) includes an EPCIS Repository, the EPCIS specifications do not cover the persistence aspect of EPCIS data. Instead, the EPCIS specification covers the interfaces between functions that generate Visibility event data and functions that consume such data either in the same or in different enterprises. EPCIS is meant to be used along with the GS1 Core Business Vocabulary (CBV)[54] which defines the different types and values of the data structures of the EPCIS related data. The role of EPCIS is to collect historical data about the supply chain and set the data into a business context in order to be shared across enterprises. The Capture layer, on the other hand, is responsible for real-time event data.

As mentioned earlier the EPCIS specification defines functionality and interfaces that belong mainly to the Share layer and partly to the Capture layer (Figure 7.19). EPCIS defines three main interfaces:

- **EPCIS Capture interface**: The Capture interface is used for pushing real-time Visibility event data to an EPCIS Repository or to external applications via the EPCIS Query interface through event push. The push of Capture events through the Query interface is implemented using the EPCIS Query Callback interface.
- **EPCIS Query Control interface**: The Query interface is used by internal as well as external enterprise applications to retrieve Visibility event data on-demand via a request/response model. The interface also offers the possibility to establish a standing request, that is, a request that is made once to an EPCIS Repository while it results in multiple periodic responses similar to a publish/notify type of request/response pattern. The periodic responses/notifications are dispatched using the EPCIS Query Callback interface. The EPCIS specification describes this interface as the third interface.
- **EPCIS Query Callback interface**: The Query Callback interface is used as a carrier of the periodic responses to a standing request or for pushing Capture events via the EPCIS Query interface. Please note that this interface is omitted from Figure 7.19.

The EPCIS specification is organized in three layers, which define the model of event data transported through the EPCIS interfaces as well as models of the interfaces themselves. The specification also covers the binding or the mapping of these data or interface models to concrete representations

---

[52]http://www.gs1.org/epcis
[53]http://www.gs1.org/epc/tag-data-standard
[54]http://www.gs1.org/epcis

of data (e.g., serialization formats) or concrete protocols for the interfaces. The three layers are the following:

- Abstract Data Model layer: This layer specifies the generic structure of EPCIS data and rules how to construct data models under the Data Definition Layer. This model cannot be changed unless the EPCIS specification is changed.
- Data Definition layer: This layer specifies the data exchanged through EPCIS, the abstract structure of the EPCIS data and the semantics of the data. The EPCIS Core Event types are specified in this layer, and there is also a binding of Core Event types to an XML Schema Definition (XSD).
- Service layer: This layer specifies the EPCIS interfaces through which EPCIS interacting entities (applications, EPCIS Repository, etc.) interact. The specification also includes a binding for the Query model into an XSD, bindings of the Capture interface to HTTP and enterprise message queues (a message queue in this context is a placeholder mechanism for point to point or publish/subscribe type of communication within an enterprise setting), bindings of the Query Control interface to SOAP[55] and AS2[56], and bindings of the Query Callback interface to HTTP, HTTPS, and AS2.

An interested reader should refer to the EPCIS specification[57] for more information on the different types of events, information models and interfaces of EPCIS.

7.10 OTHER RELEVANT STATE-OF-THE-ART

There are a lot more SDOs, alliances and community activities than the ones presented earlier in this chapter. From the previous more lengthy presentation of state-of-the-art, we have chosen to omit some groups, alliances, activities, and efforts for several reasons such as maturity and popularity; however, we briefly describe them for the interested reader since the importance of some if not all might change in the future.

7.10.1 ONEM2M

ETSI M2M (see Appendix A) produced the first release of the M2M standards in early 2012, while in the middle of 2012 seven of the leading ICT standards organizations (ARIB, TTC, ATIS, TIA, CCSA, ETSI, TTA) formed a global organization called oneM2M Partnership Project (oneM2M)[58] in order to develop M2M specifications, promote the M2M business, and ensure the global functionality of M2M systems. The ETSI M2M work was concluded after the formation of oneM2M and therefore the ETSI M2M material is included in this book as Appendix A. The oneM2M specifications reused the concepts, models, and architecture elements of ETSI M2M; however, they adopted different terminology for several functions and provided more details in the functional architecture such as bindings of the

---

[55] https://www.w3.org/TR/soap/
[56] https://doi.org/10.17487/rfc7252
[57] https://www.gs1.org/epcis
[58] http://www.onem2m.org

interfaces to HTTP, CoAP, MQTT, and OMA LwM2M interworking. Interested readers should refer to the oneM2M functional architecture (oneM2M TS 0001[59]) for more details.

7.10.2 OCF

The Open Connectivity Foundation (OCF)[60] is the result of a few mergers of industry alliances and open source activities in the IoT space. In 2014, the Open Interconnect Consortium (OIC) was formed to work on standardizing different communication models between IoT devices and between IoT devices and the cloud, such as Peer-to-Peer, bridging and forwarding, and reporting and control to/from an IoT cloud. In parallel OIC initiated a Linux Foundation[61] project called IoTivity[62] with the intention to provide a reference implementation of the specifications developed by OIC. Late in 2015 OIC acquired the Universal Plug and Play (UPnP) assets in order to develop UPnP further for IoT devices. Until then UPnP was mainly applicable for media-related devices (TVs, players, etc.). In 2016 OIC changed its name to OCF and later on merged with AllSeen Alliance while maintaining the name OCF. AllSeen Alliance developed the Alljoyn open source IoT framework which was mainly about seamless peer to peer connectivity between devices. OCF as of today supports the development of both IoTivity and AllJoyn open source projects. Similar to the OMA LwM2M and oneM2M, OCF follows a resource-oriented architecture with a client-server model. OCF specifies high-level functions and interactions based on a resource model and provides a binding to IETF CoAP for the underlying messaging between a client and a server. An OCF Entity is similar to the IoT-A [19] (Section 7.5) Virtual Entity which corresponds to a physical entity in the real world and provides OCF Resources through which a piece of software can interact with the physical entity. However, an OCF Entity seems to be an entity that has computation and communication capabilities while the notion of the Physical Entity in IoT-A is broader and can include for example objects and locations similar to the W3C WoT (Section 7.7). OCF has produced among other specifications[63] a Core specification v2.0 which includes a functional architecture and the CoAP protocol binding specification. The OCF v2.0 Core Specification is under public review at the time of writing this book. OCF also initiated oneIoTa[64] which is an open tool for defining data models which describe the OCF Resources. oneIoTa is open to everyone, and it contributes to the semantic interoperability of different IoT devices and systems.

7.10.3 IEEE

The Institute of Electrical and Electronic Engineers (IEEE) is an important organization for the standardization of several aspects of IoT such as communication protocols (IEEE 802.15.1/Bluetooth, IEEE 802.15.4, IEEE 802.11 family, etc.). IEEE maintains a comprehensive website[65] which provides an IoT definition document[66] and maintains a catalog of possible IoT activities and resources such

---

[59]http://www.onem2m.org/technical/published-documents
[60]https://openconnectivity.org
[61]https://www.linuxfoundation.org
[62]https://www.iotivity.org
[63]https://openconnectivity.org/developer/specifications/draft-specifications
[64]https://oneiota.org
[65]https://iot.ieee.org
[66]https://iot.ieee.org/definition.html

as conference and event information, educational resources, information on IoT standards, publication outlets and lists, a directory or IoT scenarios, and a start-up company directory. It is worth noting that in 2014 IEEE started the Working Group 2413[67] whose initial focus was to define an architectural framework for IoT. A draft titled "IEEE P2413 –Standard for an Architectural Framework for the Internet of Things (IoT)" is in development but has not been publicly released at the time of writing this book. The intention with P2413 is not to reinvent existing standards but rather to collect and document common functions from different application domains and present them in an architectural framework following the ISO/IEC/IEEE 42010 [106] architecture description specification. Other IoT related Working Groups and projects have also started since 2014 such as "P1451-99 –Standard for Harmonization of Internet of Things (IoT) Devices and Systems", "P1931.1 –Standard for an Architectural Framework for Real-time Onsite Operations Facilitation (ROOF) for the Internet of Things", "P2418 –Standard for the Framework of Blockchain Use in Internet of Things (IoT)", and "P2510 –Standard for Establishing Quality of Data Sensor Parameters in the Internet of Things Environment". All these Working Groups are relatively new and have not yet publicly released any draft or standards at the time of writing of this book.

7.10.4 IEC/ISO: ISO/IEC JTC 1/WG 10/SC 41

The International Standardization Organization (ISO)/International Electrotechnical Commission (ISO/IEC) Joint Technical Committee (JTC) 1, Information Technology, formed in 2015 the Working Group 10 (ISO/IEC JTC/WG10) with the aim to study and produce an IoT Reference Architecture. In 2017 the WG10 was transformed to the Subcommittee 41 (ISO/IEC JTC/SC41)[68] in order to drive the standardization of IoT and related technologies such as sensor networks as well as to provide a reference architecture. The SC41 has been working on a reference architecture (ISO/IEC CD 30141) which is in development at the time of writing of this book and not available for purchase. Public content available online provides some hints about the content of such an architecture. The architecture document contains a Conceptual Model (CM) which describes the main IoT entities and their relationships, an Entity-based Reference Model, A Domain-based Reference Model, and a few Functional Views. The CM is similar to the IoT-A [19] (Section 7.5) IoT Domain model. The Entity-based Model is a model of all the physical entities to be monitored and controlled as well as the instrumentation (sensors/actuators), communication, and computation entities (networking gateways, servers). The Domain-based reference model is a functional model with functions having the same characteristics (e.g., sensing and control domain). Similar to the IoT-A, IIC (Section 7.5), and IEEE (Section 7.10.3), the ISO/IEC CD 30141 follows the approach of describing the reference architecture with a set of architectural views.

7.10.5 AIOTI

The Alliance for the IoT Innovation (AIOTI)[69] was initiated by the European Commission (EC) in 2015, with the aim to bring together the IoT players in Europe in order to facilitate the creation of a

---

[67] http://grouper.ieee.org/groups/2413/

[68] https://www.iso.org/committee/6483279.html

[69] https://aioti.eu

European IoT ecosystem. The alliance also aimed at fostering experimentation and deployment of IoT and interoperability, mapping global and EU member states innovation activities and identifying market barriers. The IoT European Research Cluster (IERC)[70] was instrumental in the creation of AIOTI and as of September 2016 the Alliance was transformed into a nonprofit Industrial Association legal entity based in Brussels. The AIOTI counts 200 members in the General Assembly which meets twice a year. It includes thirteen Working Groups, four on horizontal aspects of IoT (IoT Research, IoT Innovation Ecosystems, IoT Standardization, IoT Policy) and nine on vertical topics (Smart Living Environment for Aging Well, Smart Farming and Food Security, Wearables, Smart Cities, Smart Mobility, Smart Water Management, Smart Manufacturing, Smart Energy, and Smart Buildings and Architecture).

The AIOTI Working Groups aim at producing deliverables most of which are public. AIOTI WG03 on IoT Standardization produced among others a High Level Architecture (HLA)[71] in which an architecture framework similar to IoT-A [19] (Section 7.5) is presented and compared with major standardization efforts such as ITU-T (Section 7.2), oneM2M (Section 7.10.1), IIC (Section 7.5, Section 8.9), and RAMI 4.0 (Section 7.6). The HLA follows the guidelines of ISO/IEC 42010 [106] for an architecture description but the current version (v3.0, June 2017) focuses on the presentation of a domain model and a functional model. The Domain model is a simplified version of the IoT-A IoT Domain Model and includes the concepts of a Thing, an IoT Device, a Virtual Entity, an IoT Service, and a User, as well as their relationships. Based on the domain model the HLA functional model is layered[72] having three main layers: Application, IoT, and Network Layer. The HLA presentation includes a set of interfaces between the different layers. The Application layer includes IoT Applications which use IoT layer and Network services. The IoT Service layer includes functions such as Thing representation (including semantic metadata), Identification, Analytics, Security, Discovery, and Device Management. The Network layer includes communication protocols, network security, Quality of Service (QoS), etc.

7.10.6 NIST CPS

The National Institute of Standards and Technology (NIST) is a US Agency which provides, among others, recommendations and best practices for a number of technology areas including Cyber-Physical Systems (CPS). NIST has a CPS Public Working Group (CPSPWG)[73] which produced in 2016 an architectural framework on CPS (CPS Framework release 1.0). Since CPS is extremely diverse with respect to application and technology domains, the core of the CPS Framework is a common vocabulary, structure, and analysis methodology for facilitating the design of a CPS system. In other words, the CPS Framework does not provide a holistic ARM per se but a common language and methodology to produce such references and model. However, the CPS Framework description document includes several appendices that describe the functional domains of CPS and several examples of applying the framework in real life.

---

[70]http://www.internet-of-things-research.eu
[71]https://aioti.eu/aioti-wg03-reports-on-iot-standards/
[72]The layers in the context of AIOTI indicate only logical grouping of functionality
[73]https://pages.nist.gov/cpspwg/

The CPS (architectural) Framework uses the ISO/IEC/IEEE 42010 guidelines [106] as well as the ISO/IEC/IEEE 15288 [107] for describing an architectural framework. It introduces a vocabulary which includes the following concepts:

- **Domains** are the different application areas of CPS such as Manufacturing, Transportation, Energy, Healthcare; the Domains typically provide the scope and grouping of the different stakeholder concerns.
- **Concerns** for an architecture are defined in a similar way as ISO/IEC/IEEE 42010 [106] and Rozanski and Woods [18] as "...requirements, objectives, intentions, or aspirations that a stakeholder has for that architecture"; Similar or related Concerns (e.g., because of belonging to the same Domain) are grouped into Aspects, which are addressed by activities within the Facets.
- **Aspects** are groups of concerns for an architecture; Example Aspects are Functional, Business, Trustworthiness, and Lifecycle.
- **Properties** are the concrete assertions that address the Concerns.
- **Facets** express the different responsibilities in the engineering process; each Facet contains a list of activities and expected outcomes; the NIST CPS Framework includes three Facets: conceptualization (activities which produce a description/design/model of a system with functional decomposition, requirements, etc.), realization (activities which include development, deployment, and operation of a system), and assurance (activities which result in assuring that the overall system behavior is as specified).

The CPS Framework includes a Conceptual Model which is a high-level model due to the diversity of the application domains of CPS. The main concept of NIST CPS, a **Device** interfaces the physical world by monitoring its state, providing a decision logic, which based on the monitored state takes an action that affects or alters the physical world state. Monitoring and actuation are realized through sensors and actuators which contain both cyber and physical parts while the decision logic lies entirely in the cyberspace. A **System** is composed of multiple devices and a **System of Systems (SoS)** includes multiple such systems which interact with each other at all levels. Humans also interact with a Device, System, or SoS.

7.11 CONCLUSIONS

This chapter provides an overview of the state-of-the-art within the reference architectures for IoT, including ITU-T, IETF, OMA, RAMI 4.0, W3C, OGC, and GS1. In Chapter 8 we cover more in-depth the IoT-A and IIC reference models and architectures. One observation is that the fragments of most architectures presented in this chapter and the IoT-A and IIC references architectures reuse similar concepts to describe an architecture (architecture views, viewpoints, stakeholders, concerns, etc.) and the Conceptual Model of several architectures distinguishes between an Entity in the physical world and the corresponding entity in the digital world (e.g., Virtual Entity, digital twin, digital shadow). The Entity in the physical world is the THING in the IoT and the rest of the entities are instrumentation, communication, and computation entities in order to enable the efficient monitoring and control of the THING. On a high level, this is the most important takeaway from all these efforts and the future efforts on IoT. An IoT architect should start with adhering to the principle "Follow the Thing ..."; in other words they should start with identifying the Things (entities of interest, assets) in a real-world system and how the stakeholders would prefer to manage them as expressed through their concerns. The rest of the system bridges this gap between the THING and the stakeholder concerns about the THING. This is a good starting point for any IoT system design.

ARCHITECTURE REFERENCE MODEL

CHAPTER OUTLINE

8.1 INTRODUCTION

This chapter consists of three main parts. The first part provides an overview of the Architecture Reference Model (ARM) for the Internet of Things (IoT), including descriptions of the domain, information, and functional models. The second part presents a generic Reference Architecture. These parts are

Internet of Things. https://doi.org/10.1016/B978-0-12-814435-0.00020-1

181

highly influenced by the IoT-A[1] [19] ARM and Reference architecture. The third part outlines the Industrial Internet Consortium (IIC) Reference Architecture (IIRA), which is a reference architecture mainly for Operational Technology (OT)[2] [108] IoT deployments, as opposed to the reference architecture of IoT-A which is mainly an Information Technology (IT)-oriented architecture and is designed for mainly consumer-oriented IoT systems. Nevertheless, there are a lot of similarities in the two reference architectures.

8.2 REFERENCE MODEL AND ARCHITECTURE

An ARM consists of two main parts: a Reference model and a Reference Architecture. For describing an IoT ARM, we have chosen to use the IoT-A ARM [19] as a guide because it currently includes the most complete model and reference architecture. However, a real system may not have all the modeled entities or architectural elements described in this chapter, or it could contain other non-IoT-related entities. This chapter serves the purpose of modeling the IoT part of a whole system and does not try to propose an all-encompassing architecture. The foundation of an IoT Reference Architecture description is an IoT reference model. A reference model describes the domain using a number of submodels (Figure 8.1).

The domain model of an architecture model captures the main concepts or entities in the domain in question, in this case, IoT. When these common language references are established, the domain model adds descriptions about the relationship between the concepts. These concepts and relationships serve as the basis for the development of an information model because a working system needs to capture and process information about its main entities and their interactions. A working system that captures and operates on the domain and information model contains concepts and entities of its own, and these need to be described in a separate model, the functional model. An IoT system contains communicating entities, and therefore the corresponding communication model needs to capture the communication interactions of these entities. These are a few examples of submodels that we use in this chapter for the IoT reference model.

Apart from the reference model, the other main component of an ARM is the Reference Architecture. A System Architecture is a communication tool for different stakeholders of the system. Developers, component and system managers, partners, suppliers, and customers have different views of a single system based on their requirements (or concerns) and their specific interactions with the system. As a result, describing an architecture for IoT systems involves the presentation of multiple facets of the systems in order to satisfy the different stakeholders [109,18]. The task becomes more complex when the architecture to be described is on a higher level of abstraction compared with the architecture of real functioning systems. The high-level abstraction is called Reference Architecture as it serves as a reference for generating concrete architectures and actual systems, as shown in Figure 8.2 [19].

Concrete architectures are instantiations of rather abstract and high-level Reference Architectures. A Reference Architecture captures the essential parts of an architecture, such as design principles,

---

[1]The IoT-A project was a Framework Program 7 (FP7) European Union Research project which was carried out between 2010–2013 by a partner consortium led by VDI/VDE Innovation + Technik GmbH. URL: https://cordis.europa.eu/project/rcn/95713_en.html

[2]https://www.gartner.com/it-glossary/operational-technology-ot/

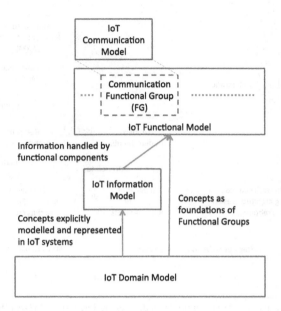

FIGURE 8.1

IoT Reference Model (redrawn from IoT-A [19]).

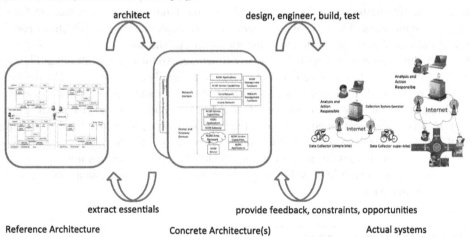

FIGURE 8.2

From reference to concrete architectures and actual systems (redrawn from IoT-A [19]).

guidelines, and required parts (such as entities), to monitor and interact with the physical world for the case of an IoT Reference Architecture. A concrete architecture can be further elaborated and mapped into real-world components by designing, building, engineering, and testing the different components of the actual system. As the figure implies, the whole process is iterative, which means that the actual

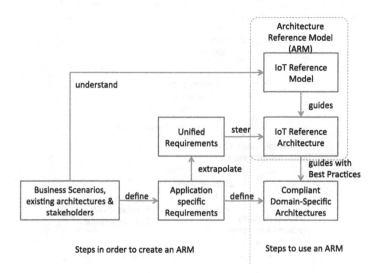

FIGURE 8.3

IoT Reference Model and Reference Architecture dependencies (redrawn from IoT-A [19]).

deployed system in the field provides invaluable feedback with respect to the design and engineering choices, current constraints of the system, and potential future opportunities that are fed back to the concrete architectures. The general essentials out of multiple concrete architectures can then be aggregated and contribute to the evolution of the Reference Architecture.

The IoT architecture model is related to the IoT Reference Architecture, as shown in Figure 8.3.

This figure shows two facets of the IoT ARM: (a) how to actually create an IoT ARM and (b) how to use it with respect to building actual systems. In this chapter we mainly focus on how to use an ARM; interested readers in the process of creation of an ARM are referred to the IoT-A ARM specification [19]. Moreover, the requirement collection, the generation process, and the Unified Requirements (Figure 8.3) are presented in [21,22]. The IoT reference model guides the process of creating an IoT Reference Architecture because it includes at least the IoT Domain Model that impacts several architecture components as seen briefly earlier (e.g., Functional Group (FGs)) and as will be seen more extensively later in the book.

8.3 IOT REFERENCE MODEL

8.3.1 IOT DOMAIN MODEL

A domain model defines the main concepts of a specific area of interest, in this case, the IoT. These concepts are expected to remain unchanged over the course of time, even if the details of an ARM may undergo continuous transformation or evolution over time. The domain model captures the basic attributes of the main concepts and the relationship between these concepts. A domain model also serves as a tool for human communication between people working in the domain in question and between people who work across different domains.

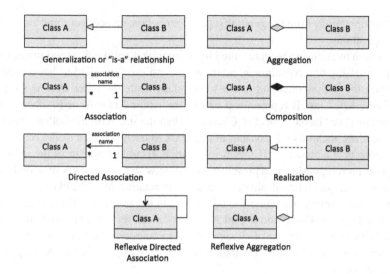

FIGURE 8.4

UML Class diagram main modeling concepts.

8.3.1.1 Model notation and semantics

For the purposes of the description of the domain model, we use the Unified Modeling Language (UML)[3] Class diagrams in order to present the relationships between the main concepts of the IoT Domain Model. The Class diagrams consist of boxes that represent the different classes of the model connected with each other through typically continuous lines or arrows, which represent relationships between the respective classes. Each class is a descriptor of a set of objects that have similar structure, behavior, and relationships. A class contains a name (e.g., Class A in Figure 8.4) and a set of attributes and operations. For the description of the IoT Domain Model, we will use only the class name and the class attributes and omit the class operations. Notation-wise this is represented as a box with two compartments, one containing the class name and the other containing the attributes. However, for the IoT Domain Model description, the attribute compartment will be empty in order not to clutter the complete domain model. The relevant and interesting attributes will be described in the text instead.

The following modeling relationships between classes (Figure 8.4) are needed for the description of the IoT Domain Model: Generalization/Specialization, Aggregation and Reflexive Aggregation, Composition, Directed Association and Reflexive Directed Association, and Realization.

The Generalization/Specialization relationship is represented by an arrow with a solid line and a hollow triangle head. Depending on the starting point of the arrow, the relationship can be viewed as a generalization or specialization. For example, in Figure 8.4, Class A is a general case of Class B or Class B is a special case or specialization of Class A. Generalization is also called an "is-a" relationship. For example, in Figure 8.4 Class B "is-a" Class A. A specialized class/subclass/child class inherits the

---

[3]Unified Modeling Language: http://www.uml.org

attributes and the operations from the general/super/parent class, respectively, and also contains its own attributes and operations.

The Aggregation relationship is represented by a line with a hollow diamond in one end, represents a whole-part relationship or a containment relationship, and is often called a "has-a" relationship. The class that touches the hollow diamond is the whole class while the other class is the part class. For example, in Figure 8.4, Class B represents a part of the whole Class A, or in other words, an object of Class A "contains" or "has-a" object of Class B. When the line with the hollow diamond starts and ends in the same class, then this relationship of one class to itself is called Reflexive Aggregation, and it denotes that objects of a class (e.g., Class A in Figure 8.4) contain objects of the same class.

The Composition relationship is represented by a line with a solid black diamond in one end and also represents a whole-part relationship or a containment relationship. The class that touches the solid black diamond is the whole class while the other class is the part class. For example, in Figure 8.4, Class B is part of Class A. Composition and Aggregation are very similar, with the difference being the coincident lifetime to the objects of classes related to composition. In other words, if an object of Class B is part of an object of Class A (composition), when the object of Class A disappears, the object of Class B also disappears.

A plain line without arrowheads or diamonds represents the Association relationship. However, in the presentation of the IoT Domain Model, we will only use the Directed Association that is represented by a line with a normal arrowhead. While all the previous relationships have implicit names represented by additional symbols (hollow triangle head, diamonds), an Association (Directed or not) contains an explicit association name. The Directed Association implies navigability from a Class B to a Class A in Figure 8.4. Navigability means that objects of Class B have the necessary attributes to know that they relate to objects of Class A while the reverse is not true: objects of Class A can exist without having references to objects of Class B. When the arrow starts and ends at the same class, then the class is associated to itself with a Reflexive Directed Association, which means that an object of this class (e.g., Class A in Figure 8.4) is associated with objects of the same class with the specifically named association.

An arrow with a hollow triangle head and a dashed line represents the Realization relationship. This relationship represents an association between the class that specifies the functionality and the class that realizes the functionality. For example, Class A in Figure 8.4 specifies the functionality while Class B realizes it.

Aggregations, Reflexive Aggregations, Associations (Directed or not), and Reflexive Associations (Directed or not) may contain multiplicity information such as numbers (e.g., "1"), ranges (e.g., "0–1", open ranges "1...*"), etc. in one or the other end of the relationship line/arrow. These multiplicities denote the potential number of class objects that are related to the other class object. For example, in Figure 8.4, a plain association called "association name" relates one object of Class B with zero or more objects from Class A. An asterisk "*" denotes zero or more while a plus "+" denotes one or more.

8.3.1.2 Main concepts

The IoT is a support infrastructure for enabling objects and places in the physical world to have a corresponding representation in the digital world. The reason why we would like to represent the physical world in the digital world is to remotely monitor and interact with the physical world using software. Let us illustrate this concept with an example (Figure 8.5).

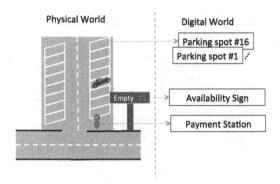

FIGURE 8.5

Physical vs. Virtual World.

Imagine that we are interested in monitoring a parking lot with 16 parking spots. The parking lot includes a payment station for drivers to pay for the parking spot after they park their cars. The parking lot also includes an electronic road sign on the side of the street that shows in real-time the number of empty spots. Frequent customers also download a smartphone application that informs them about the availability of a parking spot before they even drive on the street where the parking lot is located. In order to realize such a service, the relevant physical objects as well as their properties need to be captured and translated to digital objects such as variables, counters, or database objects so that software can operate on these objects and achieve the desired effect, i.e., detecting when someone parks without paying, informing drivers about the availability of parking spots, producing statistics about the average occupancy levels of the parking lot, etc. For these purposes, the parking lot as a place is instrumented with parking spot sensors (e.g., loops), and for each sensor, a digital representation is created (Parking spot #1–#16). In the digital world, a parking spot is a variable with a binary value ("available" or "occupied"). The parking lot payment station also needs to be represented in the digital world in order to check if a recently parked car owner actually paid the parking fee. Finally, the availability sign is represented in the digital world in order to allow notification to drivers that an empty lot is full for maintenance purposes, or even to allow maintenance personnel to detect when the sign is malfunctioning.

As seen from the example above, there is a fundamental difference between the IoT and today's Internet: today's Internet serves a rather virtual world of content and services (although these services are hosted on real physical machines), while IoT is all about interaction through the Internet with physical Things. M2M has a similar vision of representing unattended Devices accessible through a communication network in the digital world. Nevertheless, for the IoT model, the first class citizen is the Thing, and therefore Thing-oriented interaction is promoted as opposed to communication-oriented interaction for the M2M world.

As interaction with the physical world is the key for the IoT; it needs to be captured in the domain model (Figure 8.6). The first, most fundamental interaction is between a human or an application with the physical world object or place. Therefore, a User and a Physical Entity are two concepts that belong to the domain model. A User can be a Human User, and the interaction can be physical (e.g., parking the car in the parking lot). The physical interaction is the result of the intention of the human to achieve

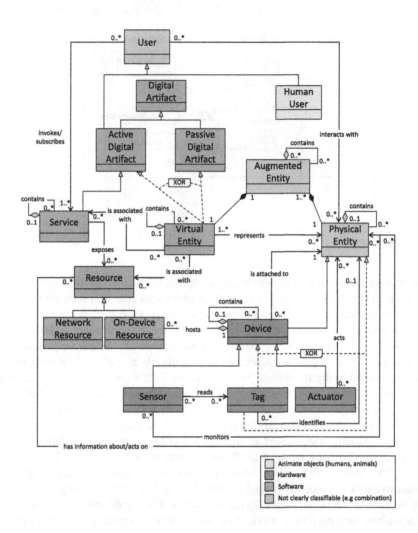

FIGURE 8.6

IoT Domain Model (adapted from IoT-A [19]).

a certain goal (e.g., park the car). In other occasions, a Human User can also choose to interact with the physical environment by means of a service or application. This application is also a User in the domain model. A Physical Entity, as the model shows, can potentially contain other physical entities; for example, a building is made up of several floors, and each floor has several rooms.

The objects, places, and things represented as Physical Entities are the same as Assets mentioned earlier in the book. According to the Oxford Dictionary, an Asset is "an item or property that is regarded as having value..."; therefore, the term Asset is more related to the business aspects of IoT. Because the domain model is a technical tool, we use the term Physical Entity instead of the term Asset. In certain cases the term Thing is also used interchangeably with the term Physical Entity.

A Physical Entity is represented in the digital world as a Virtual Entity. A Virtual Entity can be a database entry, a geographical model (mainly for places), an image or avatar, or any other Digital Artifact. One Physical Entity can be represented by multiple Virtual Entities, each serving a different purpose, e.g., a database entry of a parking spot denoting the spot availability or an (empty/full) image of a parking spot on the monitor of the parking lot management system. Each Virtual Entity also has a unique identifier for making it addressable among other Digital Artifacts. A Virtual Entity representation contains several attributes that correspond to the Physical Entity current state (e.g., the parking spot availability). The Virtual Entity representation and the Physical Entity actual state should be synchronized whenever a User operates on one or the other, if, of course, that is physically possible. For example, a remotely controlled light (Physical Entity) represented by a memory location (Virtual Entity) in an application could be switched on/off by the User by changing the Virtual Entity representation, or in other words, writing a value in the corresponding memory location. In this case, the real light should be turned on/off (Virtual to Physical Entity synchronization). On the other hand, if a Human User turns off the light by hand, then the Virtual Entity parameter that captures the state of the light should also be updated accordingly (Physical to Virtual Entity synchronization). There are cases in which synchronization can occur only one way. For example, the parking spot sensor representation in the digital world is updated whenever a car parks in the spot, but updating the digital representation does not mean that a car will magically land on that parking spot! In recent literature or product descriptions the virtual counterpart of a Physical Entity is often referred to as "Digital Twin" or "Digital Shadow".

While discussing the concept of a Virtual Entity, we also introduced another concept, that of the Digital Artifact. A Digital Artifact is an artifact of the digital world and can be passive (e.g., a database entry) or active (e.g., application software).

The model captures human-to-machine, application (active digital artifact)-to-machine, and Machine-to-Machine interaction when a digital artifact, and thus a User, interacts with a Device that is a Physical Entity. The model captures this special case of Devices being Physical and Virtual Entities as an Augmented Entity concept, which is a composition of the two constituent entities.

In order to monitor and interact with the Physical Entities through their corresponding virtual entities, the Physical Entities or their surrounding environment needs to be instrumented with certain kinds of Devices, or certain Devices need to be embedded/attached to the environment. The Devices are physical artifacts with which the physical and virtual worlds interact. Devices as mentioned before can also be Physical Entities for certain types of applications, such as management applications when the interesting entities of a system are the Devices themselves and not the surrounding environment. For the IoT Domain Model, three kinds of Device types are the most important:

1. **Sensors**: These are simple or complex Devices that typically involve a transducer that converts physical properties such as temperature into electrical signals. These Devices include the necessary conversion of analog electrical signals into digital signals, e.g., a voltage level to a 16-bit number, processing for simple calculations, potential storage for intermediate results, and potentially communication capabilities to transmit the digital representation of the physical property as well as to receive commands. A video camera can be another example of a complex sensor that could detect and recognize people.

2. **Actuators**: These are also simple or complex Devices that involve a transducer that converts electrical signals to a change in a physical property (e.g., turn on a switch or move a motor). These Devices

also include potential communication capabilities, storage of intermediate commands, processing, and conversion of digital signals to analog electrical signals.

3. **Tags**: Tags, in general, identify the Physical Entity that they are attached to. In reality, tags can be Devices or Physical Entities but not both, as the domain model shows. An example of a Tag as a Device is a Radio Frequency Identification (RFID) tag, while a tag as a Physical Entity is a paper-printed immutable barcode or Quick Response (QR) code. An electronic Device or a paper-printed entity tag contains a unique identification that can be read by radio signals (RFID tags) or optical means (barcodes or QR codes). The reader Device operating on a tag is typically a sensor, and sometimes a combined sensor and actuator in the case of writeable RFID tags.

As shown in the model, Devices can be an aggregation of other Devices, e.g., a sensor node contains a temperature sensor, a Light Emitting Diode (LED, actuator), and a buzzer (actuator). Any type of IoT Device needs to have one or more of the following options for energy supply: (a) energy reserves (e.g., a battery), (b) connection to the power grid, (c) energy scavenging capabilities (e.g., converting solar radiation to energy). The Device communication, processing and storage, and energy reserve capabilities determine several design decisions such as if the Resources should be on-Device or not, if the Device and therefore its Resources and Services will go into sleep mode or not, if the collected data can be saved locally or are transmitted as soon as acquired, etc.

Resources are software components that provide data for, or are endpoints for, controlling Physical Entities. Resources can be of two types, on-Device resources and Network Resources. An on-Device Resource is typically hosted on the Device itself and provides information, or is the control point for the Physical Entities that the Device itself is attached to. An example is a temperature sensor on a temperature node deployed in a room that hosts a software component that responds to queries about the temperature of the room. The Network Resources are software components hosted somewhere in the network or cloud. A Virtual Entity is associated with potentially several Resources that provide information or control of the Physical Entity represented by this Virtual Entity. Resources can be of several types: sensor resources that provide sensor data, actuator resources that provide actuation capabilities or actuator state (e.g., "on"/"off"), processing resources that get sensor data as input and provide processed data as output, storage resources that store data related to Physical Entities, and tag resources that provide identification data of Physical Entities.

Resources expose (monitor or control) functionality as Services with open and standardized interfaces, thus abstracting the potentially low-level implementation details of the resources. Services are therefore Digital Artifacts with which Users interact with Physical Entities through the Virtual Entities. Therefore, the Virtual Entities that are associated with Physical Entities instrumented with Devices that expose Resources are also associated with the corresponding resource Services. The associations between Virtual Entities and Services are such that a Virtual Entity may be monitored or controlled through potentially multiple redundant Resources or Services. Therefore, these associations are important to be maintained for look-up or discovery by the interested Users. It is important to note that IoT Services can be classified into three main classes according to their level of abstraction:

1. Resource-Level Services typically expose the functionality of a Device by exposing the on-Device Resources. In addition, these services typically handle quality aspects such as security, availability, and performance issues. Apart from the on-Device resources, there are also Network Resources hosted on more powerful machines or in the cloud that expose Resource-Level Services and abstract the location of the actual resources. An example of such a Network Resource is a historical database

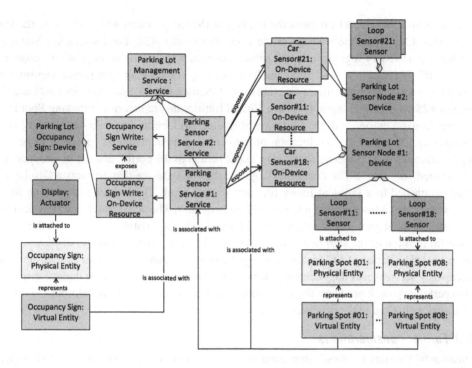

FIGURE 8.7

IoT Domain Model instantiation.

of measurements of a specific resource on a specific Device. Resource-Level Services typically include interfaces that access resource information based on the identity of the resource itself.

2. Virtual Entity-Level Services provide information or interaction capabilities about Virtual Entities, and as a result, the Service interfaces typically include an identity of the Virtual Entity.

3. Integrated Services are the compositions of Resource-Level and Virtual Entity-Level services or any combination of both service classes.

An example of an instantiation of the IoT Domain Model is shown in Figure 8.7. For this instantiation, we use the example of a simple parking lot management system presented earlier, and we model only one part of the real system. For example, the part of the model that captures the Loop Sensor #21–#28 and the associated Physical and Virtual Entities is similar to the corresponding part of the model for the Loop Sensor #11–#18 and therefore omitted. We assume that each parking spot is instrumented with a metal sensing loop (Sensor), and half of the loops are physically wired to one sensor node (Device, Sensor Node #1) while the rest are wired to another sensor node (Device, Sensor Node #2). The sensor nodes may have different identifiers (e.g., Sensor #11–Sensor #18 for the first group and Sensor #21–Sensor #28 for the second group). The loop sensors can output different impedance based on the existence or the absence of a steel object. This impedance is translated by the sensor node into binary "0" and "1" readings. Each Parking Lot Sensor Node hosts as many Car Sensor Resources as assigned parking spots. There are also two Parking Sensor Services, each running on a Sensor Node.

The Parking Sensor Service #1 provides the reading of the Loop Sensor #11–#18, while the Parking Sensor Service #2 provides the readings of the Loop Sensor #21–#28. The Parking Lot Management Service has the necessary logic to map the sensor node readings to the appropriate occupancy indicator (e.g., "0" → "free", "1" → "occupied") and to map the parking spot sensor identifier to the corresponding parking spot identifier (e.g., Sensor #11–Sensor #18 → Spot #01–Spot #08 and Sensor #21–Sensor #28 → Spot #09–Spot #16). The Virtual Entities that represent the Parking Spot Physical Entities are database entries with the following attributes: (a) identity (e.g., ID #1–#16), (b) physical dimensions (e.g., 3m×2m), (c) the location of the center of the rectangular spot with respect to the parking lot entrance (e.g., 3 meters to the west and 2 meters to the north), and (d) its occupancy level (e.g., "occupied" or "free"). The occupancy sign consists of a Device that contains a Display (Actuator) that is attached to a Physical Entity (the actual steel sign). The Device exposes one Resource with one Service that allows writing a value to the sign display. The actual steel sign (Physical Entity) is represented in the digital world with a Virtual Entity, which is a database entry with the following attributes: (a) location of the sign (e.g., GPS location), (b) status (on/off), and (c) display value (e.g., 15 free spaces). The Parking Lot Management System is a composed Service that contains the parking spot occupancy services and the occupancy sign write service. Internally given the occupancy status of all the parking spots, it produces the total number of free spots and uses this attribute to update the display actuator attached to the occupancy sign.

8.3.1.3 Further considerations

Identification of Physical Entities is important in an IoT system in order for any User to interact with the physical world through the digital world. There are at least two ways stated in [19]: (a) primary identification that uses natural features of a Physical Entity and (b) secondary identification using tags or labels attached to the Physical Entity. Both types of identification are modeled in the IoT Domain Model. Extracting natural features can be performed by a camera Device (Sensor) and relevant Resources that produce a set of features for specific Physical Entities. In addition, when it comes to physical spaces, a GPS Device or another type of location Device (e.g., an indoor location Device) can also be used to record the GPS coordinates of the space occupied by the Physical Entity. With respect to secondary identification, tags or labels attached to Physical Entities are modeled in the IoT Domain Model, and there are relevant RFID or barcode technologies to realize such identification mechanisms.

Apart from identification, location and time information are important for the annotation of the information collected for specific Physical Entities and represented in Virtual Entities. Information without one or the other (i.e., location or time) is practically useless apart from the case of Body Area Networks (BANs, networks of sensors attached to a human body for live capture of vital signals, e.g., heart rate); that location is basically fixed and associated with the identification of the Human User. Nevertheless, in such cases, sometimes the location of the whole BAN or Human User is important for correlation purposes (e.g., during the winter upon moving outdoors, the Human User heart rate increases in order to compensate for the lower temperature than indoors). Therefore, the location, and often the timestamp of location, for the Virtual Entity can be modeled as an attribute of the Virtual Entity that could be obtained by location sensing resources (e.g., GPS or indoor location systems).

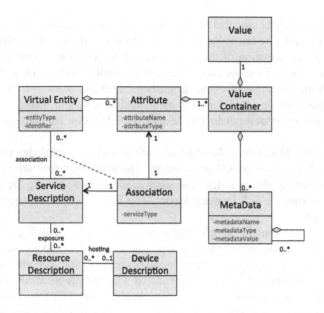

FIGURE 8.8

High-level IoT Information Model (redrawn from IoT-A [19]).

8.3.2 INFORMATION MODEL

According to the Data–Information–Knowledge–Wisdom Pyramid [110], information is defined as the enrichment of data (raw values without relevant or usable context) with the right context, so that queries about who, what, where, and when can be answered. Because the Virtual Entity in the IoT Domain Model is the "Thing" in the "Internet of Things", the IoT information model captures the details of a Virtual Entity-centric model.

Similar to the IoT Domain Model, the IoT Information Model is presented using Unified Modeling Language (UML) diagrams. As mentioned earlier, each class in a UML diagram contains zero or more attributes. These attributes are typically of simple types, such as integers or text strings, and are represented with the text under the name of the class (e.g., entityType in the Virtual Entity class in Figure 8.8).

A more complex attribute for a specific class A is represented as a class B, which is contained in class A with an aggregation relationship between class A and class B. Moreover, the UML diagram for describing the IoT Information Model contains additional notation not presented earlier. More specifically, the Association class in Figure 8.8 contains information about the specific association between a Virtual Entity and a related Service. In other words, while in the IoT Domain Model one is interested in capturing the fact that a Virtual Entity and a Service are associated, the IoT Information Model explicitly represents this association as part of the information maintained by an IoT system.

On a high level, the IoT Information Model maintains the necessary information about Virtual Entities and their properties or attributes. These properties/attributes can be static or dynamic and enter into the system in various forms, e.g., by manual data entry or reading a sensor attached to the Virtual

Entity. Virtual Entity attributes can also be digital synchronized copies of the state of an actuator as mentioned earlier: by updating the value of an Virtual Entity attribute, an action takes place in the physical world. In the presentation of the high-level IoT information model, we omit the attributes that are not updated by an IoT Device (sensor, tag) or the attributes that do not affect any IoT Device (actuator, tag), with the exception of essential attributes such as names and identifiers. Examples of omitted attributes that could exist in a real implementation are room names and floor numbers; in general, context information that is not directly related to IoT Devices, but that is nevertheless important for an actual system.

The IoT Information Model describes Virtual Entities and their attributes that have one or more values annotated with metainformation or metadata. The attribute values are updated as a result of the associated services to a Virtual Entity. The associated services, in turn, are related to Resources and Devices as seen from the IoT Domain Model. The IoT Information Model captures the above associations as follows.

A Virtual Entity object contains simple attributes/properties: (a) entityType to denote the type of entity, such as a human, car, or room (the entity type can be a reference to concepts of a domain ontology, e.g., a car ontology); (b) a unique identifier; and (c) zero or more complex attributes of the class Attributes. The class Attributes should not be confused with the simple attributes of each class. This class Attributes is used as a grouping mechanism for complex attributes of the Virtual Entity. Objects of the class Attributes, in turn, contain the simple attributes with the self-descriptive names attributeName and attributeType. As in the case of the entity type, the attribute type is the semantic type of the value (e.g., that the value is a temperature value) and can refer to an ontology such as the NASA quantities and units SWEET ontology [111]. The Attribute class also contains a complex attribute ValueContainer that is a container of the multiple values that an attribute can take. The container includes complex attributes of the class Value and the class MetaData. The container contains exactly one value such as a timestamp and metainformation (modeled as the class MetaData) describing this single value. Objects of the MetaData class can contain MetaData objects as complex attributes, as well as the simple attributes with the self-descriptive names metadataName, metadataType, and metadataValue.

As seen from the IoT Domain Model, a Virtual Entity is associated with Resources that expose Services about the specific Virtual Entity. This association between a Virtual Entity and its Services is captured in the Information Model with the explicit class called Association. Objects of this class capture the relationship between objects of the complex Attribute class (associated with a Virtual Entity) and objects of the Service Description class. The meaning of this explicit association is to link a specific attribute with the provider of the information or interaction functionality that is a Service associated with the Virtual Entity. Because the class Association describes the relationship between a Virtual Entity and Service Description through the Attribute class, there is a dashed line between the Association class and the line between the Virtual Entity and Service Description classes. The attribute serviceType can take two values: (a) "INFORMATION", if the associated service is a sensor service (i.e., allows reading of the sensor), or (b) "ACTUATION", if the associated service is an actuation service (i.e., allows an action executed on an actuator). In both cases, the eventual value of the attribute will be a result of either reading a sensor or controlling an actuator.

An example of an instantiation of the high-level information model is shown in Figure 8.9 following the parking lot example presented earlier. Here we do not show all the possible Virtual Entities, but only one corresponding to one parking spot. This Virtual Entity is described with one Attribute (among others) called hasOccupancy. This Attribute is associated with the Parking Lot Occupancy Service De-

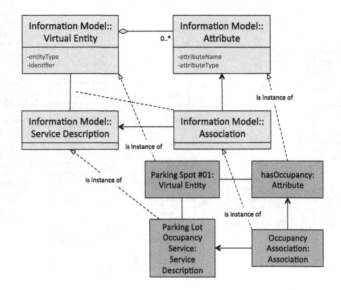

FIGURE 8.9

IoT Information Model example.

scription through the Occupancy Association. The Occupancy Association is the explicit expression of the association (line) between the Parking Spot #1 Virtual Entity and the Parking Lot Occupancy Service. Please note that the dashed arrows with hollow arrowheads represent the relationship "is instance of" for the information model, as opposed to the Realization relationship for the IoT Domain Model.

Throughout the description of the IoT Information Model, the reader might wonder about the mapping between the IoT Domain Model and the Information Model. Figure 8.10 presents the relationship between the core concepts of the IoT Domain Model and the IoT Information Model.

The Information Model captures the Virtual Entity in the Domain Model being the "Thing" in the "Internet of Things" as several associated classes (Virtual Entity, Attribute, Value, MetaData, Value Container) that capture the description of a Virtual Entity and its context. The Device, Resource, and Service in the IoT Domain Model are also captured in the IoT Information Model because they are used as representations of the instruments and the digital interfaces for interaction with the Physical Entity associated with the Virtual Entity.

The IoT Information Model is by no means complete. As this is a description of an ARM, the Information Model is a very high-level model and omits certain details that could potentially be required in a concrete architecture and an actual system. These details could be derived from specific requirements for specific use cases describing the target actual system. Because this chapter describes an ARM, we can only provide descriptions and guidelines for more information or models that could be used in real systems in conjunction with the proposed IoT Information Model. The Virtual Entity in the IoT Information Model is described with only a few simple attributes, the complex Attribute associated with sensor/actuator/tag services. As mentioned earlier, there are several other attributes or properties that could exist in a Virtual Entity description:

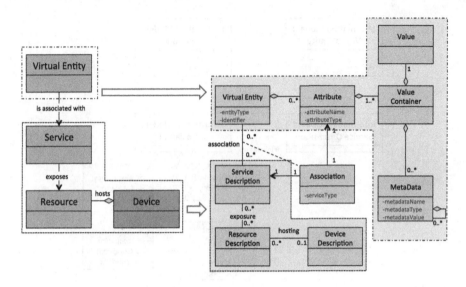

FIGURE 8.10

Relationship between core concepts of the IoT Domain Model and IoT Information Model (redrawn from IoT-A [19]).

1. Location and its temporal information are important because Physical Entities represented by Virtual Entities exist in space and time. These properties are extremely important when the Physical Entities are mobile (e.g., a moving car). Even capturing the fact that the Physical Entity is static or mobile is also a piece of useful information. A mobile Physical Entity affects the associations between Attributes and related Services, e.g., a person moving closer to a camera (sensor) is associated with the Device, Resource, and Services offered by the camera for as long as she stays within the field of view of the camera. In such cases, the temporal availability of the associations between Attributes and Services need to be captured, as availability also denotes temporal observability of the Physical Entity and therefore the associated Virtual Entity.

2. Even nonmobile Virtual Entities contain properties that are dynamic with time, and therefore their temporal variations need to be modeled and captured by an information model.

3. Information such as ownership is also important in commercial settings because it may determine access control rules or liability issues.

It is important to note that the Attribute class is general enough to capture all the interesting properties of a Physical Entity and thus provides an extensible model whose details can only be specified by the specific actual system in mind.

The Services in the IoT Domain Model are mapped to the Service Description in the IoT Information Model. The Service Description contains (among other information) the following [112,113]:

1. Service type, which denotes the type of service, such as Big Web Service or RESTful Web Service. The interfaces of a service are described based on the description language for each service type,

for example, Web Application Description Language (WADL)[4] for RESTful Web Services, Web Services Description Language (WSDL)[5] for Big Web Services, and Universal Service Description Language (USDL)[6]. The interface description includes, among other information, the invocation contact information, e.g., a Uniform Resource Locator (URL).

2. Service area and Service schedule are properties of Services used for specifying the geographical area of interest for a Service and the potential temporal availability of a Service, respectively. For sensing services, the area of interest is equivalent to the observation area, whereas for actuation services the area of interest is the area of operation or impact.
3. Associated resources that the Service exposes.
4. Metadata or semantic information used mainly for service composition. This is information such as which resource property is exposed as input or output, whether the execution of the service needs any conditions satisfied before invocation, and whether there are any effects of the service after invocation.

The IoT Information Model also contains Resource descriptions because Resources are associated with Services and Devices in the IoT Domain Model. A Resource description contains the following information:

1. Resource name and identifier for facilitating resource discovery.
2. Resource type, which specifies if the resource is (a) a sensor resource, which provides sensor readings; (b) an actuator resource, which provides actuation capabilities (to affect the physical world) and actuator state; (c) a processor resource, which provides processing of sensor data and output of processed data; (d) a storage resource, which provides storage of data about a Physical Entity; and (e) a tag resource, which provides identification data for Physical Entities.
3. Free text attributes or tags used for capturing typical manual input such as "fire alarm, ceiling".
4. Indicator of whether the resource is an on-Device resource or network resource.
5. Location information about the Device that hosts this resource in case of an on-Device resource.
6. Associated Service information.
7. Associated Device description information.

A Device is a Physical Entity that could have a sensor, actuator, or tag instantiation. An instantiation of a Device depends on the realization of it, and any information from dimensions of physical packaging to the physical placement of sensors, actuators, tags, processors, memories, batteries, cables, etc., on a printed circuit board of the Device could be captured in the Device description. A Device description should contain an identifier or name and the location of deployment, either expressed in global coordinates or local human-readable text (e.g., Auditorium).

It is important to observe that for several of these pieces of information that serve as attributes or properties of the different classes of the IoT Information Model, semantic data models or ontologies could be used in a real system implementation. For example, sensor values as Attributes could be annotated with metadata that point to the NASA SWEET ontology as already mentioned earlier. Loca-

---

[4]WADL: https://www.w3.org/Submission/wadl
[5]WSDL: http://www.w3.org/TR/wsdl
[6]USDL: http://www.w3.org/2005/Incubator/usdl/XGR-usdl/

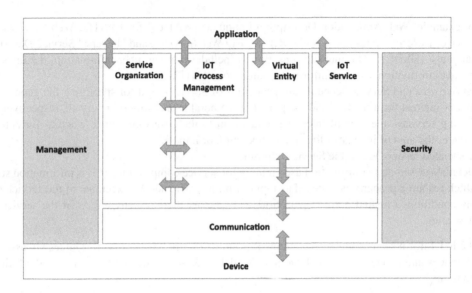

FIGURE 8.11

IoT-A Functional Model (redrawn from IoT-A [19]).

tion information could comply with an ontology such as GeoNames ontology[7], and Device description could refer to specific Device ontologies.

8.3.3 FUNCTIONAL MODEL

The IoT Functional Model aims at describing mainly the FGs and their interaction with the ARM, while the Functional View of a Reference Architecture describes the Functional Components (FCs) of an FG, interfaces, and interactions between the components. The Functional View is typically derived from the Functional Model in conjunction with high-level requirements. Readers interested in the requirement collection, generation process, and the specific Unified Requirements can refer to [21].

This section briefly describes the most important FGs, while the Reference Architecture section (Section 8.4) will elaborate on the composition of each FG. The IoT-A Functional Model is shown in Figure 8.11.

The Application, Virtual Entity, IoT Service, and Device FGs are generated by starting from the User, Virtual Entity, Resource, Service, and Device classes from the IoT Domain Model. The need for communicating Devices and Digital Artifacts was the motivation for the Communication FG. The need to compose simple IoT services in order to create more complex ones and the need to integrate IoT services (simple or complex) with existing Information and Communications Technology (ICT) infrastructure are the main drives behind the introduction of the Service Organization and IoT Process Management FGs, respectively. All the above-mentioned FGs need to be supported by management

---

[7]GeoNames Ontology: http://www.geonames.org/ontology/documentation.html

and security functionality captured by the corresponding FGs. The figure shows the flow of information between FGs apart from the cases of the Management and Security FGs that have information flowing from/to all other FGs, but these flows are omitted for clarity purposes.

8.3.3.1 Device Functional Group

The Device FG contains all the possible functionality hosted by the physical Devices that are used for instrumenting the Physical Entities. This Device functionality includes sensing, actuation, processing, storage, and identification components, the sophistication of which depends on the Device capabilities.

8.3.3.2 Communication Functional Group

The Communication FG abstracts all the possible communication mechanisms used by the relevant Devices in an actual system in order to transfer information to the digital world components or other Devices. Examples of such functions include wired bus or wireless mesh technologies through which sensor Devices are connected to Internet Gateway Devices. Communication technologies used between Applications and other functions such as functions from the IoT Service FG are out of scope because they are the typical Internet technologies. The reader is encouraged to refer to the corresponding sections in Chapter 5 related to Devices and local and Wide Area Networking (WAN) technologies.

8.3.3.3 IoT Service Functional Group

The IoT Service FG corresponds mainly to the Service class from the IoT Domain Model and contains single IoT Services exposed by Resources hosted on Devices or in the Network (e.g., processing or storage Resources). Support functions such as directory services, which allow discovery of Services and resolution of Resources, are also part of this FG.

8.3.3.4 Virtual Entity Functional Group

The Virtual Entity FG corresponds to the Virtual Entity class in the IoT Domain Model and contains the necessary functionality to manage associations between Virtual Entities with themselves as well as associations between Virtual Entities and related IoT Services, i.e., the Association objects for the IoT Information Model. Associations between Virtual Entities can be static or dynamic depending on the mobility of the Physical Entities related to the corresponding Virtual Entities. An example of a static association between Virtual Entities is the hierarchical inclusion relationship of a building, floor, room/corridor/open space, i.e., a building contains multiple floors that contain rooms, corridors, and open spaces. An example of a dynamic association between Virtual Entities is a car moving from one block of a city to another (the car is one Virtual Entity while the city block is another). A major difference between IoT Services and Virtual Entity Services is the semantics of the requests and responses to/from these services. Referring back to the parking lot example, the Parking Sensor Service provides as a response only a number "0" or "1" given the identifier of a Loop Sensor (e.g., #11). The Virtual Entity Parking Spot #01 responds to a request about its occupancy status as "free". The IoT Service provides data or information associated with specific Devices or Resources, including limited semantic information (e.g., Parking sensor #11, value = "0", units = none); the Virtual IoT Service provides information with richer semantics ("Parking spot #01 is free") and is closer to being human-readable and understandable.

8.3.3.5 IoT Service Organization Functional Group

The purpose of the IoT Service Organization FG is to host all FCs that support the composition and orchestration of IoT and Virtual Entity services. Moreover, this FG acts as a service hub between several other FGs such as the IoT Process Management FG when, for example, service requests from Applications or the IoT Process Management are directed to the Resources implementing the necessary Services. Therefore, the Service Organization FG supports the association of Virtual Entities with the related IoT Services and contains functions for discovery, composition, and choreography of services. Simple IoT or Virtual Entity Services can be composed to create more complex services, e.g., a control loop with one Sensor Service and one Actuator service with the objective to control the temperature in a building. Choreography is the brokerage of Services so that Services can subscribe to other services in a system.

8.3.3.6 IoT Process Management Functional Group

The IoT Process Management FG is a collection of functionalities that allows smooth integration of IoT-related services (IoT Services, Virtual Entity Services, Composed Services) with the Enterprise (Business) Processes.

8.3.3.7 Management Functional Group

The Management FG includes the necessary functions for enabling fault and Performance Monitoring of the system, configuration for enabling the system to be flexible to changing User demands and accounting for enabling subsequent billing for the usage of the system. Support functions such as management of ownership, administrative domain, rules and rights of FCs, and information stores are also included in the Management FG.

8.3.3.8 Security Functional Group

The Security FG contains the FCs that ensure the secure operation of the system as well as the management of privacy. The Security FG contains components for Authentication of Users (Applications, Humans), Authorization of access to Services by Users, secure communication (ensuring integrity and confidentiality of messages) between entities of the system such as Devices, Services, and Applications, and last but not least, assurance of privacy of sensitive information relating to Human Users. These include privacy mechanisms such as anonymization of collected data, anonymization of resource and Service accesses (Services cannot deduce which Human User accessed the data), and unlinkability (an outside observer cannot deduce the Human User of a service by observing multiple service requests by the same User).

8.3.3.9 Application Functional Group

The Application FG is just a placeholder that represents all the needed logic for creating an IoT application. The applications typically contain custom logic tailored to a specific domain such as a Smart Grid. An application can also be a part of a bigger ICT system that employs IoT services such as a supply chain system that uses RFID readers to track the movement of goods within a factory in order to update the Enterprise Resource Planning (ERP) system.

8.3.3.10 Modular IoT functions

It is important to note that not all the FGs are needed for a complete actual IoT system. The Functional Model, as well as the Functional View of the Reference Architecture, contains a complete map of the potential functionalities for a system realization. The functionalities that will eventually be used in an actual system are dependent on the actual system requirements. What is important to observe is that the FGs are organized in such a way that more complex functionalities can be built based on simpler ones, thus making the model modular. This is shown already in Figure 8.11, where all the bidirectional arrows show the information flow between FGs, and it is illustrated further in Figure 8.12.

The bare minimum functionalities are Device, Communication, IoT Services, Management, and Security (Figure 8.12A). With these functionalities, an actual system can provide access to sensors, actuators, and tag services for an application or backend system of a larger Enterprise. The application or larger system parts have to build the Virtual Entity functions for capturing the information about the Virtual Entities or the "Things" in the IoT architecture. Often the Virtual Entity concept is not captured in the application or a larger system with a dedicated FG, but functions for handling Virtual Entities are embedded in the application or larger system logic; therefore, in Figure 8.12A–C, the Virtual En-

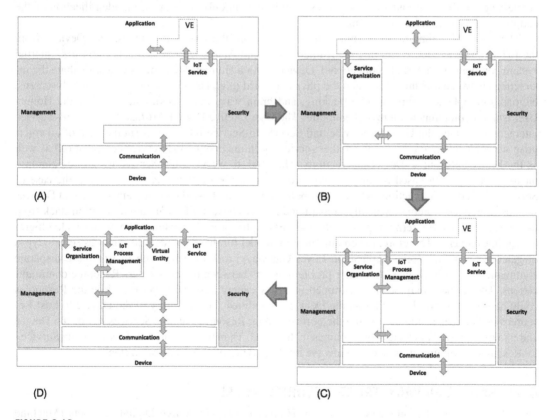

FIGURE 8.12

Building progressively complex IoT Systems.

tity is represented with dashed lines. For example, the deployment of a heating and cooling system is captured in a paper document out of which the larger system developers hardcode logic such as "if the value of sensor A is above 25°C, turn on air conditioner B". The Virtual Entity, in this case, is a room that the developer does not care to capture. The next step in complexity of an IoT system is the addition of composed services based on simpler services. This is shown in Figure 8.12B, where the Service Organization FG is added. Composed services can be used to abstract simpler services, e.g., filtering of events coming from multiple IoT Services. The following step in increasing complexity is the addition of the Business Processes functionality (Figure 8.12C) that enables Enterprise functionality to be present very close to the real IoT system, thus enabling local business control loops.

8.3.4 COMMUNICATION MODEL

The communication model for an IoT Reference Model consists of the identification of the endpoints of interactions, traffic patterns (e.g., unicast vs. multicast), and general properties of the underlying technologies used for enabling such interactions. Chapter 5 and parts of Chapter 7 (e.g., IETF architectures) describe in detail architectures, traffic patterns, and specific networking technologies used for connecting the different endpoints. As a result, this section only focuses on the identification of the endpoints of the communication paths.

The potential communicating endpoints or entities are the Users, Resources, and Devices from the IoT Domain Model. Users include Human Users and Active Digital Artifacts (Services, internal system components, external applications). Devices with a Human–Machine Interface mediate the interactions between a Human User and the physical world (e.g., keyboards, mice, pens, touch screens, buttons, microphones, cameras, eye tracking, and brain wave interfaces), and therefore the Human User is not a communication model endpoint. The User (Active Digital Artifact, Service)–to–Service interactions include the User–to–Service and Service-to-Service interactions (in the case of an enterprise service/application accessing another service, or in the case of IoT service composition) as well as the Service–Resource–Device interactions. The User–to–Service and Service-to-Service communication are typically based on Internet Protocols as described in Chapter 5, apart from the case of Service–to–Service interactions when one or both Services are hosted on constrained/low-end Devices such as embedded systems. Typically, constrained Devices have a different communication stack from the ones used for the Internet type of networks. Examples of such constrained network technologies appear in Chapter 5. Therefore, the communication model for these interactions includes several types of gateways (e.g., network, Application Layer Gateways) to bridge between two or more disparate communication technologies. A similar problem occurs between the Service–to–Resource communications. The Devices may be so constrained that they cannot host the Services, while the Resources could be hosted or not depending on the Device capabilities. This inability of the Device to host Resources or Services results in moving the corresponding Resources and/or Services out of the Device and into more powerful Devices or machines in the cloud. Then the Resource-to-Device or the Service-to–Resource communication needs to involve multiple types of communication stacks.

8.3.5 SAFETY, PRIVACY, TRUST, SECURITY MODEL

An IoT system enables interactions between Human Users and Active Digital Artifacts (Machine Users) with the physical environment. The fact that Human Users are part of a system that could potentially harm humans if malfunctioning, or could expose private information, motivates the Safety and Privacy needs of the IoT Reference Model and Architecture. The Trust and Security Model are needed in every ICT system with the objective to protect the digital world.

8.3.5.1 Safety

System safety is highly application- or application domain-specific and is typically closely related to an IoT system that includes actuators that could potentially harm animate objects (humans, animals). For example, the operation of an IoT system controlling an elevator could harm humans if it allowed the elevator doors to open with a normal user interaction when the elevator car is not behind the doors. Critical infrastructure protection is also related to safety because the loss of such infrastructure due to a malicious user attack could be detrimental to humans, e.g., attacks to a Smart Grid could result in damages ranging from simple loss of electricity in a home to electricity loss in a hospital. By not being application-specific, the IoT Reference Model can only provide IoT-related guidelines for ensuring a safe system to the extent possible and controllable by a system designer. A system designer of such critical systems typically follows an iterative process with two steps: (a) identification of hazards, followed by (b) the mitigation plan. This process is very similar to the threat model and mitigation plan that a security designer performs for an ICT system. Not all hazards or mitigation steps include IoT technology, but a system designer could add safety assertions in relevant points in the interaction between Users, Services, Resources, and Devices. For example, a Human User interaction of pressing the elevator button should result in the elevator door opening only when a Sensor Device detects the elevator car to be behind the doors. If the system designer would like to provide for a safer elevator system, the system should include mechanical safety locks that work even in the absence of the loss of electricity. However, these additional measures do not depend on an IoT system described in this book.

8.3.5.2 Privacy

Because interactions with the physical world may often include humans, protecting the User privacy is of utmost importance for an IoT system. The IoT-A Privacy Model [19,114] depends on the following FCs: Identity Management, Authentication, Authorization, and Trust and Reputation. Identity Management offers the derivation of several identities of different types for the same architectural entity with the objective to protect the original User identity for anonymization purposes. Authentication is a function that allows the verification of the identity of a User whether this is the original or some derived identity. Authorization is the function that asserts and enforces access rights when Users (Services, Human Users) interact with Services, Resources, and Devices. The Trust and Reputation FC maintains the static or dynamic trust relationships between interacting entities. These relationships can impact the behavior of interacting entities, for example, if a Device is deemed nontrusted (e.g., when its Sensor on-Device Service reports out-of-range measurements), another entity (e.g., another Sensor Device or Gateway) could contain logic to reject sensor measurements from the particular Device. The level of Trust and Reputation typically reflects the level of expected behavior of an entity. In ICT systems, trust and reputation are typically represented by a trust/reputation score used for ranking similar entities (e.g., Devices providing similar sensor measurements).

8.3.5.3 Trust

According to the Internet Engineering Task Force (IETF) Internet Security Glossary [115], "Generally, an entity is said to 'trust' a second entity when the first entity makes the assumption that the second entity will behave exactly as the first entity expects". This definition includes an "expectation" which is difficult to capture in a technical context. Nevertheless, [114] suggest that in a technical context, Trust and Reputation could be represented by a score, as seen earlier. This score could be used to impact the behavior of technical components interacting with each other. A trust model is often coupled with the notion of trust in an ICT system and represents the model of dependencies and expectations of interacting entities. The necessary aspects of a trust model according to IoT-A [114] are as follows:

- **Trust Model Domains**: Because ICT and IoT systems may include a large number of interacting entities with different properties, maintaining trust relationships for every pair of interacting entities may be prohibitive. Therefore, groups of entities with similar trust properties can define different trust domains.
- **Trust Evaluation Mechanisms**: These are well-defined mechanisms that describe how a trust score could be computed for a specific entity. The evaluation mechanism needs to take into account the source of information used for computing the trust level/score of an entity; two related aspects are the federated trust and trust anchor. A related concept is the IoT support for evaluation of the trust level of a Device, Resource, and Service.
- **Trust Behavior Policies**: These are policies that govern the behavior between interacting entities based on the trust level of these interacting entities; for example, how a User could use sensor measurements retrieved by a Sensor Service with a low trust level.
- **Trust Anchor**: This is an entity trusted by default by all other entities belonging to the same trust model, and this is typically used for the evaluation of the trust level of a third entity.
- **Federation of Trust**: A federation between two or more Trust Models includes a set of rules that specify the handling of trust relationships between entities with different Trust Models. Federation becomes important in large-scale systems.

8.3.5.4 Security

The Security Model for IoT consists of communication security that focuses mostly on the confidentiality and integrity protection of interacting entities and FCs such as Identity Management, Authentication, Authorization, and Trust and Reputation, as seen earlier.

8.4 IOT REFERENCE ARCHITECTURE

In this section we describe the IoT Reference Architecture. As mentioned earlier, the Reference Architecture is a starting point for generating concrete architectures and actual systems. A concrete architecture addresses the concerns of multiple stakeholders of the actual system, and it is typically presented as a series of views that address different stakeholder concerns [109,18]. A Reference Architecture, on the other hand, serves as a guide for one or more concrete system architects. However, the concept of views for the presentation of an architecture is also useful for the IoT Reference Architecture. Views are useful for reducing the complexity of the Reference Architecture blueprints by addressing groups of concerns one group at a time. However, since the IoT Reference Architecture does not contain details about the environment where the actual system is deployed, some views cannot be presented in detail or at all; for example, the view that shows the concrete Physical Entities and Devices for a specific scenario.

The stakeholders for a concrete IoT system are the people who use the system (Human Users); the people who design, build, and test the Resources, Services, Active Digital Artifacts, and Applications; the people who deploy Devices and attach them to Physical Entities; the people who integrate IoT capabilities of functions with an existing ICT system (e.g., of an enterprise); the people who operate, maintain, and troubleshoot the Physical and Virtual Infrastructure; and the people who buy and own an IoT system or parts thereof (e.g., city authorities).

In order to address the concerns of mainly the concrete IoT architect, and secondly the concerns of most of the above stakeholders, we have chosen to present the Reference Architecture as a set of architectural views [109,18]:

- **Functional View**: Description of what the system does and its main functions.
- **Information View**: Description of the data and information that the system handles.
- **Deployment and Operational View**: Description of the main real-world components of the system, such as devices, network routers, and servers.

The approach for this section is to describe the different views from a generic to a more specific view. It should be noted that Rozanski and Woods [18] also have the notion of a Viewpoint which is a collection of patterns, templates, and conventions for constructing one type of view. As a result, the relationship between a View and Viewpoint is the relationship between a member of a group and a group, respectively. According to Rozanski and Woods there are six viewpoints for the description of an architecture: Functional, Information, Concurrency, Development, Deployment, and Operational. We have chosen to present the Reference Architecture using the Functional, Information, Deployment, and partly the Operational Views that conform to the respective Viewpoints.

8.5 FUNCTIONAL VIEW

The functional view for the IoT Reference Architecture is presented in Figure 8.13, and its source is from IoT-A [19]. It consists of the FGs presented earlier in the IoT Functional Model, each of which includes a set of FCs. It is important to note that not all the FCs are used in a concrete IoT architecture, and therefore the actual system as explained earlier.

8.5.1 DEVICE AND APPLICATION FUNCTIONAL GROUP

The Device and Application FGs are already covered in the IoT Functional Model. For convenience, the Device FG contains the Sensing, Actuation, Tag, Processing, Storage FCs, or simply components. These components represent the resources of the device attached to the Physical Entities of interest. The Application FG contains either standalone applications (e.g., for iOS, Android, Windows) or Business Applications that connect the IoT system to an Enterprise system.

8.5.2 COMMUNICATION FUNCTIONAL GROUP

The Communication FG contains the Hop-by-Hop communication, Network Communication, and End-to-End Communication components:

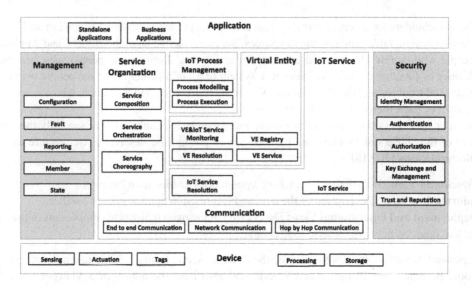

FIGURE 8.13

IoT Functional View (redrawn from IoT-A [19]).

- The Hop-by-Hop Communication is applicable in the case that devices are equipped with mesh radio networking technologies such as IEEE 802.15.4 for which messages have to traverse the mesh from node-to-node (hop-by-hop) until they reach a gateway node which forwards the message (if needed) further to the Internet. The hop-by-hop FC is responsible for transmission and reception of physical and MAC layer frames to/from other devices. This FC has two main interfaces: (a) one "southbound" to/from the actual radio on the device and (b) one "northbound" to/from the Network FC in the Communication FG.
- The Network FC is responsible for message routing and forwarding and the necessary translations of various identifiers and addresses. The translations can be (a) between network layer identifiers to MAC and/or physical network identifiers, (b) between high-level human-readable host/node identifiers to network layer addresses (e.g., Fully Qualified Domain Names (FQDN) to IP addresses, a function implemented by a Domain Name System (DNS) server), and (c) translation between node/service identifiers and network locators in case the higher layers above the networking layer use node or service identifiers that are decoupled from the node addresses in the network (e.g., Host Identity Protocol (HIP) [116] identifiers and IP addresses). Potential fragmentation and reassembly of messages due to limitations of the underlying layers is also handled by the Network FC. Finally, the Network FC is responsible for handling messages that cross different networking or MAC/PHY layer technologies, a function that is typically implemented on a network gateway type of device. An example is an IPv4 to IPv6 translation hosted in a gateway with two network interfaces, one supporting IPv4 and one supporting IPv6/6LoWPAN/IEEE 802.15.4. The Network FC interfaces the End-to-End Communication FC in the "northbound" direction, and the Hop-by-Hop Communication FC in the "southbound" direction.

- The End-to-End Communication FC is responsible for end-to-end transport of application layer messages through diverse network and MAC/PHY layers. In turn, this means that it may be responsible for end-to-end retransmissions of missing frames depending on the configuration of the FC. For example, if the End-to-End Communication FC is mapped in an actual system to a component implementing the Transmission Control Protocol (TCP) protocol, reliable transfer of frames dictates the retransmission of missing frames. Finally, this FC is responsible for hosting any necessary proxy/cache and any protocol translation between networks with different transport/application layer technologies. An example of such functionality is the HTTP-CoAP proxy, which performs transport-layer protocol translation. The End-to-End FC interfaces the Network FC in the "southbound" direction.

8.5.3 IOT SERVICE FUNCTIONAL GROUP

The IoT Service FG consists of two FCs: The IoT Service FC and the IoT Service Resolution FC:

- The IoT Service FC is a collection of service implementations, which interface the related and associated Resources. For a Sensor type of a Resource, the IoT Service FC includes Services that receive requests from a User and return the Sensor Resource value in a synchronous or asynchronous (e.g., subscription/notification) fashion. The services corresponding to Actuator Resources receive User requests for actuation, control the Actuator Resource, and may return the status of the Actuator after the action. A Tag IoT Service can behave both as a Sensor (for reading the identifier of the Tag) or as an Actuator (for writing a new identifier or information on the Tag, if possible). As mentioned earlier, Resources can also perform processing and storage (Processing or Storage Resources), and therefore their corresponding Service exposes the corresponding interfaces, for example, interfaces to provide input data and retrieve output data from a Complex Event Processing (CEP) Resource. An IoT Service for a particular Resource could also expose as a service the historical values of sensor values or actuator commands or tag identifiers.
- The IoT Service Resolution FC contains the necessary functions to realize a directory of IoT Services that allows dynamic management of IoT Service descriptions and discovery/look-up/resolution of IoT Services by other Active Digital Artifacts. The Service descriptions of IoT Services contain a number of attributes as seen earlier in the IoT Functional Model section. Dynamic management includes methods such as creation/update/deletion (CUD) of Service descriptions and can be invoked by both the IoT Services themselves and functions from the Management FG (e.g., bulk creation of IoT Service descriptions upon system start-up). The discovery/look-up and resolution functions allow other Services or Active Digital Artifacts to locate IoT Services by providing different types of information to the IoT Service Resolution FC. By providing the Service identifier (attribute of the Service description) a look-up method invocation to the IoT Service Resolution returns the Service description, while the resolution method invocation returns the contact information (attribute of the service description) of a service for direct Service invocation (e.g., URL). The discovery method, on the other hand, assumes that the Service identifier is unknown, and the discovery request contains a set of desirable Service description attributes that matching Service descriptions should contain.

8.5.4 VIRTUAL ENTITY FUNCTIONAL GROUP

The Virtual Entity FG contains functions that support the interactions between Users and Physical Things through Virtual Entity services. An example of such an interaction is the query to an IoT system of the form, "What is the temperature in the conference room Titan?" The Virtual Entity is the conference room "Titan", and the conference room attribute of interest is "temperature". Assuming that the room is actually instrumented with a temperature sensor, if the User had the knowledge of which temperature sensor is installed in the room (e.g., TempSensor #23), then the User could reformulate and retarget this query to "What is the value of TempSensor #23?" dispatched to the relevant IoT Service representing the temperature resource on the TempSensor #23. The Virtual Entity interaction paradigm requires functionality such as the discovery of IoT Services based on Virtual Entity descriptions, managing the Virtual Entity–IoT Service associations, and processing Virtual Entity-based queries. The following FCs are defined for realizing these functionalities:

- The Virtual Entity Service FC enables the interaction between Users and Virtual Entities by means of reading and writing the Virtual Entity attributes (simple or complex), which can be read or written, of course. Some attributes (e.g., the GPS coordinates of a room) are static and nonwriteable by nature, and some other attributes are nonwriteable by access control rules. In general, attributes that are associated with IoT Services, which in turn represent Sensor Resources, can only be read. There can be, of course, special Virtual Entities associated with the same Sensor Resource through another IoT Service that allow write operations. An example of such a special case is when the Virtual Entity represents the Sensor device itself (for management purposes). In general, attributes that are associated with IoT Services, which in turn represent Actuator Resources, can be read and written. A read operation returns the actuator status, while a write operation results in a command sent to the actuator. Virtual Entity attributes corresponding to Tags can be read in most of the cases by Users and can be written in special cases by other types of Users (e.g., Management applications), if possible of course, as is the case of rewriteable RFID tags. Apart from the function to operate on the Virtual Entity attributes, a Virtual Entity Service can also expose the historical variations of the attributes of a Virtual Entity.
- The Virtual Entity Registry FC maintains the Virtual Entities of interest for the specific IoT system and their associations. The component offers services such as creating/reading/updating/deleting (CRUD) Virtual Entity descriptions and associations. Certain associations can be static; for example, the entity "Room #123" is contained in the entity "Floor #7" by the building construction, while other associations are dynamic, e.g., the entity "Dog" and the entity "Living Room" are associated for a short while due to the fact that the dog moves to the living room (i.e., due to Entity mobility). Update and Deletion operations take the Virtual Entity identifier as a parameter.
- The Virtual Entity Resolution FC maintains the associations between Virtual Entities and IoT Services and offers services such as CRUD associations as well as look-up and discovery of associations. The Virtual Entity Resolution FC also provides notifications to Users about the status of the dynamic associations between a Virtual Entity and an IoT Service and finally allows the discovery of IoT Services provided the certain Virtual Entity attributes.
- The Virtual Entity and IoT Service Monitoring FC includes: (a) functionality to assert static Virtual Entity–IoT Service associations, (b) functionality to discover new associations based on existing associations or Virtual Entity attributes such as location or proximity, and (c) continuous monitoring of the dynamic associations between Virtual Entities and IoT Services and updates of their status in

case existing associations are not valid anymore. The difference between IoT Service and Resource associations and Virtual Entity and IoT-Service associations is that the former are typically static and created upon the creation of the IoT Service instantiation, while the latter are generally dynamic (without excluding static associations, of course) because of potential Virtual Entity mobility. The result of this difference is that in the IoT Service FG there is no FC to discover or monitor new IoT-Service-to-Resource associations, while in the Virtual Entity FG there is a corresponding one.

8.5.5 IOT PROCESS MANAGEMENT FUNCTIONAL GROUP

The IoT Process Management FG aims at supporting the integration of business processes with IoT-related services. It consists of two FCs:

- The Process Modeling FC provides that right tools for modeling a business process that utilizes IoT-related services.
- The Process Execution FC contains the Execution Environment of the process models created by the Process Modeling FC and executes the created processes by utilizing the Service Organization FG in order to resolve high-level application requirements to specific IoT services.

It is important to note the IoT services mentioned above are not only the services from the IoT Service FG, but also from the Virtual Entity FG and the Service Organization FG.

8.5.6 SERVICE ORGANIZATION FUNCTIONAL GROUP

The Service Organization FG acts as a coordinator between different Services offered by the system. It consists of the following FCs:

- The Service Composition FC manages the descriptions and Execution Environment of complex services consisting of simpler dependent services. An example of a complex composed service is a service offering the average of the values coming from a number of simple Sensor Services. The complex composed service descriptions can be well specified or dynamic/flexible depending on whether the constituent services are well defined and known at the execution time or discovered on-demand. The objective of a dynamic composed service can be the maximization of the Quality of Information achieved by the composition of simpler Services, as is the case with the example "average" service described earlier.
- The Service Orchestration FC resolves the requests coming from IoT Process Execution FC or User into the concrete IoT services that fulfill the requirements.
- The Service Choreography FC is a broker for facilitating communication among Services using the Publish/Subscribe pattern. Users and Services interested in specific IoT-related services subscribe to the Choreography FC, providing the desired service attributes even if the desired services do not exist. The Choreography FC notifies the Users when services fulfilling the subscription criteria are found.

It is important to note that the IoT services mentioned above are not only the services from the IoT Service FG, but also from the Virtual Entity FG and the Service Composition FC.

8.5.7 SECURITY FUNCTIONAL GROUP

The Security FG contains the necessary functions for ensuring the security and privacy of an IoT system. It consists of the following FCs:

- The Identity Management FC manages the different identities of the involved Services or Users in an IoT system in order to achieve anonymity by the use of multiple pseudonyms. The component maintains a hierarchy of identities (an identity pool), as well as group identities [114].
- The Authentication FC verifies the identity of a User and creates an assertion upon successful verification. It also verifies the validity of a given assertion.
- The Authorization FC manages and enforces access control policies. It provides services to manage policies (Create, Update, Delete, CUD), as well as taking decisions and enforcing them regarding access rights of restricted resources. The term "resource" here is used as a representation of any item in an IoT system that needs restricted access. Such an item can be a database entry (Passive Digital Artifact), a Service interface, a Virtual Entity attribute (simple or complex), a Resource/Service/Virtual Entity description, etc.
- The Key Exchange and Management FC is used for setting up the necessary security keys between two communicating entities in an IoT system. This involves a secure key distribution function between communicating entities.
- The Trust and Reputation FC manages reputation scores of different interacting entities in an IoT system and calculates the service trust levels. A more detailed description of this FC is contained in the Safety, Privacy, Trust, Security Model presented in Section 8.3.5.

8.5.8 MANAGEMENT FUNCTIONAL GROUP

The Management FG contains system-wide management functions that may use individual FC management interfaces. It is not responsible for the management of each component, but rather for the management of the system as a whole. It consists of the following FCs:

- The Configuration FC maintains the configuration of the FCs and the Devices in an IoT system (a subset of the ones included in the Functional View). The component collects the current configuration of all the FCs and devices, stores it in a historical database, and compares current and historical configurations. The component can also set the system-wide configuration (e.g., upon initialization), which in turn translates to configuration changes to individual FCs and devices.
- The Fault FC detects, logs, isolates, and corrects system-wide faults if possible. This means that individual component fault reporting triggers fault diagnosis and fault recovery procedures in the Fault FC.
- The Member FC manages membership information about the relevant entities in an IoT system. Example relevant entities are the FGs, FCs, Services, Resources, Devices, Users, and Applications. Membership information is typically stored in a database along with other useful information such as capabilities, ownership, and access rules and rights, which are used by the Identity Management and Authorization FCs.
- The State FC is similar to the Configuration FC and collects and logs state information from the current FCs, which can be used for fault diagnosis, performance analysis, and prediction, as well as billing purposes. This component can also set the state of the other FCs based on system-wide state information.

- The Reporting FC is responsible for producing compressed reports about the system state based on input from FCs.

8.6 INFORMATION VIEW

The information view consists of (a) the description of the information handled in the IoT System and (b) the way this information is handled in the system; in other words, the information lifecycle and flow (how information is created, processed, and deleted) and the information handling components. Because the information handled by an IoT system is captured mainly by the IoT Information Model described in Section 8.3.2 as part of the IoT Reference Model, we only provide a synopsis of the specific information pieces without going into details. As a second part, we describe the way some of the aforementioned pieces of information are handled in an IoT system.

8.6.1 INFORMATION DESCRIPTION

The pieces of information handled by an IoT system complying to an ARM such as the IoT-A [19] are the following:

- Virtual Entity context information, i.e., the attributes (simple or complex) as represented by parts of the IoT Information model (attributes that have values and metadata such as the temperature of a room). This is one of the most important pieces of information that should be captured by an IoT system and represents the properties of the associated Physical Entities or Things.
- IoT Service output itself is another important part of information generated by an IoT system. For example, this is the information generated by interrogating a Sensor or a Tag Service.
- Virtual Entity descriptions in general, which contain not only the attributes coming from IoT Devices (e.g., ownership information).
- Associations between Virtual Entities and related IoT Services.
- Virtual Entity Associations with other Virtual Entities (e.g., Room #123 is on Floor #7).
- IoT Service Descriptions, which contain associated Resources, interface descriptions, etc.
- Resource Descriptions, which contain the type of resource (e.g., sensor), identity, associated Services, and Devices.
- Device Descriptions such as device capabilities (e.g., sensors, actuators, radios).
- Descriptions of Composed Services, which contain the model of how a complex service is composed of simpler services.
- IoT Business Process Model, which describes the steps of a business process utilizing other IoT-related services (IoT, Virtual Entity, Composed Services).
- Security information such as keys, identity pools, policies, trust models, and reputation scores.
- Management information such as state information from operational FCs used for fault/performance purposes, configuration snapshots, reports, membership information, etc.

8.6.2 INFORMATION FLOW AND LIFECYCLE

On a high level, the flow of information in an IoT system follows two main directions. From devices that produce information such as sensors and tags, information follows a context enrichment process

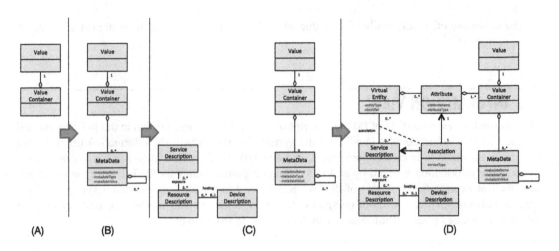

FIGURE 8.14

Information enrichment process.

until it reaches the consumer application or part of the larger system, and from the application or part of a larger system information it follows a context reduction process until it reaches the consumer types of devices (e.g., actuators). The enrichment process is shown in Figure 8.14.

Devices equipped with sensors transform changes in the physical properties of the Physical Entities of Interest into electrical signals. These electrical signals are transformed into one or multiple values (Figure 8.14A) on the device level. These values are then enriched with metadata information such as units of measurement, a timestamp, and possibly location information (Figure 8.14B). These enriched values are offered by a software component (Resource) either on the device or the network. The Resource exposes certain IoT Services to formalize access to this enriched information (Figure 8.14C). At this point, the information is annotated with simple attributes such as location and time, and often this type of metadata is sufficient for certain IoT applications or for the use in certain larger systems. This enriched information becomes context information as soon as it is further associated with certain Physical Entities in the form of Virtual Entity attributes (simple or complex, static or dynamic). Further support information such as Associations between certain attributes and IoT Services further enriches the context information of the Virtual Entity (Figure 8.14D).

Further enrichment occurs in applications or larger systems that employ, for example, data analytics, Machine Learning, and knowledge management, which produces actionable information. Parts of the context and actionable information may be stored in an information store for future use. Actionable information flows into business processes that implement an action plan. Action plans push context information about Virtual Entities to associated IoT Services, to corresponding Actuation Resources, and finally to the real actuators that perform the changes in the physical world (context information reduction flow). Actual IoT systems employ a different degree of enrichment, reduction, or storage of information. Certain IoT systems only employ enrichment while leaving the action to humans, others employ only context reduction (e.g., remote control of a heating element), or others employ the full feedback loop.

Virtual Entity context information is typically generated by data producing devices such as sensor devices and consumed either by data consumption devices such as actuators or services (IoT or other types of services such as Machine Learning processing services). Raw or enriched information and/or actionable information may be stored in caches or historical databases for later usage or processing, traceability, or accounting purposes. The historical/cache database information lifetime is often application- or regulation-specific. Typically the information maintained in a cache is ephemeral, while the information stored in historical databases can last for longer but highly application-specific durations. Certain raw sensor readings may be destroyed after fulfilling a User request, and other sensor readings may be stored for 5 years (for example) according to a data retention policy dictated by regulation. Similar rules may apply to the operation-specific information such as Device, Resource, Service Descriptions, etc. These contain the necessary information for an IoT system to operate, but not the information that human users are typically interested in. Nevertheless, this kind of information is created by FCs, is typically stored for the purpose of Fault Management, and is typically automatically destroyed, for example, when a soft state handling policy is applicable. In this context soft state information means that the subsystem that manages this kind of information destroys old information according to the time of creation and a retention policy (e.g., Service Descriptions older than one day are destroyed from the IoT Service Resolution), while the Users (FCs, management applications) that create this kind of information are responsible for refreshing it periodically in order to avoid automatic destruction.

8.6.3 INFORMATION HANDLING

An IoT system is typically deployed to monitor and control Physical Entities. Monitoring and controlling Physical Entities is in turn performed by mainly the Devices, Communication, IoT Services, and Virtual Entity FGs in the functional view. Certain FCs of these FGs, as well as the rest of the FGs (Service Organization, IoT Process Management, Management, Security FGs), play a supporting role for the main FGs in the Reference Architecture, and therefore in the flow of information. Moreover, an IoT system is typically one part of a larger system encompassing several other functions, such as Complex Event Processing (CEP), Data Collection and Processing, and Data Analytics and Knowledge Management, Machine Intelligence, as seen earlier in the book in Chapter 5. Therefore, information handling of an IoT system depends largely on the specific problem at hand. From a Reference Architecture point of view, we can only present part of the information flow space concerning the IoT Reference Model, while the technology Chapter 5 provides details on the individual and more complex information handling components and interactions. The presentation of information handling in an IoT system assumes that FCs exchange and process information. The exchange of information between FCs follows the interaction patterns below ([19]; Figure 8.15):

- Push: An FC A pushes the information to another FC B provided that the contact information of the component B is already configured in component A, and component B listens for such information push.
- Request/Response: An FC A sends a request to another FC B and receives a response from B after A serves the request. Typically the interaction is synchronous in the sense that A must wait for a response from B before proceeding to other tasks, but in practice, this limitation can be realized

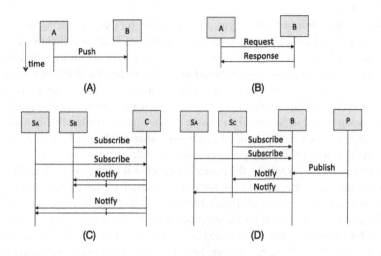

FIGURE 8.15

Information exchange patterns (redrawn from IoT-A [19]).

with parts of component A waiting and other parts performing other tasks. Component B may need to handle concurrent requests and responses from multiple components, which imposes certain requirements on the capabilities of the device or the network that hosts the FC.

- Subscribe/Notify: Multiple subscriber components (S_A, S_B) can subscribe for information to a component C, and C will notify the relevant subscribers when the requested information is ready. This is typically an asynchronous information request after which each subscriber can perform other tasks. Nevertheless, a subscriber needs to have some listening components for receiving the asynchronous response. The target component C also needs to maintain state information about which subscribers requested which information and their contact information. The Subscribe/Notify pattern is applicable when typically one component is the host of the information needed by multiple other components. Then the subscribers need only establish a Subscribe/Notify relationship with one component. If multiple components can be information producers or information hosts, the Publish/Subscribe pattern is a more scalable solution from the point of view of the subscribers.

- Publish/Subscribe: In the Publish/Subscribe (also known as a Pub/Sub pattern), there is a third component called the broker B, which mediates subscription and publications between subscribers (information consumers) and publishers (or information producers). Subscribers such as S_A and S_B subscribe to the broker B about the information they are interested in by describing the different properties of the information. A Publisher P publishes information and metadata to the broker, and the broker pushes the published information to (notification) the subscribers whose interests match the published information.

At this point, we describe a few examples of information handling by the FCs. Please note that these interaction descriptions are not complete in the sense that they do not contain all the possible ways such

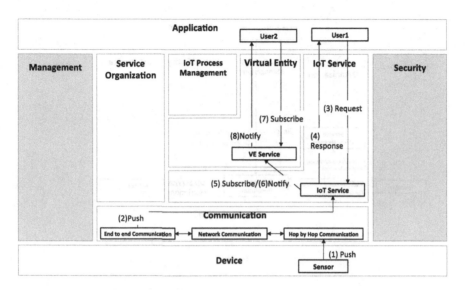

FIGURE 8.16

Device, IoT Service, and Virtual Entity Service Interactions.

interactions can take place, and they are not intended to be complete since these interactions are highly dependent on the actual IoT system requirements. In Figure 8.16 we assume that the generated sensed data is pushed by a sensor device (under Steps 1 and 2) that is part of a multihop mesh network such as IEEE 802.15.4 through the Hop-by-Hop, Network, and End-to-End communication FCs towards the Sensor Resource hosted in the network. Please note that the Sensor Resource is not shown in the figure, only the associated IoT Service. A cached version of the sensor reading on the Device is maintained on the IoT Service. When User1 (Step 3) requests the sensor reading value from the specific Sensor Device (assuming User1 provides the Sensor resource identifier), the IoT Service provides the cached copy of the sensor reading back to the User1 annotated with the appropriate metadata information about the sensor measurement, for example, a timestamp of the last known reading of the sensor, units, and location of the Sensor Device. Also, assume that the Virtual Entity Service associated with the Physical Entity (e.g., a room in a building) where the specific Sensor Device has been deployed already contains the IoT Service as a provider of the "hasTemperature" attribute of its description. The Virtual Entity Service subscribes to the IoT Service for updates of the sensor readings pushed by the Sensor Device (Step 5). Every time the Sensor Device pushes sensor readings to the IoT Service, the IoT Service notifies (Step 6) the Virtual Entity Service, which updates the value of the attribute "hasTemperature" with the sensor reading of the Sensor Device. At a later stage, a User2 subscribing (Step 7) to changes on the Virtual Entity attribute "hasTemperature" is notified every time the attribute changes value (Step 8). Please note that some of the information flow steps between the Virtual Entity and IoT Service are omitted in this figure for simplicity purposes.

Figure 8.17 depicts the information flow when utilizing the IoT Service Resolution FC. The IoT Service Resolution implements two main interfaces, one for CUD of Service Description objects in the IoT Service Resolution database/store and one for look-up/resolution/discovery of IoT Services.

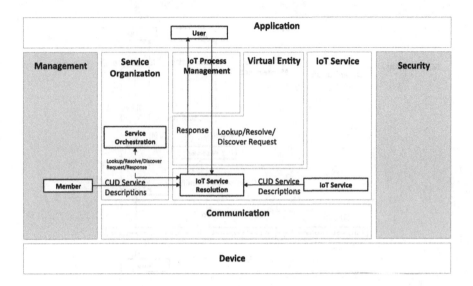

FIGURE 8.17

IoT Service Resolution.

As a reminder, the look-up and resolution operations provide the Service Description and the Service Locator, respectively, given the Service identifier and the discovery operation returns a (set of) Service Description(s) given a list of desirable attributes that matching Service Descriptions should contain. The CUD operations can be performed by the IoT Service logic itself or by a management component (e.g., Member FC in Figure 8.17). The look-up/resolution and discovery operation can be performed by a User as a standalone query or the Service Orchestration as a part of a Composed Service or an IoT Process. If a discovery operation returns multiple matching Service Descriptions, it is upon the User or the Service Orchestration component to select the most appropriate IoT Service for the specific task. Although the interactions in Figure 8.17 follow the Request/Response pattern, the look-up/resolution/discovery operations can follow the Subscribe/Notify pattern in the sense that a User or the Service Orchestration FC subscribes to changes of existing IoT Services for look-up/resolution and for the discovery of new Service Descriptions in the case of a discovery operation.

Figure 8.18 describes the information flow when the Virtual Entity Service Resolution FC is utilized. The Virtual Entity Resolution FC allows the CUD of Virtual Entity Descriptions and the look-up and discovery of Virtual Entity Descriptions. A look-up operation by a User or the Service Orchestration FC returns the Virtual Entity Description given the Virtual Entity identity, while the discovery operation returns the Virtual Entity Description(s) given a set of Virtual Entity attributes (simple or complex) that matching Virtual Entities should contain. Please note that the Virtual Entity Registry is also involved in the information flow because it is the storage component of Virtual Entity Descriptions, but it is omitted from the figure to avoid cluttering. The Virtual Entity Resolution FC mediates the requests/responses/subscriptions/notifications between Users and the Virtual Entity Registry, which has a simple CRUD interface given the Virtual Entity identity. The FCs that

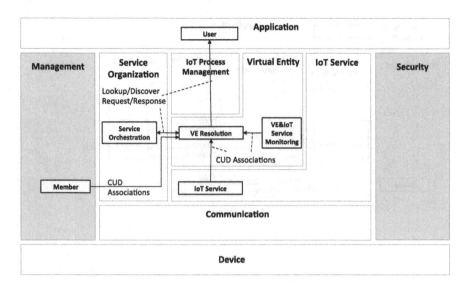

FIGURE 8.18

Virtual Entity Service Resolution.

could perform CUD operations on the Virtual Entity Resolution FC are the IoT Services themselves due to internal configuration, the Member Management FC that maintains the associations as part of the system setup, and the Virtual Entity and IoT Service Monitoring component whose purpose is to discover dynamic associations between Virtual Entities and IoT Services. It is important to note that the Subscribe/Notify interaction patterns can also apply to the look-up/discovery operations, the same as the Request/Response patterns provided the involved FCs implement Subscribe/Notify interfaces.

As a final example of the information flow, we show a Complex Event Processing (CEP) Resource mapped to an IoT Service C in Figure 8.19. The CEP Service needs the information from two IoT services, e.g., IoT Services corresponding to two Sensor Resources hosted on two Sensor Devices, and produces one output. The CEP IoT Service expects the inputs to be published/pushed to its interfaces, while the output interface conforms to a Subscribe/Notify interaction pattern. The individual IoT Services A and B expose interfaces that also comply with the Publish/Subscribe interaction pattern. The FC that can connect these three components is the Service Choreography FC that realizes a Publish/Subscribe interaction pattern. As a first step, the IoT Service C subscribes to the Service Choreography FC that it requires IoT Services A and B as inputs. In the meantime, a User subscribes to the Service Choreography FC that it needs the output of the CEP IoT Service C. When the individual IoT Services A and B publish their output to the Service Choreography FC, these outputs are published/forwarded to the IoT Service C, which needs them to produce information of type C. After performing CEP filtering, the IoT Service C publishes the output of type C to the Service Choreography FC, which publishes/forward it to the User.

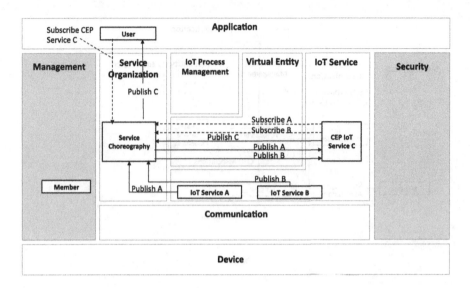

FIGURE 8.19

Service Choreography and Processing IoT Services (redrawn from IoT-A [19]).

8.7 DEPLOYMENT AND OPERATIONAL VIEW

The Deployment and Operational View depends on the specific actual use case and requirements, and therefore we present here one way of realizing the Parking Lot example seen earlier. It is by no means an exhaustive or complete example. The use case chapters presented in Part 3 of this book provide a few real-world deployment examples.

Figure 8.20 depicts the Devices view as Physical Entities deployed in the parking lot, as well as the occupancy sign. There are two sensor nodes (#1 and #2), each of which are connected to eight metal/car presence sensors. The two sensor nodes are connected to the payment station through wireless or wired communication. The payment station acts both as a User Interface for the driver to pay and get a payment receipt and as a communication gateway that connects the two sensor nodes and the payment interface physical devices (displays, credit card slots, coin/note input/output, etc.) with the Internet through WAN technology. The occupancy sign also acts as a communication gateway for the actuator node (display of free parking spots), and we assume that because of the deployment, a direct connection to the payment station is not feasible (e.g., wired connectivity is too prohibitive to be deployed or sensitive to vandalism). The physical gateway devices connect through a WAN to the Internet and to a Data Center where the parking lot management system software is hosted as one of the Virtual Machines on a Platform as a Service (PaaS; Chapter 5) configuration. The two main applications connected to this management system are human user mobile phone applications and parking operation center applications. We assume that the parking operation center manages several other parking lots using similar physical and virtual infrastructure.

Figure 8.21 shows two views superimposed, the deployment and functional views, for the parking lot example. Please note that several FGs and FCs are omitted here for simplicity purposes, and cer-

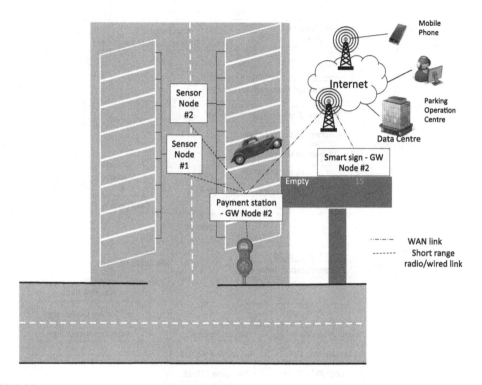

FIGURE 8.20

Parking Lot Deployment and Operational View, Devices.

tain non-IoT-specific Services appear in the figure because an IoT system is typically part of a larger system. Starting from the Sensor Devices, as seen earlier, Sensor Node #1 hosts Resource #11–#18, representing the sensors for the parking spots #01–#08, while earlier Sensor Node #2 hosts Resource #21–#28, representing the sensors for the parking spots #09–#16. We assume that the sensor nodes are powerful enough to host the IoT Services #11–#18 and #21–#28 representing the respective resources. The two sensor nodes are connected to the gateway device that also hosts the payment service with the accompanying sensors and actuators, as seen earlier. The other gateway device hosts the occupancy sign actuator resource and corresponding service. The management system for the specific parking lot, as well as others, is deployed on a Virtual Machine on a Data Center. The Virtual Machine hosts communication capabilities, Virtual Entity services for the parking spots #01–#16, the Virtual Entity services for the occupancy sign, a payment business process that involves the payment station and input from the occupancy sensor services, and the parking lot management service that provides exposure and access control to the parking lot occupancy data for the parking operation center and the consumer phone applications. As a reminder, the Virtual Entity service of the parking lot uses the IoT Services hosted on two sensor nodes and performs the mapping between the sensor node identifiers (#11–#18 and #21–#28) to parking spot identifiers (spot #01–#16). The services offered on these parking spots are to read the current state of the parking spot to see whether it is "free" or "occupied". The Virtual

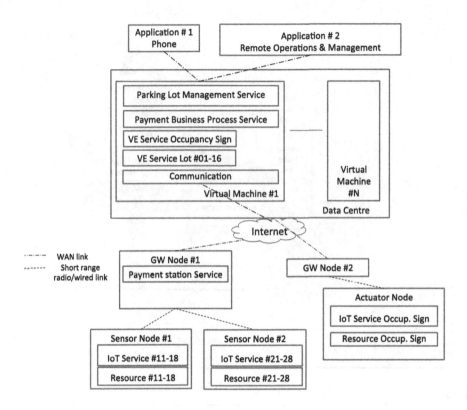

FIGURE 8.21

Parking Lot Deployment and Operational View, Resources, Services, Virtual Entities, Users.

Entity corresponding to the occupancy sign contains one writable attribute: the number of free parking spots. A User writing this Virtual Entity attribute results in an actuator command to the real actuator resource to change its display to the new value.

Of course, for the operation of the parking lot, a number of other IoT-related services would be useful, such as historical occupancy data on which Machine Learning algorithms could support the operator of the parking lot in making decisions about planning and charging. Starting from the IoT Domain Model, we attempt to perform a high-level mapping between the different classes/entities of the model and their realization. The physical sensors, actuators, tags, processors, and memory, which are parts of a Device, are deployed close to the Physical Entities of Interest, the ones whose properties are monitored or controlled.

Figure 8.22 shows an example of mapping an IoT Domain Model and Functional View to Devices with different capabilities (different alternatives) connecting to a cloud infrastructure. Alternative 1 shows devices that can host only a simple Sensor Device and a short-range wired or wireless connectivity technology (Basic Device #1). Such kind of device needs an Advanced Device of type #1 that allows the basic device to perform protocol adaptation (at least from the short-range wired or wireless connectivity technology to a WAN technology) so that the Sensor IoT service in the cloud and the

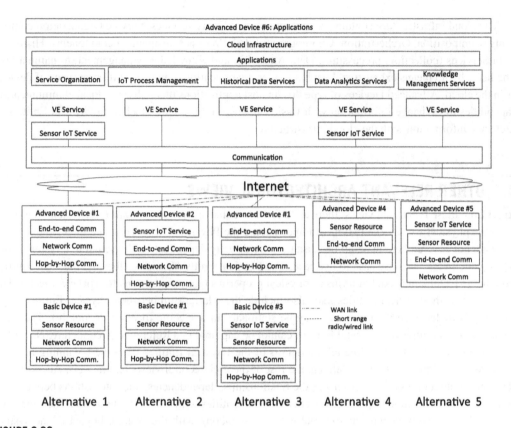

FIGURE 8.22

Mapping IoT Domain Model concepts to Deployment View.

Sensor Resource on the Basic Device #1 can exchange information. The Virtual Entity representing the Physical Entity where the Basic Device #1 is deployed is also hosted in the cloud. In alternative 2, Advanced Devices (type #2) can host the Sensor IoT Service communicating to the Sensor Resource on a Basic Device #1. The cloud infrastructure in this case only hosts the Virtual Entity Service corresponding to the Sensor IoT Service. The difference between alternative 1 and alternative 2 is that the Sensor IoT Service hosted on an Advanced Device #2 should be capable of responding to requests from Users (cloud services, Applications) with the appropriate secure mediation of course. In alternative 3, the Basic Device #3 is capable of providing the Sensor Resource and the Sensor IoT Service but still needs an Advanced Device #1 to transport IoT service requests/responses/subscriptions/notifications/publications to the Users in the cloud. According to experience, this kind of deployment scenario imposes a high burden on a Basic Device, which potentially makes the Basic Device the weakest link in the information flow. If malicious Users launch a Denial of Service (DoS) attack on the node, the probability of the node going down is very high. Alternatives 4 and 5 show Advanced Devices offering a WAN interface. In alternative 4, only the Sensor Resource is hosted on the Device, while in alternative 5, even the IoT Service is hosted on the Device. The Virtual Entity Service is hosted in the cloud.

The cloud infrastructure contains, apart from Virtual Entity services, Service Organization components (Composition, Orchestration, Choreography), IoT Process Management components, Historical Data Services (collection, processing), Data Analytics, and Knowledge Management, to name a few. The list is by no means exhaustive. Last but not least, applications can be on different types of devices or in the cloud. Advanced Devices of type #6 can host applications that either use local communication capabilities to exchange information with Basic or Advanced Devices attached to Physical Entities or exchange information with the cloud infrastructure.

8.8 OTHER RELEVANT ARCHITECTURAL VIEWS

Apart from these functional views, there are a few more that are very important for a system that interfaces the physical world. The two most important are the Physical Entity View and the Context View. They are not covered here in detail because they are directly dependent on the actual IoT system properties, which vary from use case to use case. The Physical Entity View describes the Physical Entities from the IoT Domain Model in terms of physical properties (e.g., dimensions for spaces/objects). The description of the Physical Entities may also include the relationship between Physical Entities (e.g., an entity is included in the other and may be stationary in a specific location or mobile). The large number of possibilities for Physical Entities cannot be captured in a Reference Architecture. Nevertheless, an architect needs to outline all the details of the Physical Entities from the beginning in order to assess if any Physical Property affects the rest of the architectural views and models. According to [18], the Context of a system "describes the relationships, dependencies, and interactions between the system and its environment (people, systems, external entities with which it interacts)". Therefore, the Context View should capture the external entities interacting with the system, impact of the system on its environment, external entity properties/identities, system scope, responsibilities, etc. Since the possibilities for external entities and their interactions with an IoT system depend on the assumptions on the actual system, this view is also constructed at the beginning of the design process because it sets the boundary conditions for the problem at hand. In the parking lot example above, we described briefly parts of the Physical Entity and Context View without explicit individual presentations of these views. For example, the dimensions of the parking spots are a Physical Entity property, and the fact that there are 16 parking spots physically placed in a possibly gated parking lot, the fact that there is an occupancy display near the parking lot on the roadside, and other details outline both Physical Entity properties as well as relationships between the system and its environment.

8.9 OTHER REFERENCE MODELS AND ARCHITECTURES

Apart from the IoT-A ARM which is quite generic and not dependent on the industry sector, over the past few years several organizations have produced reference architectures or reference models for specific industries. In this section, we try to describe in detail a few of these and compare them in a high level to the IoT-A.

8.9.1 INDUSTRIAL INTERNET REFERENCE ARCHITECTURE

The IIC[8] was founded in 2014 by AT&T, Cisco, General Electric, Intel, and IBM in order to promote the application of IoT technologies to industrial settings. Over the years the consortium grew from the first founding members to over 260 members. At the time of writing, the current members consist of eight founding and contributing members (Bosch, DELL EMC, General Electric, Huawei, Intel, IBM, SAP, and Schneider Electric, 3% of the total members) and large industrial (31%), small industrial (40%), and academic and governmental organizations (26%) from all over the world. IIC is an open consortium with membership eligibility based on annual fees which depend on the type and amount of revenue for each organization.

IIC is not a standardization organization; it rather evaluates existing standards and attempts to influence international Standards Development Organizations (SDOs) in the development of standards related to industrial IoT. Such evaluations and recommendations are to be captured and disseminated in specific publications that are yet to be released.

IIC also has developed several frameworks such as the Industrial Internet Reference Architecture [108] or IIRA, the Industrial Internet Connectivity Framework [117] or IICF, and the Industrial Internet Security Framework [118] or IISF. The purpose of these frameworks is to promote the use of common language among stakeholders and interoperability between different industrial IoT systems. For this purpose the developed frameworks as well as a vocabulary and interoperability guidelines are released or are to be released in the future for free as technical or white papers[9].

In addition, IIC defines industrial use cases[10] and allows members to form groups and create real-world industrial testbeds[11] applying and testing the defined frameworks and guidelines. IIC targets the following industries: (a) Energy, (b) Healthcare, (c) Manufacturing, (d) Smart Cities, and (e) Transportation. Currently, IIC has developed 25 use cases and about the same number of testbeds (in general use cases and testbeds do not correspond to each other) addressing the main focus industries as well as certain technologies such as Communication and Security.

IIC is organized in 19 Working Groups in the following six areas:

- Business Strategy and Solution Lifecycle: This Working Group supports companies with business strategy and business planning of industrial IoT opportunities, develops best practices for using and implementing the IIRA, and provides guidelines about industrial IoT project management, solution evaluation, and contractual issues.
- Liaison: This Working Group consists mainly of members that participate in Standardization Organizations, open source organizations, and alliances with the purpose of bidirectional information flow between the IIC and these organizations.
- Marketing: This Working Group develops marketing material and promotes the IIC.
- Security: This is a technical Working Group that focuses on the development of the Industrial Internet Security Framework.

---

[8]http://www.iiconsortium.org
[9]http://www.iiconsortium.org/white-papers.htm
[10]http://www.iiconsortium.org/case-studies/index.htm
[11]http://www.iiconsortium.org/test-beds.htm

- Technology: This the core Working Group that organizes and coordinates all the activities related to technical matters including the IIRA, Industrial Internet Connectivity Framework, Vocabulary, Use cases, and Liaison.
- Testbeds: This Working Group is an advisory group that evaluates testbed proposals submitted by member groups and provides guidance for new testbeds.

The IIRA description follows the standard ISO/IEC/IEEE 42010:2011 [106] which uses architecture concepts similar to Rozanski and Woods [18] such as Stakeholders and their Concerns, Viewpoints, and Views as well as other concepts such as Models and Model Kinds. The two architecture description models of Rozanski and Woods and the ISO/IEC/IEEE 42010:2011 also have similar relationships between Stakeholders, Concerns, Viewpoints, and Views with slightly differing definitions; however in substance, they capture the fact that Stakeholders have Concerns which are addressed or framed by Viewpoints. The main difference in these core concepts is that Rozanski and Woods assume that Viewpoints are collections of Views while ISO/IEC/IEEE 42010:2011 assumes that there is one to one mapping of Viewpoints and Views. However, this difference is not substantial for understanding an IoT Reference Architecture. The second major difference is the Model Kinds and Models of the ISO/IEC/IEEE 42010:2011 which are not part of the Rozanski and Woods architecture definition model. According to ISO/IEC/IEEE 42010:2011 a Viewpoint consists of multiple Model Kinds and a View consists of multiple Models. A Model is a specific application of the conventions of a Model Kind on a specific Viewpoint.

Regardless of these similarities and differences, the IIRA description is organized into a set of Viewpoints which describe the corresponding stakeholder concerns without going into explicit details about the Model Kinds, Views or Models. However, the description of each Viewpoint typically follows a structure which is specific to each Viewpoint. This structure consists of the main concepts of the Viewpoint and their relationships. The reader could notice that in the case of IoT-A the main concepts of IoT and their relationships were captured in the IoT Domain Model. As a result, these IIC structures in each viewpoint could theoretically be considered as Model Kinds or Models of the Viewpoint or the corresponding View. Since the development of the IIRA is still in progress these details are expected to be clarified in future releases.

For all practical purposes, it is not important that the IoT-A is described in terms of the architectural views while the IIRA is presented in terms of the architectural viewpoints. In both cases, the architecture descriptions contain the main pieces of information that a system architect needs to develop a concrete architecture. Both reference architecture styles also stress the fact that their descriptions are only starting points and tools to assist the system architect to capture the relevant main concepts and their relationships in a concrete architecture.

The IIRA description consists of the Business, Usage, Functional, and Implementation Viewpoints (Figure 8.23). One should note that the Viewpoints are organized in layers in order to show that the development of each layer imposes requirements and guides the layer below and the development of a layer provides feedback to the layer above with respect to the design decisions and requirements. The Business viewpoint is briefly described while the rest are described in a sufficient level of detail in order to present the important technical aspects. Briefly, the IIRA Viewpoints are the following:

- Business: The Business Viewpoint focuses on the identification of the Stakeholders, their business Vision, the Values of an IIoT system, the Key (technical and business) Objectives, and the business Fundamental Capabilities with which the business realizes the system. These concepts are related

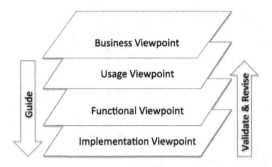

FIGURE 8.23

IIRA Viewpoints (redrawn from IIC IIRA [108]).

to each other and influence one another. The Business Viewpoint influences the Usage Viewpoint below by the usage activities and system requirements used in the Usage Viewpoint being derived by the Key Objectives and Fundamental Capabilities of the Business Viewpoint. The stakeholders interested in the Business Viewpoint are decision makers, system engineers, and product managers.

- Usage: The Usage Viewpoint describes how the system is to be used and addresses the usage concerns of the corresponding stakeholders. The description typically captures interactions between humans and software (Active Digital Artifacts in IoT-A). The typical stakeholders interested in this viewpoint are product managers, system engineers, and other stakeholders that take care of the end-user concerns.
- Functional: The Functional Viewpoint captures the FCs of the system, their interfaces, and interactions in order to support the system usage defined by the Usage Viewpoint. The typical stakeholders for this viewpoint are system architects, developers, and integrators. This viewpoint is similar to the Functional View of the IoT-A.
- Implementation: The Implementation Viewpoint describes the implementation technologies used for the realization of the FCs identified in the Functional Viewpoint. The Usage Viewpoint and the Business Viewpoint may also influence certain design choices. The stakeholders for this Viewpoint include system architects, system developers, integrators, and system operators.

8.9.1.1 IIRA Usage Viewpoint

The Usage Viewpoint describes the activities of the usage of the system, the main actors and roles that perform these activities, and the relationships between these entities (Figure 8.24). The notation is similar to the one used for the description of the IoT Domain Model (Section 8.3.1.1) with the following additions:

- In Figure 8.24 there is a bidirectional Association with two Association names ("Is-Assigned"/"Assume") and each Association name is closer to a different class: the Association name "Assume" is closer to the class "Party" while the Association name "Is-Assigned" is closer to the class "Role". The Association names are read with the closer class as the subject and the class further away as the object, i.e., the two associations are read as "A Party Assumes a Role" and "A Role is-Assigned a Party".

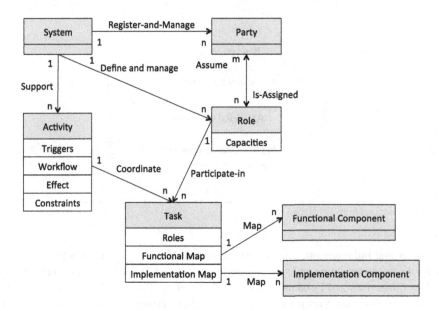

FIGURE 8.24

IIRA Usage Viewpoint main concepts and relationships (redrawn from IIC IIRA [108]).

- Some of the classes have the attributes parts of the class spelled out explicitly.

  The main concepts of this model are:

- System: The system for which the usage is described.
- Task: Basic unit of work such as the invocation of a function. A Task is similar to the Service class in the IoT-A IoT Domain Model and the services belonging to IoT Service FG in the Functional view of the IoT-A.
- Party and Roles: The different human or software entities responsible for the execution of a task; Parties can take different roles when executing tasks based on their Capacities. A Party is similar to the User class in the IoT-A IoT Domain Model.
- Activity: An activity is a coordination of tasks to achieve a higher-level goal than a task, e.g., a device on-boarding procedure. Activities have Triggers or conditions upon which the activity will be executed, a workflow which is a set of steps for achieving the higher-level goal, an Effect which is the target state of the system after the Activity execution, and system Constraints that must be attended to when executing the Activity, e.g., data integrity. An Activity is similar to the Service in the IoT-A IoT Domain Model and services belonging to the IoT Service Organization FG in the Functional view of the IoT-A.
- Functional and Implementation Components: The Functional and Implementation Components are descriptions of the respective components from the Functional and Implementation Viewpoints when these viewpoints are defined. Before the viewpoints are defined the associations in each Task are empty.

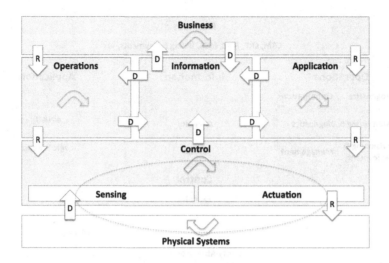

FIGURE 8.25

IIRA Functional Viewpoint main functional domains (redrawn from IIC IIRA [108]).

The Task class includes Functional and Implementation Components that may not be available at the time of developing the Usage Viewpoint. The typical development of the Usage Viewpoint starts with the initial definition of Parties, Roles, and Activities which in turn may be mapped to simpler Tasks without any associations to any Functional or Implementation Components. As the Functional and Implementation Viewpoints are developed the architect should revisit the Usage Viewpoint and refine the Tasks and their associations to different Components (Functional or Implementation).

8.9.1.2 IIRA Functional Viewpoint

The IIRA Functional Viewpoint is organized into five functional domains which are similar to the FGs in IoT-A:

- Control domain.
- Operations domain.
- Information domain.
- Application domain.
- Business domain.

Figure 8.25 shows the different functional domains and how they relate to each other with respect to information and request/control flow. The arrows with the letter "D" indicate a Data/Information flow between functional domains and the arrows with the letter "R" indicate a Request/Control flow. There are also arrows without any letter; these indicate decision flows.

Moreover, the description of the functional viewpoint includes the term "asset", which is defined in the IIC Vocabulary [119] as a "major application, general support system, high impact program, physical plant, mission-critical system, personnel, equipment or a logically related group of systems".

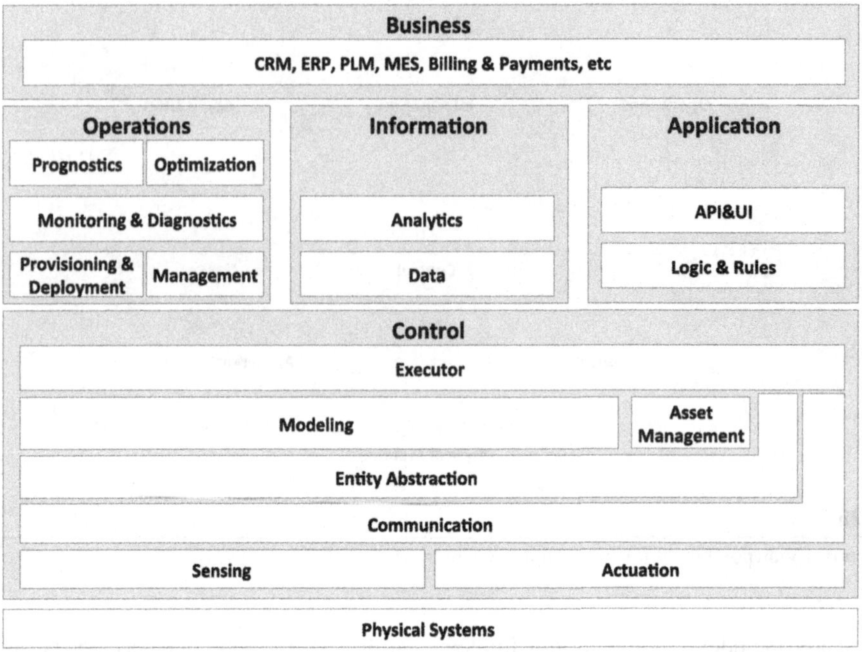

FIGURE 8.26

IIRA Functional Viewpoint functions (redrawn from IIC IIRA [108]).

The term "plant" is a common term used in automatic control, which in general means a physical object to be controlled [120]. The mathematical model of control systems in [120] shows a plant being sensed by sensors and affected by actuators, therefore the plant being a black box system with inputs and outputs. Comparing Figure 8.25 and the mathematical model of control systems the "Physical Systems" in the functional viewpoint (Figure 8.25) is similar to a plant.

Figure 8.26 shows all the functional domains in one picture for the sake of completeness. Each functional domain is described in turn.

The main characteristic of the Industrial Internet of Things is automatic control since most, if not all, industrial systems involve some form of control. As a result, the Functional Viewpoint includes Control as one of the main functional domains.

The Control functional domain includes functions that abstract simple or hierarchical control loops by means of sensors, processing logic, and actuators. Typically the implementations of these functions are sensitive to both location and time and therefore they are deployed closer to the physical assets under control and consume and generate events (reading of sensor data, applying control rules and processing, dispatching of control commands) with precise timing guarantees. As a result the Control domain is not part of a purely functional viewpoint as it includes deployment details such as proximity to the physical entities under control. This is important for the comprehension of the Control domain detailed functions described below.

The Control functional domain includes the following functions:

- The **Sensing** function includes reading sensor data from sensors and its implementation of sensing is distributed in hardware and software.
- The **Actuation** function includes dispatching commands to actuators and its implementation is also distributed in hardware and software.
- The **Communication** function represents the function with which the different sensors, actuators, and support infrastructure (gateways, controllers, routers, etc.) connect to each other for the purpose of exchanging messages. The Communication function can be an abstraction of different types of physical/link layer/networking technologies which include a variety of topologies as well (e.g., bus, mesh network, point to point). A Wi-Fi network is an example of an implementation of the Communication function. Since automatic control is an essential functional domain of the IIRA, communication characteristics such as delay and bandwidth which affect the controllability of a system are important to capture in the Communication function.
- The **Entity Abstraction** function abstracts the different underlying sensors, actuators, controllers, and systems towards the next higher layers via a virtual entity representation. The definition of a "virtual entity" from the IIC Vocabulary [119] states that a virtual entity is a "computational or data entity representing a physical entity". Based on this definition the IIRA virtual entity seems to be similar to the Virtual Entity class in IoT-A but no more information is provided in the IIRA document in order to draw more similarity conclusions.
- The **Modeling** function includes local interpretation and correlation of data from sensors into higher-level insights. A better term to describe this function would be "edge analytics" as is also stated in the description of this function in the IIRA. Edge analytics range in sophistication from simple (e.g., the average value of temperature sensor data in a building) to complex such as insights involving the application of a model of the diffusion of thermal energy across the rooms in a building floor based on the thermal properties of the physical objects on the floor. Edge analytics are typically used for two purposes: (a) as one function implementing local real-time control and (b) for reducing the amount of data transported to external systems because of financial or efficiency reasons.
- The **Asset Management** function includes Lifecycle Management (LCM) operations for the components of the underlying control systems. Examples of typical Lifecycle Management operations are on-boarding, configuration, and software/firmware updates.
- The **Executor** function executes local control logic with certain control objectives. The control objectives may be statically configured or dynamically adjusted by a local control entity or an external entity in a higher-layer tier. Similarly to the edge analytics sophistication, the control logic can be as simple as turning on a heater if the average temperature of a building floor is below a certain threshold or scheduling the activation of different heaters according to the heat diffusion model of the building floor.

The Operations functional domain includes the provisioning, management, monitoring, and optimization of collections or groups of assets in the Control domain. This is a complement to the Control domain that is responsible for the operations of a single asset (in a physical plant in the automatic control terminology). Moreover, these groups of systems may be heterogeneous in terms of assets/sensors/actuators or ownership (i.e., single systems are owned by different entities). For example, optimizing the route of a single autonomous taxi is different from optimizing a fleet of autonomous taxis.

The Operations domain consists of the following functions:

- The **Provisioning and Deployment** function includes the functionality to on-board, configure, track, deploy, and retire assets from operations, in other words, taking care of the asset Lifecycle Management.
- The **Management** function consists of the functionality for asset management centers to send commands to the assets and their respective control systems. Moreover, the management function includes the capabilities of the assets and their respective control systems to respond to such control commands from the asset management centers.
- The **Monitoring and Diagnostics** function includes the collection of asset Key Performance Indicator (KPI) data for the purpose of detection and diagnosis of potential problems.
- The **Prognostics** function refers to the predictive analytics functionality of a system. It uses historical performance data, asset properties, and models in order to predict a problematic situation before it manifests itself.
- The **Optimization** function includes functionality that aims at optimizing asset performance in terms of reliability, availability, energy, output, etc.

The Information functional domain consists of a group of functions for collecting, transforming, storing, and analyzing data from several domains, mainly the control domain. While the Modeling function in the Control Domain is responsible for Edge analytics, i.e., for collecting and analyzing data for implementing the local control loops in the edges, the data analytics in the Information domain serve the purpose of system-wide operation optimization and control (system-wide analytics vs. edge analytics).

The Information domain consists of the following functions:

- The **Data** function consists of ingesting sensor and operational states from all domains, data cleansing, format transformation, semantic transformation (e.g., association of data with the relevant context), data storage/warehousing for batch analytics, and data distribution for stream analytics.
- The **Analytics** function includes data modeling, processing, and analysis of data as well as rule engines. The analysis can be realized in two main forms, batch-oriented analysis of large amounts of stored data or stream-oriented analysis for online processing data and events as they are generated.

The Application functional domain includes functions to realize the specific application logic for which the Industrial IoT system is built. The Application domain consists of the following functions:

- The **Logics and Rules** function includes the specific functionality for the specific industry, vertical or use cases that the IIoT system was built for.
- The **APIs and UI** function includes a set of functions for exposure of the application capabilities as APIs for other application use or User Interfaces for humans use.

Finally, the Business functional domain represents the functionality for the integration of the IIoT specific functions with the typical Business Support Systems of a modern enterprise. Examples of such support systems are ERP, Customer Relationship Management (CRM), and Product Lifecycle Management (PLM).

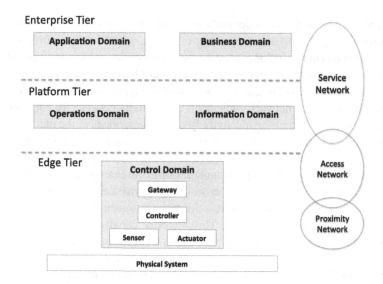

FIGURE 8.27

IIRA Implementation viewpoint Three-tier topology (redrawn from IIC IIRA [108]).

8.9.1.3 IIRA Implementation viewpoint

The Implementation viewpoint covers the representation of an IIoT architecture in terms of required technologies and components for the realization of the IIoT system. It typically includes the topology and distribution of the functions in the Functional viewpoint and their interactions including interfaces. For the case of IIRA as mentioned earlier, the Implementation viewpoint also includes a map of the activities of the Usage viewpoint to FCs and a map of the FCs to implementation components. The Business viewpoint influences the Implementation viewpoint by imposing certain technology choices, constraints in terms of costs, regulatory constraints, and business strategy constraints.

Since the choice of technologies and concrete components depends heavily on the specific industry, vertical, application, and use case, the Implementation viewpoint depends on and is specific to these parameters as well. Therefore it is not realistic to describe an implementation viewpoint in a Reference Architecture document. What is realistic, however, is to describe generic patterns that IIoT systems are expected to exhibit. According to IIRA "An architecture pattern is a simplified and abstracted view of a subset of an IIoT system implementation that is recurrent across many IIoT systems, yet allowing for variants". In other words, architecture patterns are common and recurring implementation mechanisms across industries, verticals, and applications.

The IIRA has identified three architectural patterns, the first two of which were also identified in the IoT-A:

- The **Three-tier** architecture pattern is a system topology pattern and it consists of three tiers, i.e., the edge, platform, and enterprise tiers (Figure 8.27). Each of these tiers also includes different types of connectivity technologies. The edge tier includes mainly proximity types of networks (e.g., ZigBee, Bluetooth Low Energy, Wi-Fi) and access networks (e.g., wired, mobile wireless such as 2G, 3G, 4G, 5G) and the platform tier includes access types of networks and service types of networks. The proximity network connects the sensors, actuators, controllers, and assets collectively called edge nodes. The edge nodes connect to gateways with multiple types of networks such as proximity and access networks. The access network implements connectivity between the edge and platform tiers. The service network implements connectivity between the services in the platform tier and the enterprise tier and the services within those two tiers. The edge tier collects data from the edge nodes and distributes control commands to the edge nodes using the proximity network. The platform tier receives processes and forwards control commands from the enterprise tier and receives, processes, and forwards sensor data from the edge tier to the enterprise tier. The platform tier also includes device and asset management. The enterprise tier receives sensor data and dispatches control commands from/to the edge and platform tiers. The enterprise tier also hosts application logic, business support logic, and User Interfaces. Figure 8.27 shows the three tiers and the Functional domains of the Functional viewpoint of IIRA. Typically the edge tier implements most of the Control domain, the platform tier implements the Operations and Information domain functionality, while the enterprise tier implements the Application and Business domain functionality. However this mapping is only a rough correspondence as in a real system, functions from a functional domain mapped to the platform tier may be implemented in the edge tier. An example being analytics in the edge nodes enabling edge computing.

- The **Gateway-Mediated Edge Connectivity and Management** architecture pattern consists of local connectivity technology to connect all the edge nodes and a gateway to enable the edge nodes to communicate to the entities far from the edge via WAN technology. The topology of the local network depends on the connectivity choice and it can be star, bus, or mesh. The gateway plays the following roles: (a) the gateway bridges physical, link layer, networking, and in some cases application layers between the edge and the wide area, (b) the gateway plays the role of local sensor data processing and feedback control loop host, (c) the gateway is a host of a local Device Management functionality, (d) the gateway hosts application-specific logic which is needed to be located close to the edges, and (e) the gateway acts as a security boundary protecting the edge nodes from security attacks from the WAN.

- The **Layered Databus** pattern consists of multiple databuses in layers. A databus is a collection of endpoints that exchange data following a common schema and common data model allowing interoperability between the endpoints. These schemas and models may be different for different layers in the architecture. Since control is the main characteristic of an IIoT system and since control can be hierarchical, the control endpoints (sensors, actuators in the wider sense, controllers, etc) need to communicate with a common data schema, or over a common databus. Therefore hierarchical control results in hierarchical or layered databuses. Between the layers, there may be adapters of different schemas and data models. A recommendation for the IIRA is the implementation of a databus to be based on the publish-subscribe communication model, i.e., data producing endpoints publish to a publish-subscribe component, while data consuming endpoints subscribe to the publish-subscribe component and receive notifications from the publishing endpoints.

8.10 **BEST PRACTICES**

Now that at least two approaches of an IoT reference architecture have been described and fragments of reference architectures were presented in Chapter 7, some natural questions to ask are how an architect or a system designer can use all this knowledge and what the first step is towards producing a more concrete architecture. Both the IoT-A and IIRA and other complete or partial architectures outline the main concepts and their relationships or at least a designer should extract this knowledge from different reference model and architecture documents. The main concepts, the first class citizens, typically show the main system components that need to be included in the final architecture.

During the design process, there are more components and support functions that need to be placed in the architecture but these are discovered in the process. A good design practice is to identify the main assets or entities of interest and collect the main requirements or concerns based on use case collection sessions with the main stakeholders. The main concerns should be functional or nonfunctional. Functional concerns typically focus on desired information or desired action, often associated with the relevant desired information in a control loop. A typical example is a room (asset/entity of interest) and a functional requirement is the ability to measure temperature in a room with the aim to eventually control the temperature within a desirable range. An example of a nonfunctional requirement is that all the aforementioned interactions should take place within a specific amount of time and in a secure manner.

The stakeholder concerns should contain information about assets of entities of interest (room in the previous example), information about real-world phenomena characteristics (temperature in the example above), information about the entities of interest, and any constraints or limitations. Based on the desirable characteristics to be monitored or controlled, appropriate instruments (sensors and actuators) need to be identified to realize the stakeholder concern. If a piece of information about an entity cannot be directly provided by an available sensor or cannot be directly controlled by an available actuator, the necessary gap in functionality needs to be introduced in the functional architecture. For example, estimating the location of a person in an indoor environment cannot be readily provided by a specific sensor but rather by a combination of sensors and software. For example, there may be cameras that can uniquely identify and localize a person in a room but since people are concerned about being recorded by cameras, this sensor is not desirable in the system. A person could be localized by the use of their mobile phones through tracking of a short-range radio on the mobile phone (e.g., Bluetooth, Wi-Fi) but the accuracy may not be so high. Radio tomographic [121] techniques could be used to localize people without the people being identified or without any devices on the people themselves. However, the room needs to be equipped with an array of short-range radio transceivers and the data collected by them to be processed in a server locally or in the cloud. This solution introduces hardware, sensors, and FCs in the architecture. After identifying the assets/entities of interest, the sensors or actuators, and the necessary functionality to bridge the gap between the desired entity information or control capabilities and the available capabilities, the information model and view of the architecture can be designed to follow the previous design steps. For example, if a radio tomography solution is chosen for a person localization, the information view includes the radio transceivers as information sources and the analytics or sensor fusion function as information processors in the information flow. The location of a person can be used to populate a database, which is required to maintain the location of persons for all the rooms and floors in a building. The database is an information sink in the information view of the architecture. By iterating this approach one identifies the FCs in a functional view and information flows in an information view of an architecture and the next step is to consolidate and merge any of the functions that perform the same or similar operations for different assets or entities of interest.

The next steps after the functional and information views are generated are the deployment views in which the designer needs to be meticulous about the physical world deployment of the chosen sensors, actuators, and computation elements.

The steps above provide some initial approach to an IoT System design taking into account the different reference architectures relevant for the domain of application.

8.11 CONCLUSIONS

This chapter described two main approaches with respect to an IoT Reference Model and Architecture, the IoT-A and IIRA approaches. The IoT-A is mainly an IT-based architecture focusing on simple monitoring and actuation while more complex sensor data analytics and control loops are abstracted in the IoT-A Reference Architecture as IoT Services. On the other hand, the IIRA, coming from an industrial perspective, focuses more on the OT world, which emphasizes the control aspects of IoT, i.e., control loops in several levels in topological hierarchy (e.g., on the device, close to the device, on the cloud) and potentially loops that operate in real-time. Data analytics and actions on the analytics insights are used for controlling the physical world and are therefore explicitly shown and described in the IIRA Analytic functions. Moreover, related to analytics are optimization and prediction as well as action planning. Some of the common aspects that both approaches share are the abstraction of real-world Entities, the communication considerations, sensing and actuation, and the integration to the business process systems.

DESIGNING THE INTERNET OF THINGS FOR THE REAL WORLD

9

CHAPTER OUTLINE

9.1 INTRODUCTION

This chapter outlines the technical design constraints to illustrate the questions that need to be taken into account when developing and implementing Internet of Things (IoT) solutions in the real world. This provides some background and thoughts for the use cases outlined in Part 3 of this book.

9.2 TECHNICAL DESIGN CONSTRAINTS – HARDWARE IS POPULAR AGAIN

The IoT will see additional circuitry built into a number of existing products and machines – from washing machines to meters. Giving these things an identity and the ability to represent themselves online and communicate with applications and other things represents a significant, widely recognized opportunity.

For manufacturers of products that typically contain electronic components, this process will be relatively straightforward. Selection of appropriate communication technologies that can be integrated with legacy designs (e.g., motherboards) will be relatively painless. The operational environments and the criticality of the information transmitted to and from these products, however, will present some unconventional challenges and design considerations. These are discussed later in the context of new and potential applications.

The IoT will, on the other hand, allow for the development of novel applications in all imaginable scenarios. Emerging applications of M2M and wireless sensor and actuator networks have seen deployment of sensing capabilities in the wild that allow stakeholders to optimize their businesses, glean new insight into relevant physical and environmental processes, and understand and control situations that would have previously been inaccessible.

The technical design of any IoT solution requires a fundamental understanding of the specificity of the intended application and business proposition, in addition to heterogeneity of existing solutions. Developing an end-to-end instance of an IoT solution requires the careful selection and, in most cases, development of a number of complementary technologies. This can be both a difficult conceptual problem and an integration challenge, and it requires the involvement of the key stakeholder(s) on a number of conceptual and technological levels. Typically, it can be considered to be a combinatorial optimization problem – where the optimal solution is the one that satisfies all functional and nonfunctional requirements, whilst simultaneously delivering a satisfactory cost-benefit ratio. This is particularly relevant for organizations wishing to compete with existing offerings, or for start-up ventures in novel application areas. Typically, capital costs in terms of "commissioning" and operational costs in "maintenance" must be considered. These may be balanced by resultant optimizations.

Typical M2M or IoT applications conform to the general functional architecture presented in Chapters 4, 7, and 8. Assuming that the system designer has selected the appropriate communication technologies to bridge the device and application domains (likely standard Internet Protocol (IP)-based methods as described in Chapter 5), they must consider the application at several levels: the device (or M2M Area Network; i.e., hardware), representation (i.e., data and visualization thereof), and interaction (i.e., local or remote control).

9.2.1 DEVICES AND NETWORKS

As introduced in Chapter 5, devices that form networks in the capillary or M2M Area Network domain must be selected, or designed, with certain functionality in mind. At a minimum, they must have an energy source (e.g., batteries, increasingly EH [122]), computational capability (e.g., an MCU), an appropriate communication interface (e.g., a Radio Frequency Integrated Circuit (RFIC) and front-end RF circuitry), memory (program and data), and sensing (and/or actuation) capability.

These must be integrated in such a way that the functional requirements of the desired application can be satisfied, in addition to a number of nonfunctional requirements that will exist in all cases.

9.2.1.1 Functional requirements

Specific sensing and actuating capabilities are basic functional requirements. In every case – with the exception of devices that might be deployed as a routing device in the case of range issues between sensing and/or actuating devices – the device must be capable of sensing or perceiving something interesting from the environment. This is the basis of the application. Sensors, broadly speaking, are difficult to categorize effectively. Selecting a sensor that is capable of detecting a particular phenomenon of interest is essential. The sensor may directly measure the phenomenon of interest (e.g., temperature), or may be used to derive data or information about the phenomenon of interest, based on additional knowledge (e.g., a level of comfort). Sensors may sense a phenomenon that is local (e.g., a meter detecting total electricity consumption of a space) or distributed (e.g., the weather).

In many cases, sensing may be prohibitively expensive or unjustifiable at scale, thus motivating the derivation of models that can reason over the sensor readings that are available. Air and water quality monitoring systems are typical of this type of problem.

Given a particular phenomenon of interest, there are often numerous sensors capable of detecting the same phenomenon (e.g., types of temperature sensors), but have widely varying characteristics. These characteristics relate to the accuracy of the sensor, its susceptibility to changing environmental conditions, its power requirements, its signal conditioning requirements, and so forth.

In some cases, for example, a complementary (e.g., temperature) sensor is required in addition to the primary sensor such that variations in readings of the primary sensor that are caused by variation in temperature can be understood in context.

Sensing principle and data requirements are also of essence when considering the real-world application. Consider a continuously sampling sensor, such as an accelerometer, vs. a displacement transducer. Displacement can be sampled intermittently, whereas if an accelerometer is duty-cycled, it is likely that data points of interest (i.e., real-world events) may be missed. Furthermore, the data requirements of the stakeholder must be taken into account. If all data points are required to be transmitted (which is the case in many scenarios, irrespective of the ability to reason locally within an M2M Area Network or WSN), this implies higher network throughput, data loss, energy use, etc. These requirements tend to change on a case-by-case basis.

9.2.1.2 Sensing and communications field

The sensing field is of importance when considering both the phenomenon to be sensed (i.e., whether it is local or distributed) and the distance between sensing points. The physical environment has an implication on the communications technologies selected and the reliability of the system in operation thereafter. Devices must be placed in close enough proximity to communicate. Where the distance is too great, routing devices may be necessary. Devices may become intermittently disconnected due to the time-varying, stochastic nature of the wireless medium. Certain environments may be fundamentally more suited to wireless propagation than others. For example, studies have shown that tunnels are excellent environments for wireless propagation, whereas, where RF shielding can occur (e.g., in a heavy construction environment), the communication range of devices can be significantly reduced. The recent explosion of LPWAN technologies should allow applications with much larger geographical distributions to thrive.

9.2.1.3 Programming and embedded intelligence

Devices in the IoT are fundamentally heterogeneous. There are, and will continue to be, various computational architectures, including MCUs (8-, 16-, 32-bit, ARM, 8051, RISC, Intel, etc.), signal conditioning (e.g., ADC), and memory (ROM, (S/F/D)RAM, etc.), in addition to communication media, peripheral components (sensors, actuators, buttons, screens, LEDs), etc. In some applications, where it would previously have been typical to have homogeneous devices, a variety of sensors and actuators can actually exist, working collaboratively, but constituting a heterogeneous network in reality.

In every case, an application programmer must consider the hardware selected or designed and its capabilities. Typically, applications may be thought of cyclically and logically. Application-level logic dictates the sampling rate of the sensor, the local processing performed on sensor readings, the transmission schedule (or reporting rate), and the management of the communications protocol stack, among other things. Careful implementation of the (embedded) software is required to ensure that the device operates as desired. This continues to be nontrivial and highly specialized. For heterogeneous devices, the embedded software will vary by device.

The ability to reconfigure and reprogram devices is still an unresolved issue for the research community in sensor networks, M2M, and the IoT. It relates to the physical composition of devices, the logical construction of the embedded software, and the addressability of individual devices and security, to name a few. Operating Systems are typically used to make programming simpler and modular for embedded system designers, but each comes with conceptual and implementation differences that impact the ability to handle certain desirable features.

9.2.1.4 Power

Power is essential for any embedded or IoT device. Depending on the application, power may be provided by the mains, batteries, or conversion from energy scavengers (often implemented as hybrid power sources). The power source has a significant implication on the design of the entire system. If a finite power supply is used, such as a battery, then the hardware selected, in addition to the application level logic and communication technology, collectively has a major impact on the longevity of the application. This results in short-lived applications or increased maintenance costs. In most cases, it should be possible to analytically model the power requirements of the application prior to deployment. This allows the designer to estimate the cost of maintenance over time.

9.2.1.5 Gateway

The Gateway, described in Chapter 5, is typically more straightforward to design if it usually acts as a proxy; however, there are few effective M2M or IoT Gateway devices available on the market today. Depending on the application, power considerations must be taken into account. It is also thought that the Gateway device can be exploited for performing some level of analytics on data transitioning to and from capillary networks.

9.2.1.6 Nonfunctional requirements

There are a number of nonfunctional requirements that need to be satisfied for every application. These are technical and nontechnical:

- Regulations.
  - For applications that require placing nodes in public places, planning permission often becomes an issue.
  - Radio Frequency (RF) regulations limit the power with which transmitters can broadcast. This varies by region and frequency band.
- Ease of use, installation, maintenance, accessibility.
  - Simplification of installation and configuration of IoT applications is as yet unresolved beyond well-known, off-the-shelf systems. It is difficult to conceive a general solution to this problem. This relates to positioning, placement, site surveying, programming, and physical accessibility of devices for maintenance purposes.
- Physical constraints (from several perspectives).
  - Can the additional electronics be easily integrated into the existing system?
  - Are there physical size limitations on the device as a result of the deployment scenario?
  - What kind of packaging is most suitable (e.g., IP-rated enclosures for outdoor deployment)?
  - What kind and size of antenna can I use?
  - What kind of power supply can I use given the size restrictions (relates to harvesting, batteries, and alternative storage, e.g., supercapacitors)?

9.2.1.7 Financial cost

Financial cost considerations are as follows:

- **Component Selection:** Typically, the use of devices in the M2M Area Network domain is seen to reduce the overall cost burden by using nonleased communication infrastructure. However, there are research and development costs likely to be incurred for each individual application in the IoT that requires device development or integration. Developing devices in small quantities is expensive. Given the recent proliferation of LPWAN technologies seeking to exploit existing cellular infrastructure networks, subscription costs are likely to apply.
- **Integrated Device Design:** Once the energy, sensors, actuators, computation, memory, power, connectivity, physical, and other functional and nonfunctional requirements are considered, it is likely that an integrated device must be produced. This is essentially going to be an exercise in Printed Circuit Board (PCB) design, but will in many cases require some consideration to be paid to the RF front-end design. This means that the PCB design will require specific attention to be paid to the reference designs of the RFIC manufacturer during development, or potentially the integration of an additional Integrated Circuit (IC) that deals with the balun and matching network required.

9.3 DATA REPRESENTATION AND VISUALIZATION

Each IoT application has an optimal visual representation of the data and the system. Data that is generated from heterogeneous systems has heterogeneous visualization requirements. There are currently no satisfactory standard data representation and storage methods that satisfy all of the potential IoT applications.

Data-derivative products will have further ad hoc visualization requirements. A derivative in these terms exists once a function has been performed on an initial dataset – which may or may not be raw data. These can be further integrated at various levels of abstraction, depending on the logic of the integrator. New information sources, such as those derived from integrated data streams from various logically correlated IoT applications, will present interesting representation and visualization challenges.

9.4 INTERACTION AND REMOTE CONTROL

To exploit remote interaction and control over IoT applications, connectivity that spans the traditional Internet (i.e., from anywhere) on the side of the application manager, or other authorized entity, to the endpoint (i.e., an embedded device), continues to be a challenging problem. Aside from authentication and availability challenges, for most constrained devices, heterogeneous software architectures, such as event-based Operating Systems running on devices with significantly varying concurrency models, continue to pose significant challenges from a remote management perspective.

Elements of Device Management, specifically reprogramming and reconfiguration of deeply embedded devices, will be required, particularly for devices deployed in inaccessible locations. This requires, among others, reliability, availability, security, energy efficiency, and latency performance to be satisfactory whilst communicating across complex distributed systems.

Another significantly underresearched topic is the definition and delivery of end-to-end Quality of Service (QoS) metrics and mechanisms in IoT-type applications. These will be necessary if Service

Agreements (SAs) or Service Level Agreements (SLAs) are to be defined in the case of service provisions for IoT applications – which may or may not be desirable to the application owner. This will be situation-specific. End-to-end latency, security, reliability, availability, times between failure and repair, responsibility, etc. are all likely to feature in such agreements.

IOT USE CASES

3

The following chapters provide an overview of some prominent and diverse use cases for the Internet of Things (IoT). IoT applications can manage different assets, or manage entire complex physical infrastructures. They rely on different tools to process data, generate insights, and automate different business- and society-related processes. The use cases show where the different technology standards and architecture practices come together to realize the concrete solutions that create value for organizations and across value chains.

IoT systems constitute the backbone of modern society as they use control systems and Information Technology (IT) to guarantee a multitude of tasks, such as the production of goods and delivery of services, and generally monitor and control much of the processes upon which the global economy is based.

IoT is expected to play a key role and provide the ability to realize new approaches that are cost- and energy-efficient, while in parallel they are flexible and open to future innovations. Integration of IoT for monitoring and control is not expected to be trivial, especially in an environment that used to be isolated with a tight control of the infrastructure and a much slower pace of evolution. Industrial systems pose significantly different hard requirements that need to be tackled, especially when related to critical infrastructures. We will investigate here how the transition of existing factories to future ones with the use of service-oriented technologies and the emerging cloud can be realized.

Future factories are expected to be Systems of Systems (SoS) that will empower a new generation of applications and services that are today hardly realizable or too costly. New sophisticated enterprise-wide monitoring and control approaches will be possible due to the prevalence of Cyber-Physical

Systems (CPS), which have made IoT interactions a key competitive advantage and market differentiator. Similar systems based on largely identical technologies will be at the core of Smart Cities and the Smart Grid, as well as other Smart "Utilities". These systems will increasingly be coupled with people engagement such as from Participatory Sensing systems.

Let us see how some of these use cases are implemented in the following chapters.

ASSET MANAGEMENT

10

CHAPTER OUTLINE

10.1 INTRODUCTION

The emergence of the Internet of Things (IoT) with its billions of envisioned devices poses a clear challenge to the management of these. Existing asset management practices consider the operations applicable to various physical assets and, in their majority, refer to monitoring of their operations and to some degree to the adjustment (control) of their behavior. However, such operations have been up to now strongly coupled with what the devices and underlying systems are capable of, bound with (mostly) proprietary protocols in order to cover the largest possible spectrum of functionalities and guarantee results, and are mostly static. Only the last couple of years, a significant movement has been witnessed towards open protocols, mostly over IP, that are slowly finding their way into widespread industrial settings [123].

Current practices are undergoing a transformation as the Operational Technologies (OT) and Informational Technologies (IT) are merging. The reasons include the explosion of the heterogeneity of devices that are now deployed in modern infrastructures, which can deliver high-quality data that have clear business relevance at a fraction of the cost compared to some years ago. The usage of open standards is one promising way to go to achieve large-scale manageability; however, it is not going to be enough. The reasons lie in the increasing complexity not only of the devices themselves, but also of the constellations they take part in, and how their functionalities are used in modern applications. On top of this, applications are not monolithic from one provider, but on the contrary, depend on multiple layers that are developed from other stakeholders. Hence, managing different devices requires significant overhead in order to be able to effectively integrate, monitor, and control/reconfigure them.

A typical manageability nightmare scenario includes the configuration of assets in modern enterprises in order to enable data collection according to enterprise service needs [55]. For example, today employees use several computing devices (e.g., laptops, desktops, smartphones, tablets, access cards, security tokens) in conjunction with the shop-floor OT such as sensors and PLCs; all of these need to be accounted for in backend systems, to be integrated with enterprise-wide monitoring solutions and to comply with the organization's policies and requirements. To achieve that is challenging; for example, how does one protect company assets and the data they contain from unauthorized access or usage? The latter also comes with, for example, a trade-off between security and functionality in order to enhance the employees' performance and benefits from using the devices, while still adhering to the overall enterprise constraints.

Internet of Things. https://doi.org/10.1016/B978-0-12-814435-0.00023-7

10.2 EXPECTED BENEFITS

The IoT era is dominated by interactions among the devices, access to their data, and dynamic configuration/management of them. The prevalence of modern IT concepts and adoption of Internet-based technologies is slowly penetrating traditional domains such as those of energy, manufacturing, etc. Hence, management in the IoT era, although challenging [124], may yield significant benefits and enable the mastering of the vast device-based infrastructure.

Several benefits [65] are expected with IoT in asset management, for instance:

- Reduced costs, e.g., because of remote telemetry without the need of field personnel to be engaged.
- Increased quality, e.g., because of the fine-grained monitoring of data that could be done even in near real-time.
- Increased resilience, e.g., because of analysis of a device's status, can lead to Predictive Maintenance, which minimizes, apart from costs, also downtime and unexpected failures.
- Increased performance and security: remote updates may enhance the operational capabilities of the assets.
- Increased security: updates of the asset's software can help correct its behavior and security holes.
- Asset location tracking, e.g., easier recovery and theft prevention of assets.
- Operation optimization, e.g., fleet management optimizing for journeys on the fly.
- New services, e.g., energy awareness via smart metering, location-based services, and e-ticketing.

On the downside, if all of the benefits rely on the correct tackling of complex issues, then only partially fulfilling them might quickly turn an advantage into a disadvantage. As an example, remote access to the devices lowers integration costs, but also potentially opens the door for third parties to tamper with them and their functionalities, which may lead to increased security risks, etc. Asset management is considered as a significant challenge, but with pivotal benefits in the IoT era, where real-time monitoring, Predictive Maintenance [125], and intelligent infrastructure management are expected. Its applications are expected to empower the next generation of innovation in multiple domains, such as residential homes, healthcare, buildings, cities, transportation networks, energy grids, manufacturing, supply chains, homeland security, workplace safety, and environmental monitoring, just to name a few.

10.3 E-MAINTENANCE IN THE IOT ERA

The recent focus is on e-maintenance, i.e., "maintenance support which includes the resources, services, and management necessary to enable proactive decision process execution" [126], especially due to the expected benefits due to IoT that can be harnessed at multiple levels. However, in order to effectively implement e-maintenance applications for asset Lifecycle Management, several requirements need to be met, and this is a challenging task. Apart from interoperability, a major roadblock is the absence of sophisticated IT platforms [127] that can bring added value [128] and master the new challenges and can fully support e-maintenance practices.

Key strategies of e-maintenance include [126,127]:

- **Remote Maintenance**: The capability empowered by Information and Communication Technologies (ICT) to provide maintenance practices from anywhere without being physically present (e.g., third-party entities outside the enterprise borders). This approach enables far more effective reaction to maintenance and dramatically affects business models having maintenance services as part of their functionalities.

- **Predictive Maintenance**: Here the adoption of models and methodologies is implied that analyzes the operational performance of assets and attempts to predict malfunctions and failures, or determine on-demand when maintenance checks are needed (and not as usually done at fixed intervals). Here, apart from the immediate benefit of asset reliability, one can also witness optimization of maintenance schedules carried out by field personnel, enhanced planning for replacement of assets, increased customer satisfaction, etc.
- **Real-Time Maintenance**: Current failures on the shop floor, although evident, might require significant time until they are assessed, repaired, and replanned with respect to the operations the enterprise systems had scheduled. With real-time notifications up to enterprise systems, immediate assessment of the operational factory or enterprise-wide processes can be achieved and addressed.
- **Collaborative Maintenance**: This capability enables traditional maintenance concepts to integrate collaboration among different areas of the enterprise that may lead to leaner processes, simpler landscapes, and more effective management.

To achieve these objectives, we have to harness the new capabilities that IoT brings. IoT approaches integrate Internet technologies that have been proven to be open enough (compared to many practices in individual industry domains) and scalable, it is expected that integration overall will be easier considering the large heterogeneity of assets. Since the assets are no longer passively monitored entities within an infrastructure, they can take advantage of the event-driven methods present in IoT and on-demand report their health status, failures, and/or other information. This migration from a traditional information pull to an event-driven approach is expected to minimize unnecessary traffic and empower the targeted information dissemination within the infrastructure.

Another key aspect of IoT that e-maintenance can benefit from is the dynamic discovery of assets and their "surrounding context" information such as location or real-world sensing information of the process they monitor or control. Integration of protocols that allow a simple "plug and play" approach mean that assets can be connected, be immediately recognized by the network services, get automatically configured, and operate within the organization's policies.

However, apart from these example asset-related enhancements that IoT brings, collaborative e-maintenance is a promising approach that integrates multienterprise experts that can together effectively address any asset issues. This is of critical importance and paves the way for new business models deploying subject matter experts when and where needed remotely, which leads to more effective use of the expertise and better solutions to the maintenance problems [127].

Cross-company communication is already a reality but constrained at enterprise-level operations without in-depth information that could trigger sophisticated replanning scenarios. However, with IoT real-time connection to the devices, malfunctions can be quickly analyzed and resolved with the help of multiple experts such as those with knowledge of the specific process, those responsible for the hardware/software (HW/SW) of the device, and potentially the field personnel on-site (as depicted in Figure 10.1). All these can now collaborate using multiple Internet-based technologies with seamless data, voice, and video integration over a (future) e-maintenance platform.

Communication done directly (e.g., via common trusted third-party service providers) may simply couple the two companies for the specific business case and remove the overhead of costly home-grown solutions by propagating all the necessary information to the engaged stakeholders. Synergies can be

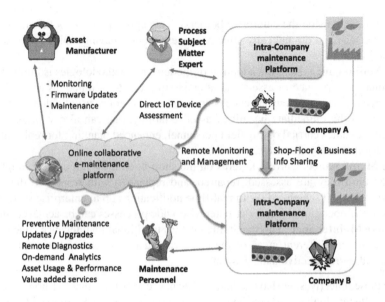

FIGURE 10.1

Outsourced remote continuous cross-company e-maintenance.

identified, and information that up to now was too costly to be obtained in a timely manner can flow into cross-company applications and services. This approach is very well suited for dynamic and short-lived interactions that can be set up, exploited, and removed as easy as a simple composite service.

Cross-company collaboration allows us to realize new functionality and innovate on services offered [127]. Especially in the case of outsourcing of maintenance, specialized partners can now bring in their expertise and monitor remotely the devices on the shop floor and maintain them. Assets on which the company operates may in the future not be owned by the company as such, but instead be provided to them over specific Service Level Agreements (SLAs), e.g., a production line with an uptime of 99%. How this is achieved and maintained is the responsibility of the respective service provider. As a result, companies can now focus more on their core business, while SLAs can regulate shop floor performance that better matches the business process goals, and how this is achieved is now the responsibility of the e-maintenance partner. This can facilitate the development of new business models based on remote maintenance service delivery through e-maintenance platforms.

10.4 HAZARDOUS GOODS MANAGEMENT IN THE IOT ERA

An example of asset management that goes beyond the traditional monitoring approaches is that realized within the research project Collaborative Business Items [129]. There, an IoT platform was used to monitor hazardous goods (i.e., chemicals) and guarantee their safe storage. Proximity of stored incompatible goods, i.e., goods that could be flammable if brought together, or exceeding the compliance

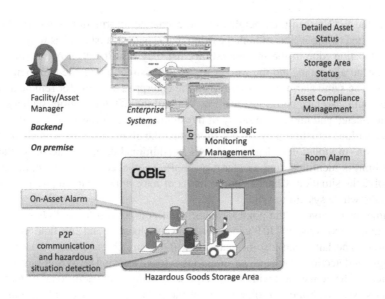

FIGURE 10.2

IoT-enhanced Hazardous Goods Management in CoBIs.

guidelines would raise alarms locally and at the enterprise system that would need to be resolved [129, 130].

Traditionally, the monitoring was done with passive Radio Frequency Identification (RFID) tags. However, CoBIs deployed Wireless Sensor Networks (WSNs) that could execute business logic locally and communicate among themselves as well as with the enterprise systems [130]. WSNs are seen as one of the most promising technologies that will bridge the physical and virtual worlds, enabling them to measure, assess, and actuate real-world environments [131,65].

A typical demonstration of how IoT at large helps with the hazardous goods management scenario is shown in Figure 10.2. All drums are equipped with WSNs that have internally stored information about the chemical within the drum, an "incompatible goods" list indicating which chemicals are dangerous to be in near proximity with it as corrosion might lead to explosions, etc., as well as the max limit according to current regulations of this chemical that can be stored within a limited space. The WSNs are able to beacon other WSNs nearby and exchange information about the chemicals within the drum. A nearby WSN bridge interacts with the WSNs of the drums, has a connection to an enterprise system, and hence mediates all information from the field (the drum storage facility).

The WSN bridge is also capable of not only providing the backend with WSN information (bottom-up communication) but also providing other WSNs with information from the backend (top-down communication). The latter is handy, as information provided initially to WSNs may change; for example, the chemical incompatibility list may be extended or the max storage limits may be increased or reduced. The WSNs and the bridge also feature visible and audible alarms.

In a typical safety-critical situation, a worker might transfer by mistake a drum and position it on the wrong side of the storage room (i.e., near another drum whose potential combination may cause explosions). As the WSNs scan their neighbors and exchange information, the dangerous situation is immediately detected by the WSNs of both the drum that is being transferred as well as the other drums already in that location (due to the incompatibility list they host locally). An alarm goes off on the drums and in the storage room, and the worker resolves the situation on-site by moving it away to the correct location. The same alarm is also visible on the enterprise systems for remote asset monitoring managers who could also take additional actions. Without the IoT interaction, such a situation would probably go undetected and impact the safety of the storage room workers.

A similar hazardous situation may arise if more than the allowed drums with chemicals are stored in a single location (in violation of legislation). As an additional drum is placed, similarly the excess of the storage limit is identified, and an alarm goes off. The backend, of course, has additional info on how to resolve the situation. Changes in the limits of max storage can be pushed wirelessly to the WSNs to keep the whole system in sync.

The real-time interaction at "the point of action" (i.e., on-premise by IoT) has profound benefits. The infrastructure empowered by IoT can react to events locally, and the field personnel can take corrective actions. The latter can also be realized without any connection to the enterprise systems because the logic of detection of hazardous situations is locally hosted on WSNs (and updatable by the enterprise system). The connection with the enterprise systems offers additional benefits linking asset management processes with the real-time status of the field, effectively enabling remote monitoring and actuation scenarios related to asset management.

10.5 CONCLUSIONS

Asset management is an area that can hugely benefit from machine-to-machine communication and especially the proliferation of cross-layer IoT interactions. New innovative solutions can be realized that take advantage of the networked embedded devices on-premise, the information they provide, and the collaboration with enterprise systems [56]. However, for such solutions to be adopted, several challenges [55], such as complexity management, interoperability, security, and Quality of Service (QoS)-guaranteed communication, have to be tackled, especially when scenarios involving critical infrastructures are involved. Nevertheless, there is a huge potential for many diverse domains where new business models and opportunities will arise with asset management.

INDUSTRIAL AUTOMATION

11

CHAPTER OUTLINE

11.1 SOA-BASED DEVICE INTEGRATION

The trend in industrial environments is to create system intelligence by a large population of intelligent, small, networked, embedded devices at a high level of granularity, as opposed to the traditional approach of focusing intelligence on a few large and monolithic applications. This increased granularity of intelligence distributed among loosely coupled intelligent physical objects facilitates the need for more adaptability and reconfigurability of the system, allowing it to meet business demands not foreseen at the time of design and providing real business benefits [132].

The Service-Oriented Architecture (SOA) paradigm can act as a unifying technology that spans several layers, from sensors and actuators used for monitoring and control at shop floor level, up to enterprise and engineering systems as well as their processes, as shown in Figure 11.1. This common "backbone" means that machine-to-machine communication is not limited to direct (e.g., proximity) device interaction, but includes a wide range of interactions in a cross-layer way with a variety of heterogeneous devices, as well as systems and their IoT services. This yields multiple benefits for all stakeholders involved. Such visions have been proposed [133] and realized, demonstrating the benefits and challenges involved [134,132,135,136]. There are several patterns on how such integration can be realized [137], and technologies play a key role in such orchestration of devices and systems.

Internet Protocol (IP)-based technologies, and more specifically web technologies and protocols (e.g., OPC-UA, DPWS, REST, and Web Services (WS)), constitute a promising approach for integration [66], including the Factory of the Future [138,123,136]. IP technologies in industrial automation act as an enabler towards the fundamental goal of enabling easy integration of device-level services with enterprise systems overcoming the heterogeneity and specific implementation of hardware and software of the device. In addition, industry-specific requirements for security, resilience, and availability of near real-time event information need to be effectively addressed. The latter are also seen as key enablers for a more real-time approach towards interaction with enterprise systems and applications such as real-time business activity monitoring, overall equipment effectiveness optimization, and maintenance optimization.

The SOA-everywhere vision is not expected to be realized overnight, but may take a considerable time depending on the lifecycle processes of the specific industry, and may be impacted by micro- and macroeconomic aspects. Hence, it is important that migration capabilities are provided so that we can harvest some of the benefits today and provide a step-wise process towards achieving the

Internet of Things. https://doi.org/10.1016/B978-0-12-814435-0.00024-9

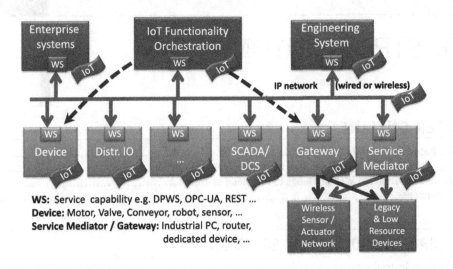

FIGURE 11.1

IoT SOA-based integration.

vision. The concepts of gateway and service mediator [135], as depicted in Figure 11.2, can help towards this direction by enabling new functionalities while also assisting with the migration of legacy systems. Dynamic device discovery is a key functionality in the future IoT-enabled infrastructure. As an example, Figure 11.3 depicts how Windows 7 can discover dynamically heterogeneous devices that are SOA-ready (i.e., equipped with web services; Devices Profile for Web Services, DPWS).

Figure 11.2 demonstrates two different approaches of integrating devices via a gateway and a service mediator.

- **Gateway**: It is a device that controls a set of lower-level nonservice-enabled devices, each of which is exposed by the Gateway as a service-enabled device. This approach allows gradually replacing limited-resource devices or legacy devices by natively WS-enabled devices without impacting the applications using these devices. This is possible since the same WS interface is offered this time by the WS-enabled device and not by the Gateway. This approach is used when each of the controlled devices needs to be known and addressed individually by higher-level services or applications.
- **Service Mediator**: Originally meant to aggregate various data sources (e.g., databases and log files), the Mediator components evolved and are now used to not only aggregate various services but possibly also compute/process the data they acquire before exposing it as a service. Service Mediators aggregate, manage, and eventually represent services based on some semantics (e.g., using ontologies). In our case, the Service Mediator could be used to aggregate various non-WS-enabled devices. This way, higher-level applications could communicate to Service Mediators offering WS instead of communicating to devices with proprietary interfaces. The benefits are clear, as we do not have the hassle of (proprietary) driver integration. Furthermore, now processing of data can be done at Service Mediator level, and more complex behavior can be created which was not possible before from the standalone devices.

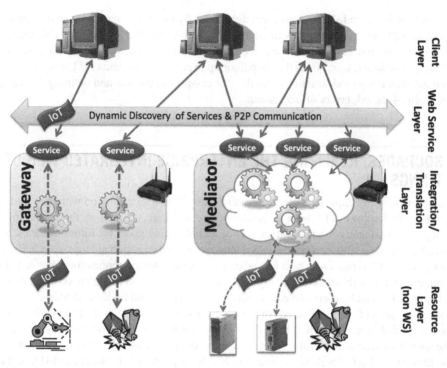

FIGURE 11.2

Device integration: Gateway vs. Service Mediator.

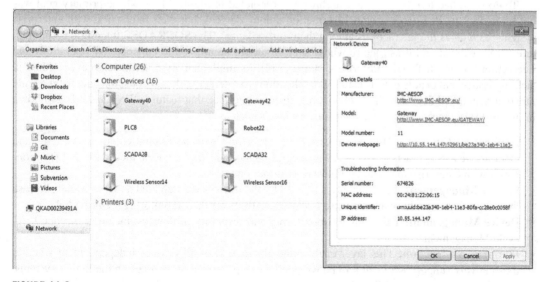

FIGURE 11.3

Dynamic device discovery via DPWS in Windows 7.

As we can see in future IoT infrastructures dominated by billions of devices with different capabilities and needs, we have to consider how these integrate with each other [139] and enable the realization of new innovative approaches. This assumes increased integration and collaboration among the various layers existing in industries, i.e., from the shop floor up to enterprise systems [55]. Several concepts and efforts are directed towards abstracting from the device-specific aspects and defining device-agnostic, but functionality-focused, layers of integration.

11.2 SOCRADES: REALIZING THE ENTERPRISE INTEGRATED WEB OF THINGS

Agility and flexibility are required from modern factories. This, in conjunction with the rapid advances in Information Technology (IT), both in hardware and software, as well as the increasing level of dependency on cross-factory functionalities, sets new challenging goals for future factories. The latter are expected to rely on a large ecoSystem of Systems where collaboration at large scale will take place. Mashing up services has proven to be a key advantage in the Internet application area; and if now the devices can either host web services natively or be represented as such in higher systems, then existing tools and approaches can be used to create mash-up apps that depend on these devices.

A visionary project that followed this line of thinking was the Industry-driven European Commission-funded project SOCRADES [140]. Driven by the key need for cross-layer collaboration, i.e., at the shop floor level among various heterogeneous devices as well as among systems and services up to the Enterprise (ERP) level, an architecture had been proposed, prototyped [141], and assessed [142]. SOCRADES proposed and realized SOA-based integration, as shown in Figure 11.2, including migration of existing infrastructure via gateways and service mediators, as shown in Figure 11.3.

To do so, it had to rely on an IoT architecture (depicted in Figure 11.3), whose primary goal was not limited to the Peer-to-Peer (P2P) interaction among devices, but their interaction with enterprise systems, as well as the interactions among them (cross-layer). The SOCRADES Integration Architecture (SIA) realized an infrastructure whose services could be utilized to enhance IoT operations and interactions among IoT and enterprise systems [142]. Its implementation and assessment resulted in tackling issues that enable enterprise-level applications to interact with and consume data from a wide range of networked devices using a high-level, abstract interface that features WS standards (as shown in Figure 11.4). One can distinguish the various levels such as:

- **Application Interface:** This part enables the interaction with traditional enterprise systems and other applications. It acts as the glue for integrating the industrial devices, their data, and functionalities with enterprise repos and traditional information stores.
- **Service Management:** Functionalities offered by the devices are depicted as services here to ease the integration in traditional enterprise landscapes. Tools for their monitoring are provided.
- **Device Management:** This includes monitoring and inventory of devices, including service Lifecycle Management.
- **Platform Abstraction:** This layer enables the abstraction of all devices independent of whether they natively support WS or not, to be wrapped and represented as services on the higher systems. In addition to service-enabling the communication with devices, this layer also provides a unified view of remotely installing or updating the software that runs on devices.

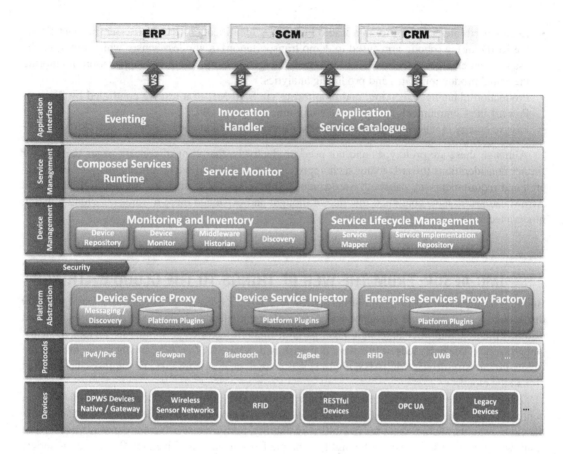

FIGURE 11.4

The SOCRADES Integration Architecture (SIA).

- **Devices and Protocols:** These layers include the actual devices that connect over multiple protocols to the infrastructure. The respective plugins of course need to be in place so that they can be seamlessly integrated to SIA.

To realize discovery and interaction in a P2P way, a local gateway/service-mediator is implemented. This prototype, named Local Discovery Unit (LDU), enables the dynamic discovery of devices on-premise and their coupling with the SIA. SIA has been used in several scenarios as proof of concept for the integration among different devices, both locally and with enterprise systems. Examples include [142]:

- Integration between a Programmable Logic Controller (PLC), a robotic gripper, and SunSPOT wireless sensor nodes, while these are monitored by the SAP Manufacturing Integration and Intelligence software (SAP MII [143]), which is also responsible for the execution of the business logic.

- Event-based interaction between a Radio Frequency Identification (RFID) Reader (product ID via the RFID tag), a robotic arm (used to demo transportation), a wireless sensor which monitors the usage of an emergency button, an IP-plugged emergency lamp, and a web application monitoring the actual production status and producing analytics.
- Production planning, execution, and monitoring via SAP MII of a test rig controlled by Siemens Power Line Communication (PLC) and communicating over OPC.
- Passive energy monitoring via the usage of sensors (Ploggs) and gateways.

Although these are prototypes and not used productively, the proof of concept paves the way for further considerations towards using such approaches in real-world environments once the exact operational requirements are satisfied. Subsequent follow-up projects, such as IMC-AESOP [144] and Arrowhead [145], have continued this line of thought and expanded more towards the cloud utilizing modern technologies.

11.3 IMC-AESOP: FROM THE WEB OF THINGS TO THE CLOUD OF THINGS

Building on the experiences acquired during the SOCRADES project [140], it was evident by 2009 that cloud technologies and concepts would not only be applicable to the industrial automation but also had the potential to change significant parts of modern factories pertaining to devices, systems, and processes. The first seeds of this were demonstrated in the SOCRADES, and therefore in 2010 the follow-up industry-driven IMC-AESOP [144] was started, dealing with visionary approaches bundled around the cloud.

Considering the rapid advances in hardware and software, as well as IT concepts ([146–148]) in conjunction with the cloud capabilities, visions were set and prototypes were demonstrated that harness the cloud benefits, such as resource flexibility, scalability, performance, efficiency, and resilience. The result was in line with the vision of a highly dynamic flat information-driven infrastructure (as shown in Figure 11.5) that will empower the rapid development of better and more efficient next-generation industrial applications, while in parallel satisfying the agility required by modern enterprises.

This vision [147] is only realizable due to the distributed, autonomous, intelligent, proactive, fault-tolerant, reusable (intelligent) systems, which expose their capabilities, functionalities, and structural characteristics as services located in a "service cloud". The infrastructure links many components (devices, systems, services, etc.) of a wide variety of scales, from individual groups of sensors and mechatronic components, to, for example, control, monitoring, and supervisory control systems, performing SCADA, DCS, and MES functions (for example).

Although today factories are composed and structured by several systems viewed and interacting in a hierarchical fashion following mainly the specifications of standard enterprise architectures [148], there is an increasing trend to move towards information-driven interaction that goes beyond traditional hierarchical deployments and can coexist with them. With the empowerment offered by modern service-oriented architectures, the functionalities of each system (or even device) can be offered as one or more services of varying complexity, which may be hosted in the cloud and composed by other (potentially cross-layer) services, as depicted in Figure 11.5.

This transition marks a paradigm shift towards a cross-layer and information-driven interaction among the different systems, applications, and users. Although the traditional hierarchical view co-

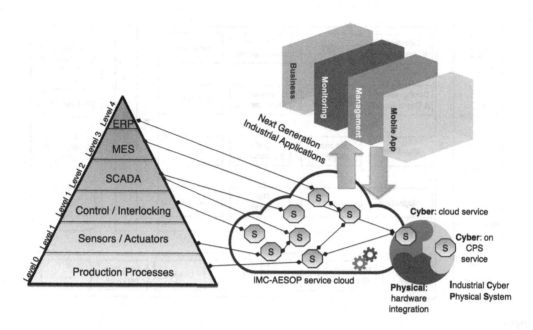

FIGURE 11.5

Cloud-based composition of CPS services.

exists, there is now a flat information-based architecture that depends on a wide variety of services exposed by the Cyber-Physical Systems (CPS) and their composition [149]. Next-generation industrial applications can now rapidly be composed by selecting and combining the new information and capabilities offered (as services in the cloud) to realize their goals. The envisioned transition to the future cloud-based industrial systems [147,150,151] is depicted in Figure 11.6.

Several "user roles" will interact with the envisioned architecture (shown in Figure 11.6), either directly or indirectly as part of their participation in a process plant. The roles define actions performed by staff and management and simplify grouping of tasks into categories such as business, operations, engineering, maintenance, and engineering training.

As shown in Figure 11.6, it is possible to distinguish several service groups for which there have also been defined some initial services [151]. All of the services are considered essential, with varying degrees of importance for next-generation, cloud-based, collaborative automation systems. The services are to provide key enabling functionalities to all stakeholders (i.e., other services), as well as CPS populating the infrastructure. As such, all these systems can be seen as entities that may have a physical part realized in on-premise hardware, as well as a virtual part realized in software potentially on-device and in-cloud. This emerging "Cloud of Things" [67] has the potential to transform the way we design, deploy, and use applications and CPS.

Typical functionalities include alarms, configuration, and deployment, some control, Data Management, data processing, discovery, Lifecycle Management, HMI, integration, simulation, mobility support, monitoring, and security. It is clear that this is a proposal that will need to be further refined in real-world scenarios. However, it clearly depicts a step towards a highly flexible IoT-enabled infras-

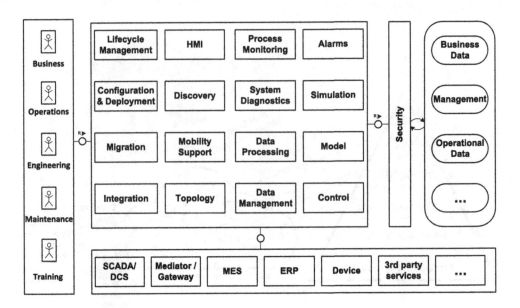

FIGURE 11.6

IMC-AESOP cloud-based architecture.

tructure for the automation domain that abstracts from devices and focuses on functionalities that can reside on-device or in-network, and harness the power of the cloud.

Outsourcing key functionalities to the cloud is challenging, and further research needs to be done towards the applicability of it for several scenarios. For instance, monitoring scenarios have been successfully realized demonstrating the benefits of the IT concept prevalence to traditional industrial system design and operation. However, several aspects with relation to real-time interactions, reliability, and resilience, as well as control (especially closed-loop control), are still at an early stage. Constructing such complex systems additionally bears challenges with respect to safety, maintainability, and security.

11.4 CONCLUSIONS

The prevalence of advanced embedded devices, in conjunction with increased processing and communication capabilities, is transforming industrial automation. Service-oriented architecture approaches are now possible at the device level and bring with them benefits previously only available to the designers and managers of enterprise systems. We see a clear trend towards information-driven integration that empowers IoT interactions and integration with enterprise systems and the respective business processes. The latest visions depict a fully dynamically customizable infrastructure that harnesses the best breed of functionalities at the device and at the cloud level in order to empower the next generation of applications and services. Surely several challenges are open and need to be effectively tackled; however, on-going research will shed more light on the real-world applicability of such visions.

SMART GRID

12

CHAPTER OUTLINE

12.1 INTRODUCTION

A revolution is currently underway in the electricity system, which has remained largely unchanged for more than 100 years. Rapid advances in Information and Communication Technologies (ICT) are increasingly integrated into several infrastructure layers covering all aspects of the electricity grid and its associated operations. In addition, intelligent networked devices are emerging whose Internet of Things (IoT) interactions create new capabilities in the monitoring and management of the electricity grid and the interaction between its stakeholders. IT-empowered innovations integrated with the electricity network and the stakeholders' interactions have paved the way [152–156] towards a "Smart Grid" that takes advantage of sophisticated bidirectional interactions.

The US National Institute of Standards and Technology (NIST) Smart Grid Conceptual Model [154] defines a framework that outlines seven domains: Bulk Generation, Transmission, Distribution, Customers, Operations, and Markets and Service Providers. Complementary to that, the IEEE views the Smart Grid as "a large System of Systems" where each NIST Smart Grid domain is expanded into three Smart Grid foundational layers: (i) the Power and Energy Layer, (ii) the Communication Layer, and (iii) the IT/Computer Layer. The interplay of these layers via a highly sophisticated ICT infrastructure brings intelligence to the grid and enables it to provide new added-value services to its stakeholders [157]. This view is also exemplified in Figure 12.1, where different stakeholders are interconnected, providing and consuming energy-related services, that are based on concepts of data analytics, market integration, real-time communication, and intelligent asset management.

The deregulation of the energy market and the subsequent new operational context will permanently change the structures associated with the production and distribution of energy [158]. Deregulation or reregulation of the energy market is designed to break up the value-added chain, with the production, transfer, and distribution of electric power forming separate segments. The objective is to establish a more open market designed to promote increased competition and flexibility for consumers. A much more decentralized and diversified energy production system will emerge as a result. New energy technologies for cogenerated heat and power, and increased use of renewable energies such as biomass, solar energy, and wind power, will introduce a considerable number of diversified systems into the power grid in addition to traditional large-scale plants. Consequently, the share of decentralized generated power – produced by industrial or private producers – will increase significantly. This will create

Internet of Things. https://doi.org/10.1016/B978-0-12-814435-0.00025-0
Copyright © 2019 Elsevier Ltd. All rights reserved.

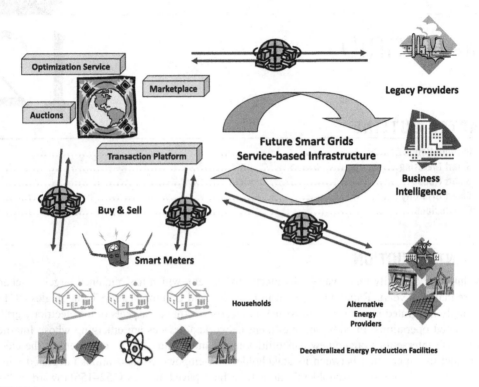

FIGURE 12.1

The future ICT-empowered interaction-rich Smart Grid.

a new infrastructure where the future user may not be a simple passive consumer as today, but also a producer (commonly referred as a prosumer).

The integration of small, highly distributed energy production sources and their coupling with advanced, information-driven services will give rise to a new infrastructure, i.e., an Internet of Energy [158,152], whose key technology building blocks are the IoT and Internet of Services (IoS). As shown in Figure 12.2, IoT and IoS make possible diverse interactions locally and over the Internet, empowering traditional business relationships with fine-grained monitoring and control of the large energy infrastructure [157]. The resulting Internet of Energy is the vision of the Smart Grid, spanning not only the technical grid infrastructure but covering all energy-related aspects of the grid, its devices (including appliances), and their interactions, as well as high-level applications and systems that depend on their data.

The Smart Grid is one of the key areas where the economic benefits of monitoring and control are likely to be realized [153,155,159]. It is driven by "the degree of decentralization of the system components and their interrelation with electricity networks, the variability of renewable generation, the increased distance between electricity generation and consumption, the intelligence level of the involved systems created by smart products and associated smart services, the legal framework, the associated regulation of market-based product and service choices versus natural monopoly products and services

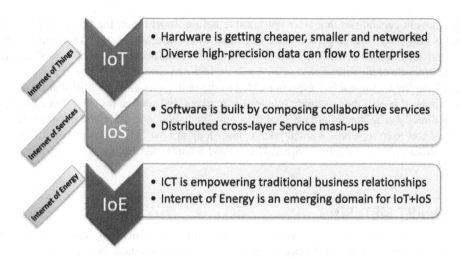

FIGURE 12.2

The Smart Grid as a combination of IoT and IoS.

and the business roles for actors involved in all aspects of networks and intelligent electric systems"
[159]. Currently, several projects [155,156] with varying degrees of maturity are addressing several
aspects of the Smart Grid value chain, investigating the benefits for the diverse stakeholders involved.

Due to the complexity of the multiple Smart Grid stakeholder interactions at several layers, it is
mandatory to look at the Smart Grid from the network viewpoint, as an ecosystem where collaboration
and information-driven interactions characterize it. It is expected that in such networks, all distributed
energy producers and consuming entities will be highly interconnected via information flows, many
of which will depend on IoT interactions from smart meters to energy management systems [158]. In
that sense, a paradigm changes from existing passive and information-poor to active information-rich
energy networks, which reverses the trend of one-way flow because the electricity networks that were
initiated are underway. Such a future infrastructure is expected to be service-oriented and give rise to
new innovative applications that will drastically change our everyday environment. The bidirectional
information exchange will put the basis for cooperation among the different entities [160], as they will
be able to access and correlate information that up to now either was only available in a limited fashion
(and thus unusable on a large scale) or extremely costly to integrate. Examples of such emerging
capabilities are demand-side management and local energy trading [161–163].

IoT-capable assets that are increasingly introduced not only in the industrial domain such as sensors
and SCADA/DCS systems, but also on the end-user side, for example, as Electric Vehicles (EVs), white
label appliances, and even light bulbs, lie at the heart of the Smart Grid. One example of this trend is the
IoT-fication of many everyday objects, from white appliances (offered by Samsung, GE, etc.) to light
bulbs (offered by Philips, GE, etc.). Such objects can be customized and controlled usually via com-
panion apps running on the user's smartphone. Their capability of offering fine-grained information
about a device's energy signature as well as control capabilities depicts a shift towards a highly observ-
able and controllable infrastructure that can dynamically adjust to the real-world conditions and create
intelligent living environments. The latter empowers the Smart Grid vision with massive monitoring

and management capabilities that were not feasible before. In addition, the fine-grained information provided is also used for high-performance analytics, which benefits a variety of existing approaches ranging from technical monitoring to Predictive Maintenance and accurate economic model assessment during the lifecycle of the assets.

Apart from the traditional IoT aspects tackled in the bulk generation and distribution domains, of particular interest is also the customer, operations, markets, and service provider domains, especially those that directly or indirectly are affected by a previously uncontrollable and passive infrastructure (i.e., that of end-users). Hence, we will take a closer look at how IoT empowers end-users and what new capabilities they can exercise in Smart Houses and Smart Grid Cities.

12.2 SMART METERING

Traditional utility management processes involve the collection of metering data from a centralized point (the meter) in the household once or twice a year. With the emergence of smart metering, the collection of data from households has increased, with the aim to have measurements at 15-minute intervals (or less). This coupling of smart meters to an infrastructure that can monitor at such a fine-grained form the electricity consumption for residential households has profound implications for the grid itself as well as other value-added services, if coupled with other tools such as IoT data analytics (covered in Section 5.3).

Many utility providers today strive towards offering new energy services that will enable customers to get a better grasp of their energy consumption by seeing the amount of energy they consume in short intervals. The aim is to move from the 15-minute interval, towards 1-minute resolution or as near as possible to real-time. Coupled with variable tariffs [164], this may provide customers with a view on the upcoming costs and assist end-users in adjusting their energy consumption behavior. Such services may lead to more sustainable energy management and avoid cases such as the "bill shock", where the users get shocked when they discover the high electricity costs in the electricity bill months after the actual consumptions. To realize such services, key IoT aspects such as integration, remote monitoring, remote management, and real-time analytics are needed.

Figure 12.3 depicts the paradigm change from an infrastructure delivering metering data for billing purposes towards a general purpose monitoring infrastructure that acts as an enabler for multiple stake-holders and value-added services. Data generated include not only the energy metering data such as consumption (or production of energy) but also other data (e.g., related to power quality, device status), which may provide additional added value towards asset management. Such smart metering infrastructures are currently under investigation for their performance [165,166], scalability, and the added value they can deliver beyond smart metering, e.g., towards energy management, asset management, and integration of heterogeneous systems [161].

The new smart meters, or their extensions (such as dedicated devices depicting energy consumption and other info), are used as communication media where the utilities push information, e.g., about the current energy tariff, upcoming maintenance, or costs. By additionally adopting varying tariffs and projecting that to the end-users, the aim is to have soft control of the infrastructure; more precisely, the expectation is that due to higher prices at key times during the day, the users will shift part of their activities to noncritical times. When the latter is realized successfully from a large number of users (critical mass), it can have a significant impact on the infrastructure resources and help tackle aspects such as

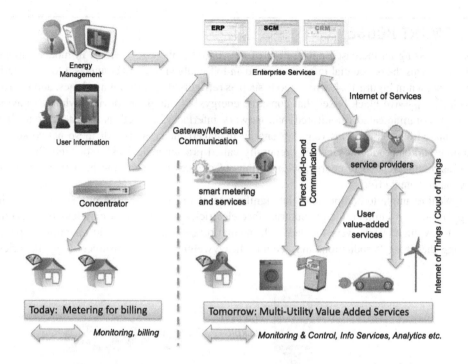

FIGURE 12.3

Smart Metering enabling multiutility value-added services.

peaks of consumption, which would be too costly. On a similar train of thought, lowering the price implies shifting energy consumption to those intervals where the energy is available (usually coming from intermittent resources such as wind parks and photovoltaic farms), but not enough consumption is present, for example. Although the industrial infrastructures have more fine-grained control, this is the first step towards introducing financial-driven energy management, a notion of control, to residential households (which were previously passive, unmanageable consumers of energy only).

IoT is key to this bidirectional communication between the utilities and their customers. The millions of smart meters required to realize the vision of fine-grained monitoring (and potentially control) are currently under assessment and deployment worldwide. Indicatively, it is expected that around 200 million electricity smart meters will be deployed in Europe by 2020, which translates to approximately 72% of European consumers [167].

Smart metering customer behavior trials [155,156] indicate that there might be benefits for all involved stakeholders, including reduction of electricity bills and better usage of electricity. At an installation cost of approximately €200–€250, this poses a significant investment, while "on average, smart meters provide savings of €309 for electricity per metering point (distributed amongst consumers, suppliers, distribution system operators, etc.) as well as an average energy saving of 3%" [167]. Apart from direct benefits, more importantly, other benefits may also exist, such as improvement of competition in retail markets, better demand–response, and future coupling with energy management and automation systems.

12.3 SMART HOUSE

We are witnessing an increased penetration of networked embedded systems in modern appliances, which transforms the residential infrastructure to an IoT-enabled one and connects it to the Smart Grid. Hence, as shown in Figure 12.4, white goods such as refrigerators, washing machines, and microwaves are no longer passive black boxes that consume energy, but intelligent devices whose behavior may be actively communicated (monitored) via network interfaces, as well as adjusted (controlled) via interactions with other systems (e.g., an energy management system). All these networked energy consuming and/or producing devices (generally called prosumer devices) exponentially expand the IoT infrastructure related to energy, with billions of devices that can now be active participants in energy management efforts.

A typical example to comprehend the significance of IoT in residential infrastructures is the following. During the summer, many countries face electricity shortages that may lead to blackouts. Up to now, many such events were tackled with manual processes, i.e., the grid operator (in parallel to increasing the energy production) may contact heavyweight energy consumers (e.g., factories) and

FIGURE 12.4

IoT interactions in the context of a Smart House.

ask them to reduce their load. However, blackouts still occur and are costly for all stakeholders. On the one hand, this can clearly be done only with large industrial energy consumers; however, such an approach does not scale and also has other drawbacks. On the other hand, if peaks coming from the residential consumers could be reduced (or shifted), that would greatly ease energy management and make it more efficient. Hence, in the above scenario, instead of contacting only a limited number of industrial consumers, now thousands of residential consumers could be contacted and assist by adjusting (also proactively) their energy consumption. For instance, all unnecessary devices could be turned off momentarily in case of urgency. Prioritization on the device level (e.g., do not turn off the fridge or reschedule the washing machine) as well as on process or location (e.g., do not cut off power in hospitals and emergency infrastructures) could be applied. In any case, the key message here is that similar results can now be achieved on a large scale if a critical mass of residential consumers can be reached. Also since more intelligent measures can be taken, also proactively, the grid stability is enhanced, while highly dynamic situations may be better tackled.

Several research projects [155,156] focus on the integration of the Smart House and its appliances to the Smart Grid and investigate how energy could be better managed in order to increase efficiency, without noticeably impacting the quality of life. One example of such a project is SmartHouse/Smart-Grid [168,169], which focused on the in-house as well as the Smart House with enterprise systems integration and trialed various approaches, as depicted in Figure 12.5.

It is unlikely that a "one-size-fits-all approach" will work in such cases, but rather many and diverse approaches for managing residential areas and connecting Smart Houses and devices to the Smart Grid; systems integration will be a key issue for ensuring that benefits from such solutions are delivered. SmartHouse/SmartGrids build on the following elements:

- in-house energy management based on user feedback, real-time tariffs, intelligent control of appliances, and provision of (technical and commercial) services to grid operators and energy suppliers,
- aggregation software architecture based on agent technology for service delivery by clusters of Smart Houses to wholesale market parties and grid operators,
- usage of Service Oriented Architecture (SOA) and strong bidirectional coupling with the enterprise systems for system-level coordination goals and handling of real-time tariff metering data [165].

Within the household, appliances and devices are integrated via some form of gateway or concentrator that connects to the Smart Grid (as depicted in Figure 12.5). Several integration approaches were experimented with, e.g., the PowerMatcher, the Bi-directional Energy Management Interface (BEMI), the Magic system, as well as direct web service integration [168,169]. These represent not only different technical integration approaches, but also different models of management. The integration via DPWS/REST enabled the direct integration of SOA-ready devices (i.e., devices hosting native web services) or at gateway level with enterprise systems, easing their management. Similarly, in the mediated interactions, various approaches such as multiagent systems or middleware were used, which offered delegation of the intelligence and decision making near to the actual infrastructure (the devices).

What is also required for viable business cases with respect to Smart Houses as part of the Smart Grid is the integration of in-house services with enterprise-level services. The last include typical business-to-customer (B2C) services such as billing, but also other business-to-business (B2B) services such as the interaction among different players such as the Distributed Generation (DG) operator, energy retailer, and wholesale market.

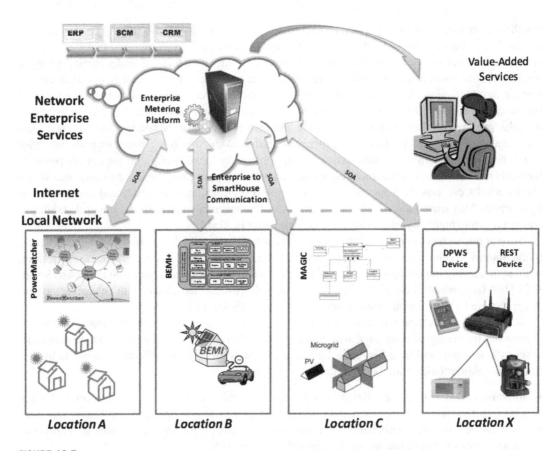

FIGURE 12.5

Smart House integration with enterprise services.

12.4 SMART GRID CITY

With an increasing amount of the grid core infrastructure embracing IT technologies, the Smart Grid is advancing and is being complemented with similar functionality coming from smart houses and smart buildings. The same principles apply increasingly to other parts of the modern city energy infrastructure, including buildings, traffic systems, renewable energy parks with PV and wind turbines, the public lighting system, etc., just to name a few. Hence, we witness the metamorphosis of cities into smart grid cities where innovative approaches towards energy efficiency can be applied at a new level and enable their better energy footprint management.

Future smart cities are expected to provide superior quality of life to its citizens, and as explained in Chapter 14, IoT will greatly boost such efforts. However, in a similar way, new innovative services and applications are expected to empower better understanding and tackling of energy-related issues. For instance, monitoring applications could offer a real-time view on a city's Key Performance Indicators,

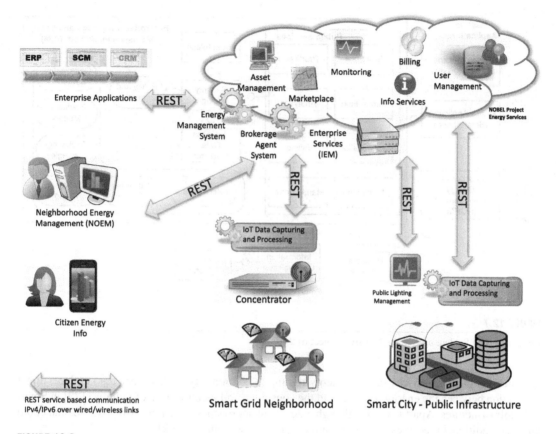

FIGURE 12.6

Smart City Energy Services.

such as CO_2 footprint, energy consumption increase, and penetration of renewables. For example, citizens may now be able to calculate with smart apps the environmental impact of their in-city traveling options. Alternatively, public authorities may be able to better identify energy-hungry processes at a city-wide level and plan how to tackle those as well as a more sustainable city expansion.

Monitoring, as well as being able to exercise control capabilities, plays a pivotal role in order to not only extract the necessary information for understanding key energy processes, but also to be able to apply control when decisions are taken. Once this is a reality on a large scale, new innovative applications that depend on IoT data as well as their control capabilities can be realized, with impact on individual citizens as well as a city-wide impact as shown in Figure 12.6.

As an example, the NOBEL project [170] dealt with integrating information coming from various energy aspects of the city infrastructure, such as energy producing/consuming customers (e.g., residential houses, buildings) as well as the public lighting infrastructure. By tapping into the extended information of smart metering, it was possible to enhance existing services, e.g., real-time monitoring of energy consumption, real-time billing, and asset management [171]; it was also possible to pro-

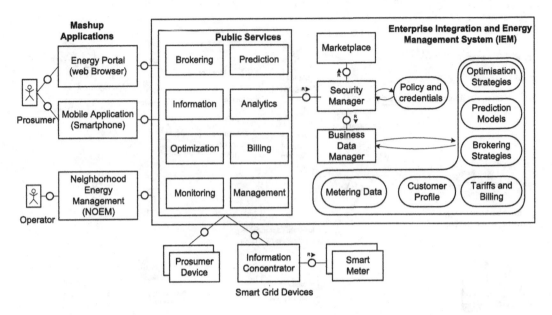

FIGURE 12.7

Enterprise IEM System used in the NOBEL project trial.

vide new innovative services such as energy trading [172] in a city-wide energy marketplace, direct interaction between the energy provider and the consumer, and identification of failures [173] in Smart Grid infrastructures. Apart from that though, additional approaches were now also possible, that could enhance traditional processes such as prosumer forecasting accuracy [162].

To be able to provide value-added energy services, an Integration and Energy Management System (IEM) was created [171] which eased the IoT interactions over IPv4 and IPv6 [174] and especially data collection, which was assessed and provided to the respective energy services. The architecture of the IEM is depicted in Figure 12.7. The IEM itself and several applications depending on it [175] have been extensively tested and used operationally in the second half of 2012 as part of the NOBEL pilot project that took place in the city of Alginet in Spain. During the trial, data in 15-minute resolution, of approximately 5,000 meters, were streamed over a period of several months to the IEM, while the IEM services were making available several functionalities ranging from traditional energy monitoring up to futuristic energy trading [171,175].

As depicted in Figure 12.7, we can distinguish the following key functionalities commonly offered via mobile (e.g., smartphones, tablets) and traditional (e.g., desktops, web browsers) devices, other more complex services, and end-user applications:

- **Energy Monitoring**: For acquiring and delivering data related to the energy consumption and/or production of prosumer devices.
- **Energy Prediction**: For forecasting consumption and production based on historical data acquired by IEM and other third-party services (e.g., weather data).
- **Management**: For handling the asset, user, and configuration issues in the infrastructure.

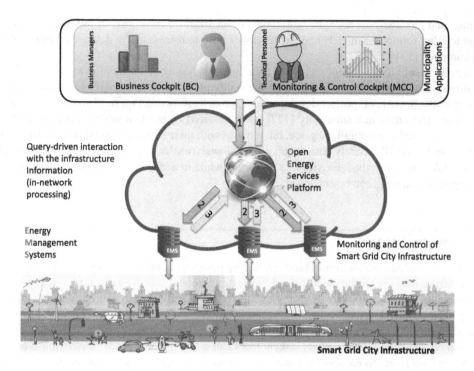

FIGURE 12.8

The SmartKYE query-driven integration and management.

- **Energy Optimization**: For interacting with existing assets of the Smart City; as a proof of concept the public lighting system was used for energy balancing.
- **Brokerage**: Offering energy trading to all prosumer citizens who in a stock-exchange manner could interact via the platform and buy potentially cheaper energy or sell excess production from their PV panels.
- **Billing**: Offering a real-time view of the energy costs and benefits (from transactions on the Smart City energy market), hence avoiding "bill shock" scenarios.
- **Other**: Value-added services offering bidirectional interaction between the users and the energy provider, such as notification for extraordinary events.

The hands-on experiences [171,175] within the NOBEL project revealed several benefits for Smart Grid Cities, as well as several aspects that are crucial when dealing with IoT infrastructures. Designing open services is a must, and the needs of various stakeholders need to be considered, especially when they intersect. With the enhanced monitoring capabilities at the IoT layer, as well as the increased access and correlation with the available enterprise data, the era of IoT "Big Data" reaches the Smart Cities. This has profound implications on the apps and services, depending not only on the processed outcome of the data but also on their quality and timely acquisition. To this end, e.g., validation of data values and syntax based on model semantics, correct timestamping, duplicate detection, security validation and risk analysis, anonymization, data normalization, estimation of missing data, and conversion to other formats or models may be necessary prior to release of the data for further processing or consumption [171].

Once the capability to collect city-wide data and to run queries on them is available [176], new capabilities towards decision making, as well as the interplay of different stakeholders, are evident. As an example, the SmartKYE project developed an approach (shown in Figure 12.8) for monitoring Key Performance Indicators in a smart city [177] and visualized them via a business cockpit [176]. The data is not collected in a centralized place, but remains with their owners, while distributed queries and analytics can be run. The timely data acquisition and analytics for city-wide data can enable decision makers to take more informed decisions, in real-time, while in addition support for existing processes, e.g., long-term planning, can be enhanced.

12.5 CONCLUSIONS

IoT and IoS, coupled with the Smart Grid, are a very powerful combination. IoT has the capability of revolutionizing the core of the electrical grid infrastructure and transforming it to a truly smart one. In combination with the IoS and the IoT, new applications and services can be realized, effectively tackling older problems and providing innovative solutions. The ongoing efforts in smart metering are only a small part of what the future energy infrastructure will be able to do, and we have already shown that based on smart grid data, various stakeholders can enjoy benefits at the Smart House as well as at the Smart City level. To harness the benefits, there are both technological and social challenges that need to be addressed. The future Smart Cities will need to be built on principles of cooperation, openness/interoperability, and trust. Extracting and understanding the business-relevant information under temporal constraints, and being able to effectively integrate it in solutions that utilize the monitor–analyze–decide–manage approach for a multitude of domains, is challenging [171].

COMMERCIAL BUILDING AUTOMATION

13

CHAPTER OUTLINE

13.1 INTRODUCTION

A Building Automation System (BAS) is a computerized, intelligent system that controls and measures lighting, climate, security, and other mechanical and electrical systems in a building. The purpose of a BAS is typically to reduce energy and maintenance costs, as well as to increase control, comfort, reliability, and ease of use for maintenance staff and tenants.

Some example use cases include:

- Control of heating, cooling, and ventilation based on time of day, outside temperature, and occupancy (e.g., Morning Warm-up).
- Automatic control of air handlers to optimize the mixture of outside air in ventilation based on, for example, inside temperature, pressure, and time of day.
- Supervisory control and monitoring to allow maintenance staff to quickly detect problems and perform adjustments.
- Outsourcing of monitoring and operations to a remote operations center.
- Data collection to provide statistics and facilitate efficiency improvements.
- Alarms for high CO and CO_2 levels.
- Individual metering per apartment (to give incentive to save energy in multitenant buildings).
- Intrusion and fire detection.
- Building access control.

A BAS is normally distributed by nature to allow every subsystem to continue operation in case of failure in another system. A BAS consists of the following components (Figure 13.1):

- Sensors (i.e., devices that measure, such as thermometers, motion sensors, and air pressure sensors).

Internet of Things. https://doi.org/10.1016/B978-0-12-814435-0.00026-2

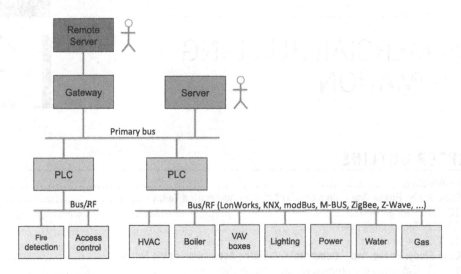

FIGURE 13.1

Central parts in a BAS.

- Actuators (i.e., controllable devices, such as power switches, thermostats, and valves).
- Programmable Logic Controllers (PLCs) that can handle multiple inputs and outputs in real-time and perform regulating functions, for example.
- A server which monitors and automatically adjusts the parameters of the system, while allowing an operator to observe and perform supervisory control.
- One or more network buses (e.g., KNX, LonWorks, or BACnet). We have divided the case study into two phases. In Phase One we give an example of what is commonly available today in regard to building automation. In Phase Two we explore new opportunities for building automation, such as the Smart Grid and the Internet of Things (IoT).

13.2 CASE STUDY: PHASE ONE – COMMERCIAL BUILDING AUTOMATION TODAY

13.2.1 BACKGROUND

Company A wants to improve energy efficiency in their buildings and become GreenBuilding Partner-certified[1], which requires lowering their energy consumption by at least 25%.

After discussions with a building automation company (Company B), they have come to understand that this is a very good investment that will quickly justify itself in terms of reduced energy costs. They agree on a five-step plan that starts with collecting data from the buildings, followed by analysis,

---

[1] https://www.sgbc.se/in-english

adjustments, and connecting the systems in the buildings to a local server, and finally connecting the buildings to a remote operations center.

They can now start with collecting data from existing systems. In some cases this requires new meters to be installed. Everything from water usage to heat and electricity consumption is logged continuously, as well as performance of the ventilation and room temperatures.

By comparing the Key Performance Indicators with comparative figures, the need for corrective actions is assessed and used as a basis for an action plan that consists of adjusting the existing systems and installing new software. These adjustments quickly increase the efficiency of the systems and are continuously optimized during the project. Examples of adjustments are hot water temperature, improved control of indoor temperatures, as well as better control of fan and pump operation to avoid unnecessary operation.

One of the most central features of the improved system is the new web-based E-report. It provides information about current energy consumption and other key parameters from the buildings. This information is used to make both short-term decisions as well as long-term planning. Everyone has access to the web portal because it is not only important for the maintenance staff, but also needed to create awareness among everyone in the company.

The next phase of the project consists of connecting the systems in the buildings and analyzing the dynamics to be able to perform intelligent control. This both improves performance and reduces maintenance costs.

The final step to completion involves setting up a web-based Supervisory Control and Data Acquisition (SCADA) system for remote monitoring of the building systems. Through the web portal, the users can access information from the buildings in a coherent manner. Company A decided to outsource the operations and daily maintenance of the systems to Company B by utilizing their cloud-based offering. Company B's remote operations center is continuously monitoring the building systems. When building system operations deviate from their expected behavior, Company A's maintenance staff and their supervisors are notified by SMS and email. Typical events that can trigger a notification are, for example, mechanical failures or undesirable temperature deviations. Apart from notifications, Company B can also assist with equipment operation and adjustments remotely. For Company A, this arrangement is perfect because their in-house maintenance staff can respond to an alert 24 hours a day.

For Company A, the most important improvement has been the 35% energy reduction after the completion of the project. Another critical aspect has been the knowledge transfer from the experts at Company B that allows Company A to maintain the efficiency of the systems as well as the ability to continuously improve the operations of them.

13.2.2 TECHNOLOGY OVERVIEW

Figure 13.2 depicts the setup for Company A.

Each building is equipped with a set of meters and sensors to measure temperature, water consumption, and power consumption, as well as one or more Programmable Logic Controllers (PLCs).

As seen in Figure 13.2, the PLCs perform real-time monitoring and control of the devices in the building. They also feature a User Interface for configuration and calibration of (for example) the regulators, curves, and time relays. It is possible to remotely configure the PLCs from the Operations Center using the PLC Control system, which is connected to the PLC via a 3G modem and an Internet Protocol (IP) modem that converts between RS-485 networks and Transmission Control Protocol/Internet

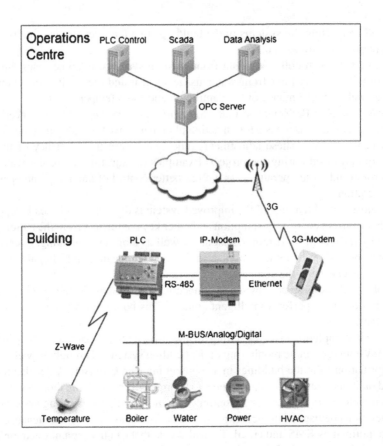

FIGURE 13.2

Illustration of the BAS.

Protocol (TCP/IP) networks. The PLCs communicate with the devices using several protocols, such as M-BUS, analog, digital, and Z-Wave, which is a low-power radio mesh-network technology. All logic necessary to operate the buildings is contained within the PLCs, allowing for minimal bandwidth requirements on the connection towards the Operations Center as well. It also means that the building systems can remain fully operational during periods of network outage.

The OLE for Process Control (OPC) server provides access to data, alarms, and statistics from the PLCs. When a value is requested from a user, a request is sent from the user's OPC Client to the OPC Server, containing an OPC Tag that identifies which PLC to contact and which value to ask for. The type of OPC communication used is called OPC Data Access. The OPC server then contacts the PLC in question and asks for the value using a protocol supported by the PLC (LonWorks or ModBus).

The SCADA system is used for operational monitoring of the buildings and provides information from all the relevant building systems. It uses the open and standardized OPC protocol, which enables

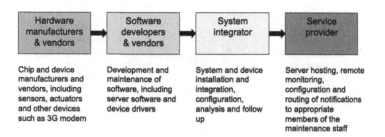

FIGURE 13.3

Applied value chain for Company A's system.

integration with devices from many different vendors. The maintenance and operations staff can connect to the system using a web browser with a username and password to access dynamic flowcharts, drawing tools, timers, set points, actual values, historic readings, alarm management, and event logs, as well as configuration for notifications over email, fax, or SMS.

The Data Analysis server logs all historical readings from the buildings and makes it possible to follow up on different aspects of the energy and resource consumption, satisfying the varying needs of the tenants, economy department, and landlord. Through the OPC server it is possible to gather readings from all the building systems, regardless of vendor. Typical reports include trends, cost, budget, prognosis, environment, and consumption of electricity, heating, water, and cooling.

13.2.3 VALUE CHAIN

Figure 13.3 shows an applied value chain.

13.3 CASE STUDY: PHASE TWO – COMMERCIAL BUILDING AUTOMATION IN THE FUTURE

13.3.1 EVOLUTION OF COMMERCIAL BUILDING AUTOMATION

Two major factors will drive the evolution of Building Automation: information and legislation (Figure 13.4).

Access to well-packaged information will provide the basis needed for decisions and behavioral changes. This can (for example) be electricity prices or where and when energy is used, and this will allow for well-founded decisions that provide the best results.

Legislation, and taxes or tax credits to some degree, will provide the second driver. Legislative demands on green buildings and the Smart Grid will give rise to new opportunities, such as Demand/Response, Micro Generation, and Time-of-Day Metering.

Market growth will result in economies of scale, standardization and commoditization, driving down prices and increasing availability of devices and services. It will be possible to buy advanced devices off-the-shelf, perform installation, and connect them directly to service providers on the Internet.

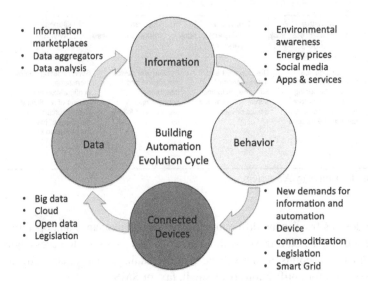

FIGURE 13.4

Building automation evolution cycle.

13.3.2 BACKGROUND

A few years have passed and Company A has decided to outsource the maintenance of its buildings to a local contractor who provides services to several other customers in the neighborhood. This will save money since that will enable them to utilize a shared caretaker pool.

At the same time they plan to upgrade their buildings to become fully automated with, for example, occupancy sensors, automated lighting, and integrated access control. To do this cost-efficiently, they intend to make use of the existing IP infrastructure in the buildings, which also saves on Operating Expenditures as the network administrators can also manage the BAS infrastructure. According to studies, a converged IP and BAS network can reduce maintenance costs by around 30% while also lowering the initial investment for installation and integration by around 20% (according to studies performed by Cisco). A shared infrastructure also leads to increased energy efficiency.

New political incentives in regard to energy efficiency have increased the development pace in the building automation area. Many neighboring buildings in Company A's area are now fitted with building automation, which allows for sharing of information and resources. The increased customer base has also enabled new niches in the value chain, which has been split up to a large degree. Where before the rule was to have one single integrator and service provider, we now see a multitude of new actors, such as specialized service providers for remote monitoring, security, optimization, data collection, and data analytics. This allows Company A to choose freely what combination of service providers to use, while also providing a smooth transition when moving to a new provider. This is made possible by a new niche in the value chain: the Cloud Service Broker (Figure 13.5).

The process of integrating with the maintenance contractor's systems is simplified by the service broker because it provides immediate access to Company A's BAS. The caretakers can use their own

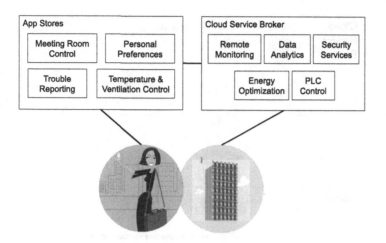

FIGURE 13.5

Cloud Service Broker.

specialized software as the service broker provides a bridge that can convert between several common protocols used for building automation.

When it comes to selection of devices, Company A opts for using standardized protocols to avoid vendor lock-in. They also decide to keep certain parts of the old system, as these would be too costly to replace. To still benefit from a fully integrated system, they also invest in a Constrained Application Protocol (CoAP) gateway that translates between legacy devices and the new system.

By exporting historical data and configuration parameters from the old OPC Server it is possible for Company A to choose new service providers that can replace the old systems for PLC control, SCADA, and data analytics with a minimum of manual effort.

As an added service, the new platform also provides data brokering. This gives access to a multitude of data sources, such as the following:

- Historical and current KPIs to similar buildings.
- Integration with local government facilities.
- Weather forecast information.
- Utility prices, both current and future.

Apart from providing access to new service providers, the cloud broker also hosts a client API that enables third-party app developers to create smartphone applications. A number of users have purchased apps that allow them to do the following, for example:

- Control HVAC settings in meeting rooms.
- Report problems and service requests.
- Integrate with Outlook to adjust meeting rooms in advance.
- Create personal profiles to automatically adjust room settings.
- View instant and historical personal energy consumption and compare to others using social media.

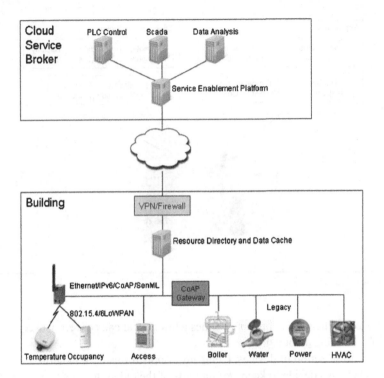

FIGURE 13.6

Architectural overview of the upgraded system.

13.3.3 TECHNOLOGY OVERVIEW

Thanks to the rapid development of technologies for IP Smart Objects, it is now possible to use IP for both constrained devices, such as battery-powered sensors and actuators. The new system is to a large degree based on IP technology (Figure 13.6). There are several IP-based protocols to select from, but in this case CoAP and Sensor Markup Language (SenML) were selected. CoAP provides both automatic discovery as well as a semantic description of the services the device provides. This drastically reduces installation costs, as much less configuration is needed. CoAP is similar to Hypertext Transfer Protocol (HTTP), but is binary to reduce the size of the messages. It also defines a Representational State Transfer (REST)-like Application Programming Interface (API) optimized for IoT applications. As with HTTP, a format for the content is also needed, in this case SenML, which is used as a format for sensor measurements and device parameters.

As mentioned, there are still a few legacy devices, and these need a gateway to enable communication with the IP-based systems.

A local Resource Directory and data cache is also installed to keep track of all the devices in the company network. This allows for local look-ups of devices and data and serves as a safeguard in case of failure. To protect the system from intruders, a normal network firewall is used. For the connection

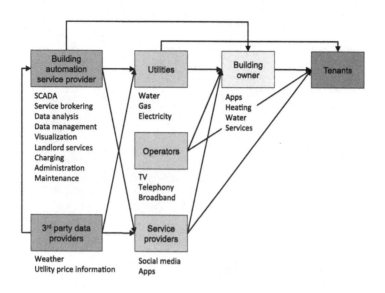

FIGURE 13.7

Evolved value chain for building automation.

towards the service broker and the service providers, a permanent Virtual Private Network (VPN) connection is established.

Historical data is exported from the OPC server to the Cloud service broker's data storage to make it available to new service providers. Apart from data storage, the Cloud service broker also provides management functionality for control and data access, as well as access to specific service provider's management portals. It also offers a global Resource Directory with semantic resource and data description, along with a contextual model that covers schematics, geospatial information, and indoor location.

13.3.4 EVOLVED VALUE CHAIN FOR COMMERCIAL BUILDING AUTOMATION

As the demand for M2M services grows, new niches in the value chain will emerge, such as information brokers, service brokers, and service enablement providers (Figure 13.7). These will enable new use case areas by allowing vertical domains to be integrated (e.g., security, energy, waste management, police, and public transport). It will also provide the openness needed for third-party service providers to create apps and social media integration, to allow for (for example) comparisons with neighboring buildings, competitions, and end-user involvement. Privacy and security will, however, be essential to build up the trust needed for this ecosystem to develop.

In addition to new use case areas, major improvements can be expected within the areas of efficiency, convenience, comfort, reliability, and safety.

SMART CITIES

CHAPTER OUTLINE

14.1 INTRODUCTION – WHAT IS A SMART CITY?

"Smart City" is a phrase that has many meanings and is used in many different contexts. It is a phrase that is used to cover technical solutions that cover the Internet of Things (IoT), big data, analytics, and Machine Learning. At the other end of the spectrum, it is a phrase used to represent technical interactions with citizens and enabling a better integration of social, environmental, economic, and political aims in different regions. In short, the definition of the smart city is as complex as the definition of a city itself – every country and region in the world has different types of cities and these differences reflect the different backgrounds of the inhabitants and the different environments within which cities developed and spread in different areas of the world. There is no unifying definition of a city and as a result, there is no unifying definition of a smart one.

For the purposes of this chapter, however, we must settle on one that will allow a relevant discussion of our work and the role of IoT within city contexts. While all cities are different, we take the stance that city leadership across the globe are working to secure the economic success of their citizens, a reasonable level of environmental protection in order to attract people to live within the city, and a thriving culture that attracts tourists and business visitors. From this perspective, the role of IoT takes a new meaning and is less about standardization and the in-depth radio and silicon discussion. The IoT of smart cities is the IoT of technology placed within a broad social context. While this may seem a simplified definition to some readers, within it we are able to place IoT in a proper context – within the complex interactions a city and its citizens need to handle on a daily basis.

14.2 SMART CITIES – A TECHNICAL PERSPECTIVE

From a technical perspective, smart city concepts are built on the idea of sensors installed in various parts of city infrastructure – e.g., roads, cars, CCTV, buildings, public transport, and citizens' smart phone data. The use of smart phones and personal sensing is covered more in Chapter 15 and in this chapter, we therefore focus on the use of IoT in city infrastructure. See Figure 14.1.

Internet of Things. https://doi.org/10.1016/B978-0-12-814435-0.00027-4
279

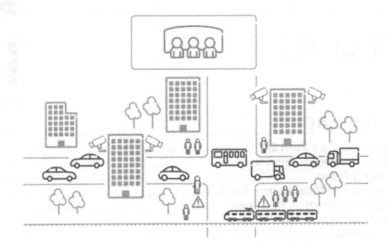

FIGURE 14.1

A Smart, IoT-Enabled City (source: Ericsson 2016).

Table 14.1 Data Supply Chains (source: Gurguc and Mulligan, 2018)

Supply Chain	Characteristics	Data Supply Chain	Emergent areas
Flow of physical artefacts (materials, products, services) from initial source(s) to final customer.	Content	Flow of multisource, multiform data artefacts (or even processed data, information or knowledge) from inbound and outbound activities of the firm	Data heterogeneity Data quality Data privacy and security
Demand-led supply chain (only produce what is pulled through), targeting in production maximisation, revenue and value creation, quality, service, safety, etc. Price-driven (strategically decoupled and price driven)	Strategy	Innovation-led (through ideas, practices, and business models; value for DSC is not solely created from an information product/ service but also through the disruption of the existing business and operational models). Outcome-driven (strategically coupled and value driven)	Data generation and exploitation Innovation (business model and product/service development)
Shared information across the whole chain (end to end pipeline visibility). Collaboration and partnership (mutual gains and added value for all)	Integration	Integration of multiple data sources (internal and external to the focal firm). Collaboration, interconnection and value co-creation (value through business model innovation)	Multisource data Interconnectivity
IT enabled; Physical manufacturing systems; agile and lean; mass customization methods	Tools/ Methods	Analytics-enabled; Cyber-physical manufacturing systems; agile, lean and real-time; tailored customization methods	Data collection, processing, storage methods/ tools and provenance Data Analytics

Within a city, IoT is often used in conjunction with other data sources including Big Data and also advance analysis techniques such as Machine Learning and increasingly Artificial Intelligence; these have all been covered in Chapter 5. Often, however, smart city solutions also include "Open Data". According to OpenDefinition, "Open data is data that can be freely used, re-used and redistributed by anyone – subject only, at most, to the requirement to attribute and sharealike". A critical aspect of smart cities, therefore, is the ability to combine data from very different sources. This raises a number of questions about the tracking and tracing of data, ensuring the data comes from the device that it claims to come from and to ensure that it has not been changed or manipulated in transit. Effectively, the use of IoT in a smart city develops the need to properly manage and control data in a data supply chain. Table 14.1 illustrates the differences between a physical supply chain and a data supply chain, including the new characteristics demanded by data supply chains, including data heterogeneity, data quality, privacy, and security. In addition, it illustrates that the products in question are related to data generation and exploitation.

14.3 IOT DATA SUPPLY CHAINS
Building on the work outlined in Chapter 3, we here illustrate a data supply chain for a smart city installation. In this case, we have a system that is combining data from a number of IoT installations as well as other data sources. In Figure 14.2, a solution for assessing the status of the city infrastructure

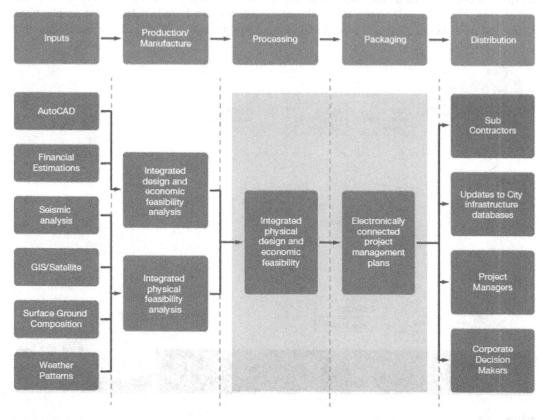

FIGURE 14.2

Data Supply Chain for IoT data in a Smart City installation.

and for forecasting the need and ability of a city to develop and deliver new infrastructure is presented. On the left-hand side of the diagram, we see the inputs to the data supply chain – namely input from seismic sensors, satellites, weather sensors, and also AutoCAD and financial data from the corporate databases. The data streams from these sources are combined in the production/manufacture, processing, and packaging sections of the data supply chain.

14.4 IOT DATA AND CONTEXT MANAGEMENT IN SMART CITIES

One of the main issues associated with smart cities is how to manage data from a large variety of sources, in particular the context within which the data needs to be understood. For example, I may have a dataset from a sensor about air quality, but the context that it is in a particular park in a particular city needs to be understood. These contextual pieces of data assist in the proper processing of the data in question. In addition, we can often find that in a city the context of the information may change over time; as a person moves through the city, they change from being an employee at a firm, to a commuter on a bus. This contextual data is often tagged within different systems in different ways. Figure 14.3 illustrates this in the city space; here we have IoT data about people, trees, buses, and roads being combined in different systems in different ways. In addition, much data in a

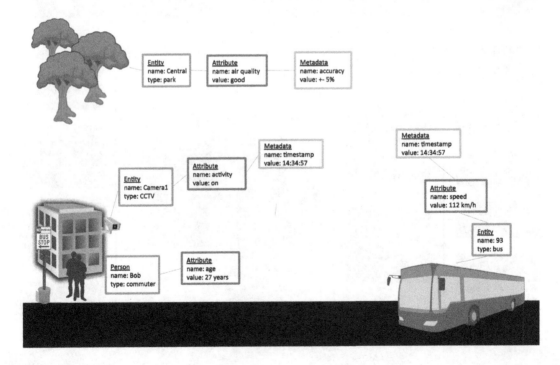

FIGURE 14.3

Data Context in a Smart City.

city space is only useful for a short period of time – its usefulness is relevant until the next reading is taken. Examples are where a bus is on a particular route at a particular point in time or the air quality reading taken in the morning or the afternoon. A final piece of contextual information may be the accuracy of the sensors – e.g., an air quality sensor may have a variation of $+/-$ 5 percent.

This contextual information management can create a number of problems when managing such IoT data. One body that has attempted to create guidelines for such issues is the ETSI Industry Specification Group (ISG) on Context Information Management (CIM).

14.5 ETSI ISC CONTEXT INFORMATION MANAGEMENT

The ETSI ISG for CIM has outlined the issue associated with the broad number of stakeholders working within a Smart City space and how this large ecosystem requires context information management APIs and models in order to fully ensure solutions that are able to manage such data effectively, as illustrated in Figure 14.4.

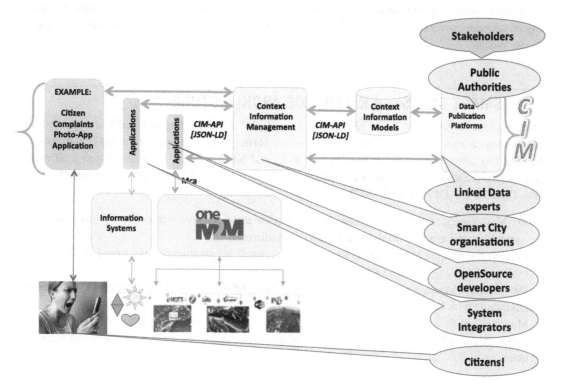

FIGURE 14.4

ETSI ISG CIM Context Information Management Layer (source: ETSI ISG CIM).

Within "smart cities", there are a large number of use cases and an emerging number of standards that need to overlap/interact with one another. Many of these standards are from areas completely outside the normal communication and telecommunication standards arenas and include smart grid standards, citizen data standards, building standards, and construction standards. This is a completely different approach than what the technical industries are used to – in many cases the technical standards from, e.g., ETSI or similar bodies need to act as the fabric/glue for disparate construction and city standards. Another complexity that emerges is the regulatory frameworks that cover these systems – from personal data management (European Union General Data Protection Regulation, GDPR) to health and safety regulations, IoT solutions within the smart city space need to be able to manage and appropriately handle all. This is one of the reasons that the monetization of smart city projects has often been so difficult. Standards such as ETSI ISG CIM are required in order to:

- ensure vendor neutrality for users such as Cities,
- reduce technological barriers to development and deployment,
- enable a community of entrepreneurs to build innovative services.

In addition to context management, it is also necessary to construct reference architectures for smart city installations; several attempts have been made to do this – due to space limitations we have selected one to look at.

14.6 SMART CITIES – A REFERENCE ARCHITECTURE

The Synchronicity Project[1] is an EU-funded activity that is developing detailed learnings and understandings across a variety of smart city projects. Through combining existing work from its own baseline city areas and OASC, FIWARE, EIP-SCC, and NIST IES-CF, it has developed a reference architecture for smart cities, which is illustrated in Figure 14.5. It has aligned with standards in the ITU-T SG20 / FG-DPM and ISO TC268.

- IoT Management: to interact with the devices that use different standards or protocols making them compatible and available to the SynchroniCity platform.
- Context Information Management: to manage the context information coming from IoT devices and other public and private data sources.
- Data Storage Management: to provide functionalities related to the data storage and data quality interacting with heterogeneous sources.
- Marketplace: to implement a hub to enable digital data exchange for urban data and IoT capabilities providing features in order to manage asset catalogues, orders, and revenue management.

---

[1] https://synchronicity-iot.eu

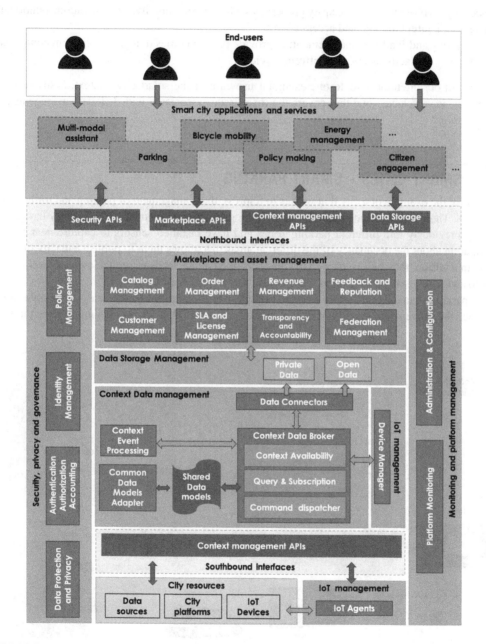

FIGURE 14.5

Synchronicity Smart City Reference Architecture (source: https://synchronicity-iot.eu/about).

- Security: to provide crucial security properties such as confidentiality, authentication, authorization, integrity, nonrepudiation, and access control.
- Monitoring and Platform management: to provide functionalities to manage platform configuration and to monitor activities of the platform services.

We turn our attention now to one exemplar use case of the smart city context. Many others exist, but we have selected one that covers the main issues associated with IoT data management and monetization.

14.7 SMART CITIES – SMART PARKING

Let us take the example of a city that wishes to improve the management of parking and traffic flow on its streets. Their assets include a number of parking garages and on-street parking. They have divided the city previously into different parking zones and have installed sensors to capture near real-time information about the occupancy of the different areas that allows them to publish capacity of parking areas. In addition, they have sensors and cameras aimed at the roads to see the current level of traffic flow to and from these parking areas. This is illustrated in Figure 14.6.

As parking demand and supply rises and drops over the entire day, the sensors are able to track and trace the movements in and out of the parking areas and plan for improvements over the longer term.

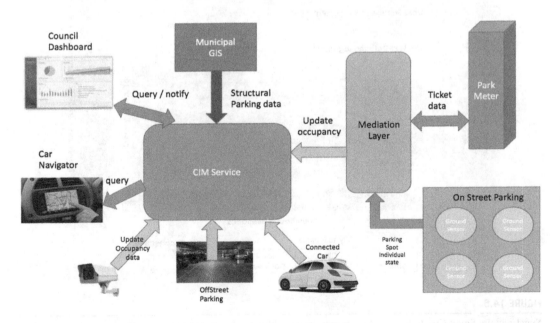

FIGURE 14.6

Smart Parking Use Case (source: ETSI ISG CIM, 2017).

Drivers that are looking for a parking space, meanwhile, are able to be directed to the place where there is the highest probability of finding a spot. In addition, car navigation systems will be able to integrate the traffic flows to and from parking areas and include that in the algorithm for directing drivers to parking. This will reduce the overall flow of traffic around the city and allow a city to have more efficient use of assets.

PARTICIPATORY SENSING

15

CHAPTER OUTLINE

15.1 INTRODUCTION

Participatory Sensing (PS) [178], also known in the literature as Urban, Citizen, or People-Centric Sensing, is a form of citizen engagement for the purpose of capturing the city surrounding environment and daily life. The first PS projects appeared in the early 2000s and focused more on city dweller campaigns to capture problematic situations (such as road faults, air pollution, low-lit parts of the city), daily routines (such as commuting by bike/car), personal individual health, or even combinations thereof. More recently, the concept has gone through a transformation in terms of branding, people engagement practices, the target audience, and business models. Nevertheless, the main constituent of any PS activity is the city inhabitants acting as human sensors, and often as end-users of an end PS product.

The purpose of early PS was to empower citizens to transform their cities into a collective optimum living environment. With today's technology, citizens have several means of capturing and sharing their daily life and their view of the quality of living in the city. Most citizens today have mobile phones with which they can capture at least pictures, movies, and sounds and share them over the Internet. This information can be analyzed and interpreted by individuals, groups, or city officials, and conclusions could be drawn which would result in actions. The information collection can be initiated by a specific sensing/collection campaign organized by any interested actor. This chapter describes the traditional PS concept along with an example case study, as well as the more recent and future trends around the general idea of people sensing. We have chosen the term *Participatory Sensing* for the description of both the early and the present concepts for the rest of the chapter.

Internet of Things. https://doi.org/10.1016/B978-0-12-814435-0.00028-6

15.2 ROLES, ACTORS, ENGAGEMENT

The main potential actors in a PS system are "individuals" and "city authorities". The city inhabitants have access to sensing devices and can use them to capture their environment. Individuals can be one group of the receivers of the collected information and the target for environmental changes based on PS. The city authorities can be the receiver of the collected information and potentially the organizer of the sensing/collection campaigns. City authorities can also analyze the collected information, potentially with individual citizens, set plans of actions, and follow-up on the actions.

There are several roles for the citizens or city authorities in a PS system. An individual can potentially act as the Data Collector using their mobile phone to collect sensor data. An individual or city authority can play the role of the Collection System Operator who owns and operates the collection system from multiple data collectors. An Analysis Provider that processes, stores, and analyzes the collected data can be assumed by any party (individuals, cities) that could also prepare plans of, and execute actions based on, the conclusions of the campaign (Action Responsible). As a general observation, the city authorities typically assume the nondata collector roles (Collection System Operator, the Analysis Provider, and the Action Responsible) while the citizen's major role is that of a Data Collector, not excluding, however, any other combination of assignments of roles to actors. Depending on the degree of engagement of the different actors, there are three main models of participation in a PS activity, described below.

15.2.1 COLLECTIVE DESIGN AND INVESTIGATION

In this model individual citizens design the sensing campaign, participate in the data collection, and analyze and interpret the collected data. Therefore, the citizens are fully empowered to contribute to the change that they would like to see in their living environment.

15.2.2 PUBLIC CONTRIBUTION

Individual citizens only take active participation in the data collection phase organized by another individual or organization (e.g., city authorities), but they do not necessarily analyze or interpret the results.

15.2.3 PERSONAL USE AND REFLECTION

Individual citizens monitor and record their daily lives for themselves without any organized campaign. A person may choose to refrain from sharing personal information and details or may choose to share certain specific information or aggregation of collected information. Therefore, the collected data are used mainly for personal reflection or sharing and reflection within a very small and private group (e.g., individual's relatives).

15.3 PARTICIPATORY SENSING PROCESS

The basic steps of a typical PS process are shown in Figure 15.1.

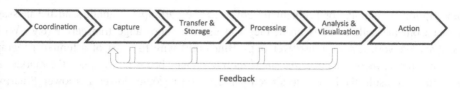

FIGURE 15.1

Typical Participatory Process steps.

During the Coordination phase the participants need to either organize themselves, or be recruited by some other entity (e.g., city authorities) within the context of a sensing campaign, and the objective of the campaign needs to be communicated among all of them. Then the participants spend some predetermined amount of time to capture (Capture phase) the desired sensing modalities using their mobile phone applications or custom designed applications for the sensing campaign. The data entering a PS system does not need to originate only from the data collectors. Several other publicly available sources, such as weather, air quality, and traffic reports, could be used for drawing richer conclusions. The collected data are transferred (Transfer and Storage phase) to the data collection system through the phone connectivity options and stored in Internet servers (private or public). The data are then subject to preprocessing (Process phase) so that the privacy of the data collectors is preserved, and access control rules are added so that the data can be accessed anytime by only authorized individuals or services. The collected data are analyzed by relevant analysis tools, aggregated (if possible), correlated with each other in order to detect patterns, and in the end visualized for better understanding for the target group of the campaign (Analysis and Visualization phase). Last but not least, certain actions (Action phase) may be taken by individuals or city authorities. Feedback is present throughout the whole process and typically assists the Capture phase of the processes. If, for example, the captured data transfer and storage fails, then the participant may be notified to retransmit or recapture the target environment. If Processing and Analysis & Visualization result in very little or ambiguous information (e.g., when the processed picture has a bad quality), or participants enter in an area of interest, then they may be notified to (re)capture the situation if possible (Figure 15.1).

15.4 TECHNOLOGY OVERVIEW

One of the main pieces of technology for a PS campaign is the mobile phone. It encompasses both basic and complex sensors, input and output hardware, one or more processors, one or more modes of communication, location sensing capabilities, and a software Execution Environment that can potentially allow the execution of third-party software. The minimum capability for a mobile phone in order to be useful for a PS activity is the communication capability to allow transmission and reception of messages and potentially sensor data. Active participants can always send text messages with their own sensed data (e.g., a pothole on the corner of Main and 3rd Street). The minimum communication capabilities include a cellular (2G, 3G, LTE, WiMAX, etc.), Wi-Fi, or Bluetooth transceiver for the participant to potentially receive the campaign start signal (e.g., phone call or short message) and to send the message(s) with captured information. Of course, if for reasons of cost, subscription limitations, or problematic coverage the participant cannot send the messages to the Internet from the cellular transceiver, they can use the short-range transceiver (e.g., Wi-Fi) or Personal Area Network (PAN) transceiver (e.g., Bluetooth) with the cost of a limited geographical reach and, of course, inconvenience. As smart mobile phones begin to dominate the market, inconvenience and, as a result, the barrier to citizen participation becomes lower and lower. Smartphones are equipped with all kinds of sensors, such as high-resolution photo/video sensors that can measure light intensity, accelerometers, gyroscopes, compasses, location sensors (GPS), infrared sensors, and of course the microphone that could measure a part of the caller's voice and the ambient noise. They are also equipped with some actuators such as displays (e.g., for receiving campaign messages), vibrators, or speakers (e.g., for notifying campaign participants that their location is ideal for a capture).

An important requirement for any sensor system, and therefore for a mobile phone as a means for sensing, is that it should have the ability to annotate the collected sensor data with some time and location information. Otherwise, the collected data for a single participant cannot be correlated with the other publicly available city sources such as weather reports. Moreover, sensed data missing time or location information cannot be correlated with the corresponding data from other participants and of course cannot be aggregated to produce useful statistics.

The time and location information is not listed as a minimum requirement on the phone capabilities because the participant can annotate the collected data him/herself, for example, by writing the time and location in the message that conveys the sensed data; but of course, the level of inconvenience is high. Fortunately, location sensors are standard in most modern smartphones, and it is often the case that images and videos are automatically annotated with location information without any intervention from the owner. Timestamping images and videos are also supported in the majority of modern phones. In the case of other sensor modalities useful for a campaign (e.g., pothole detection by means of a phone accelerometer), a special application is often designed and disseminated as part of the campaign in order to make it convenient for people to participate. For example, there might be an application that always records the sensor data, time, and location, forms the PS message, and dispatches to the data collection system when the time, cost, and network availability conditions are favorable.

Because the sensed data from mobile phones need to be collected in one or multiple places for further processing, there is a need for one or more dedicated servers from either a server farm in the premises of the Collection System Operator or from a commercially available centralized or distributed cloud provider platform (described in Chapter 5). The machines host the appropriate applications for supporting most of the steps in a PS process. While the simplest of the campaigns may need only a simple content database to store sensed data (e.g., ambient noise recordings or photos of the city), more sophisticated campaigns may include preprocessing of individual data, such as filtering, validation whether or not the collected data is meaningful or not, anonymization of the source of information, removal of people's faces from images/videos, etc. Further actions can also take place such as annotation with city addresses based on location information, compression, storage, etc.

After the necessary data is collected, the Analysis Provider is responsible for analyzing the data. Analysis can be as simple as citizens sifting through the data and discussing with other citizens their conclusions; for example, looking at hundreds of photos of poorly maintained neighborhoods. Or the analysis could be as sophisticated and automatic as determining the light or noise levels of the city according to neighborhoods and time of day, annotating a city map with such information, or creating videos showing the variation of sensed data on a city map. The level of sophistication is clearly up to the imagination of the campaign responsible.

Of course, each PS scenario requires different sensor modalities and different collection processing, analysis, and visualization capabilities. The more organized the campaign in terms of software and hardware capabilities, the lower the barrier to participation and the higher the number of participants.

15.5 AN EARLY SCENARIO

In a modern developed city, people move from place to place in order to commute to home, work, school, and extracurricular activities. They walk, drive, or ride private or public means of transportation to get from one point of the city to another, and their mobility is the perfect solution that can ensure as close to complete sensing coverage as possible. The specific use case that we present here involves bikers moving in the city (e.g., commuting between home and work), carrying their mobile phones on themselves [178] and possibly several other sensor devices on their bikes (Figure 15.2) [179]. The bikers can have simple bikes with the only sensing device being their mobile phone, or superbikes equipped with sensors such as microphones, magnetometers, GPS, CO_2 meters, and speedometers. The mobile phone plays the role of the sensor and the communication device. As bikers move in the city, all the data from the sensors are transported to dedicated servers, stored, and preprocessed. Individuals or city authorities retrieve anonymized raw data and analyze them, while letting automatic analysis tools produce useful statistics. Raw data are transformed into statistics such as preferred bike routes, traffic problems, road faults, air quality reports, ambient noise levels, and evening light levels, and they can be correlated with map and city infrastructure information (e.g., road intersections). Both types of data (raw and analyzed) can be presented to individuals or

FIGURE 15.2

CycleSense/BikeNet use case [179].

action-responsible city authorities to remedy problems or contribute to (re)planning of the city infrastructure.

15.6 RECENT TRENDS

More recently, the concept of PS has gone through a transformation in terms of branding, people engagement practices, target audience, and business model. The terms used today such as Urban Sensing, Social Sensing, or Citizen Sensing [180,181] do not stress the participatory aspect anymore, although people can still participate in sensing campaigns. The term "campaign" itself implies active participation for the design and execution of the sensing session and therefore active participation in the changes citizen would like to see. However, today's Citizen Sensing focuses more on analyzing and visualizing any data coming from people, regardless of whether people actively design and participate in the on-going research or not. Today's PS activities are mainly uncoordinated, and they are either active or passive. This shift in the engagement practice can be expressed with the introduction of two new participation models, i.e., the spontaneous participation or citizen journalism and the passive or unaware participation.

15.6.1 CITIZEN JOURNALISM

Individuals monitor and record their lives similarly to the personal use and reflection model of early PS models without an organized sensing campaign. This means that they report their findings through social media (e.g., blogs, twitter feeds, social networking websites), and typically these reports are open to the public, unlike reports that target personal use and reflection. Citizen journalism is an active sensing engagement. However, the target audience for the sensing campaign is the journalist's own followers/readers/viewers. This type of model is typically used by individuals or authorities during exceptional circumstances such as disasters, without excluding the citizen journalists who regularly post their version of the news. Examples include pictures posted on Facebook about a disaster scene in the city, or tweets providing very terse descriptions of the situation from the individual's point of view. The value of these citizen journalist reports is the freshness of the reports because the individual witnesses are on scene before any city authorities or news correspondents arrive.

15.6.2 PASSIVE PARTICIPATION

The behavior of citizens is captured, stored, and analyzed, with actual citizens often unaware of the fact that their behavior data could be used for the public or the private sector. In such cases, the data are (or should be) anonymized, and possibly aggregated, in order to preserve the privacy of the citizens. Examples include traffic cameras, electricity metering of certain neighborhoods, and credit card transactions. Either city authorities or private companies collect data about people's behavior and use them for their purposes, such as city planning or targeted marketing. Therefore, the target audience for the analyzed data is not necessarily citizens or city authorities, but also private sector employees.

The most recent Citizen Sensing activities put more weight on making the collection, analysis, and visualization more automated before people can receive the product of the research in order to take action. A human can still be in the loop, assisting the automated processes to analyze the collected data better, and of course as an end-user of the produced information. The incentive for shifting the focus from the manual processing and analysis to more automated methods is due to the emergence of Data Management, Machine Intelligence, and Knowledge Management technologies (all covered in Chapter 5), which promise fast analysis and semantic interpretation of vast amounts of data from both streaming sources (e.g., active or passive citizens) and fixed sources (e.g., city's open data). However, these technologies need know-how that volunteers or city authorities do not necessarily have available. Therefore, private companies seize the opportunity to collect data from citizen behavior, analyze the data, semantically annotate and correlate the data with other sources, and create information and knowledge. This information and knowledge can be sold in an information marketplace. Examples are the analysis of Twitter messages for detecting market trends or consumer satisfaction.

15.6.3 SOCIAL SENSING

Social Sensing [182] refers to sensing of physical phenomena by humans or devices used by humans (e.g., mobile phones or sensors). In Social Sensing, the data reported by humans or the utilized devices are a combination of sensor data collected by sensors, natural language text, and user-captured media such as photos, audio, and video. The main challenge of Social Sensing is the overall data reliability and Quality of Information (QoI) for assessing the state of the physical world based on human reporting [183]. There two main groups of challenges: (a) cyber-physical and (b) linguistic and social challenges. The cyber-physical challenges are about the modeling of human reporting in a similar way as a typical sensor which has a well-specified reliability and accuracy model and a well-specified model of the observed phenomena. Human reporting represents the perception of an event and not the actual event and thus it is susceptible to distortion coming from intentional or unintentional mistakes, mistakes in conclusion drawing, social network bias (a human report based on a friend's report), exaggeration, etc. The linguistic and social challenges include the vagueness and ambiguity of human language when reporting an event and the implicit context that humans often omit and which is important for even a human to draw a correct conclusion.

15.7 A MODERN EXAMPLE

More recent examples of PS activities focus on exploiting citizen journalist reports from a disaster site in order to produce richer information content and potentially help others who are close to the disaster area. The use case that we describe here consists of three citizen journalists that observe a fire on the corner of Oak and Birch Street near Oak Park (Figure 15.3). At different time instances, each of the participants posts a short message (Tweet) with a different description, different intention, and probably spelling and language errors. For example, CityJour#1 names the location of the fire as "Oak Park", CityJour2# as "oak parl", and the third Tweeter as "corner of Oak and Birch street", while there are typos or omissions such as "wins" instead of "winds". As the Twitter feeds allow only short messages, there are typically links to more content for the more interested Tweeter followers/readers; for example, a photo taken from the site annotated with the GPS location. The two Tweeters include

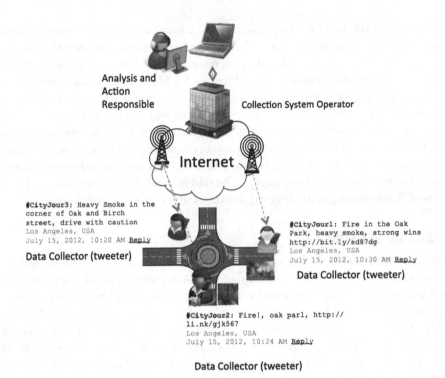

Analysis and Action Responsible

Collection System Operator

Internet

#CityJour3: Heavy Smoke in the corner of Oak and Birch street, drive with caution
Los Angeles, USA
July 15, 2012, 10:20 AM Reply

Data Collector (tweeter)

#CityJour1: Fire in the Oak Park, heavy smoke, strong wins
http://bit.ly/sd87dg
Los Angeles, USA
July 15, 2012, 10:30 AM Reply

Data Collector (tweeter)

#CityJour2: Fire!, oak parl, http://li.nk/gjk567
Los Angeles, USA
July 15, 2012, 10:24 AM Reply

Data Collector (tweeter)

FIGURE 15.3

Citizen Sensing use case.

such links in their Tweet with the links redirecting to another site where photos from the fire are stored.

In terms of the PS process (Figure 15.4), all or most of the spontaneous Tweets by the citizen journalists need to be discovered and collected, which is not an easy task. For identifying the related report's analysis on the actual text, metadata and linked content at off-site servers may need to take place in order to select the relevant tweets. Moreover, third-party, publicly available data sources such as maps or city authority buildings and addresses are used to enable the transformation of location information from one format to another (e.g., from GPS coordinates to road names or building addresses). In addition, weather information provides a potential for prediction of the course of the fire. All these sources of information are subject to correlation and analysis before any visualization and action steps are taken. An example of further analysis is the semantic analysis of all the sources, extraction of the thematic context, as well as the temporal and spatial context, and the fusion of all these pieces of information into one coherent report. In this case, the report is that the location of the fire is in Oak Park, specifically on the corner of Oak and Birch Street, and because there are strong winds in the area, the fire is intense, and there is heavy smoke that obstructs driving, especially in the intersection. Based on weather reports, the report could be enhanced with fire movement prediction and warnings to drivers and citizens with nearby homes.

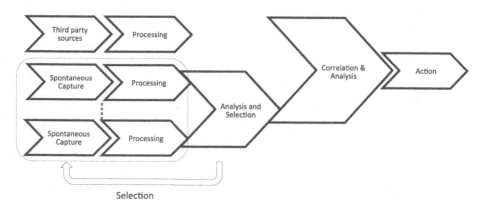

FIGURE 15.4

Citizen Sensing process.

15.8 **CONCLUSIONS**

This chapter focused on the use case of participatory, or urban or citizen or recently emerged social sensing. Despite the different terms used in the past and present, this form of sensing assumes that humans, with the help of devices such as mobile phones and peripheral sensors, monitor their environment and report sensor data or text describing a phenomenon. In the past, most of the focus of research and development for this kind of sensing was on the applications and coordination of groups of people by means of campaigns. Recently research and development focus more on the reliability of the data collection process, data provenance, data independence (data free of influence from social networks), relevant context, and extraction of useful information from human assessment of a situation or a phenomenon by written text. In any case, this kind of sensing is still interesting since the infrastructure is widely available (mobile phones and nearby short-range sensors) and humans often report interesting events anyway without any obligation.

AUTONOMOUS VEHICLES AND SYSTEMS OF CYBER-PHYSICAL SYSTEMS

16

CHAPTER OUTLINE

16.1 INTRODUCTION

Autonomous vehicles are not necessarily an Internet of Things (IoT) "use case". Rather, while they constitute connected *things*, they are a consequence of sustained progress in the development of sensor, actuator, control, energy, communication, and computational technologies, among others, integrated with vehicles and enabling systems more generally. This section describes how the IoT is facilitating progress in interconnecting a number of evolving subsystems, including microinstances of Cyber-Physical Systems such as autonomous vehicles [184], broadly defined, with macrosystems like the transportation networks within which they reside and increasingly interact, e.g., within so-called Smart Cities (Chapter 14).

A number of examples of autonomous vehicles and systems are described in the following subsections, beginning with the automotive industry and considering recent advances in complementary areas, such as Unmanned Aerial Vehicles (UAVs) and intelligent infrastructure, and how their convergence is destined to create complex systems of Cyber-Physical Systems (Section 7.10.6), wherein new relationships and opportunities are created through the near real-time exchange of information, e.g., multimodal point-to-point transportation [85] or improved logistics (Chapter 17).

Internet of Things. https://doi.org/10.1016/B978-0-12-814435-0.00029-8

16.2 AUTONOMOUS CARS

Self-driving cars are, at the time of writing, a very hot topic in research, industry, governments, and society at large. Several of the world's largest companies, like Google (which spun out its self-driving car project as Waymo, to be precise), the majority of the automotive industry, and many newer companies like Uber and Tesla, already have the technology to demonstrate fully autonomous driving capabilities. As described in Section 6.6, there are varying levels of autonomous driving capabilities defined, and one of the key considerations is safety.

16.2.1 A VERY BRIEF HISTORY OF AUTONOMOUS CARS

This section gives a brief overview of some of the key milestones in the history and evolution of autonomous or self-driving cars, which stretches back as far as the 1920s, when radio-controlled cars were demonstrated in New York. Over the following decades, several notable advances in terms of guided driving were made using a variety of technologies. In the late 1960s, Stanford Professor John McCarthy's essay[1] on Computer Controlled Cars described a similar vision as what today's autonomous cars resemble, but it was not until the 1980s that fully autonomous cars were prototyped and demonstrated. This was led by a number of groups and projects, including Carnegie Mellon University's Navlab and ALV and Bundeswehr University Munich's Eureka Prometheus Project. These were followed by a number of similar initiatives in industry and academia.

The next major initiatives were spurred by DARPA's Grand Challenges, which began by offering major cash incentives to research organizations capable of developing an autonomous car that could navigate more than 140 miles in the Mojave Desert. On the first occasion, a maximum of 8 miles over several hours was achieved. Three competitions were held, in 2004, 2005, and 2007. The latter was an *urban* challenge, with several successful entries, six finishing the course, and was won by the Tartan Racing team – a collaboration between Carnegie Mellon University and General Motors.

These challenges showed how much additional effort would be needed to realize the dream of fully autonomous cars. However, in the 2000s, a number of auto makers added certain autonomous capabilities to their cars, beginning with self-parking systems. Toyota, Lexus, Ford, and BMW were among those to offer this feature by the end of the 2000s. Since then, the major manufacturers have all included many more self-driving features, and technology companies like Uber, Waymo, and Tesla have all contributed to advancing the state-of-the-art towards full autonomy. This has led to several firsts, including the state of Nevada issuing a driver's license to a Google autonomous vehicle (2012), and the first deaths caused by self driving vehicles; in 2016, when a Tesla driver was killed while the vehicle was in *autopilot* mode, and in 2018, when an Uber killed a pedestrian. The latter has led to a lot of speculation as to whether a LIDAR blind spot or a software fault was to blame.

16.2.2 ENABLING TECHNOLOGIES

Autonomous cars have arrived thanks to advances in many complementary engineering disciplines. Local (i.e., on-board) decision making leverages advances in Machine Learning technology and associated algorithms to interpret and react to a variety of sensor data in close to real-time. The primary

---

[1] http://www-formal.stanford.edu/jmc/progress/cars/cars.html

sensors that are used to interpret the vehicle's immediate surroundings include Radar, LIDAR, and camera sensors to detect objects and orientation [185,186]. Route planning and location information are more often than not retrieved from the Internet based on destination and present location (e.g., GPS coordinates triangulated from the vehicle or occupant's cell phone taken together with inertial sensor readings [187]), trajectory, and speed. These parameters are often communicated back from the vehicle to an associated service (e.g., Uber) to enable tracking of the vehicle. This is often also very useful for logistics applications (Chapter 17).

A significant amount of computing power is typically required to enable an autonomous car. Recent advances in GPU technology, such as NVIDIA's DRIVE PX AI platform, designed specifically to enable deep learning in-vehicle, combined with continued improvements to cars' electronic control units (of which there are now hundreds per vehicle) and improved communications and standards, such as automotive Ethernet [188] and AUTOSAR [189], respectively, are rapidly simplifying the development of autonomous cars. Perhaps the other major technical improvements worth mentioning with regard to autonomous vehicles relate to battery design and recharging capabilities for Electric Vehicles. Tesla's Supercharger and Megacharger networks are prime examples. The ability to rapidly refuel an EV is essential to ensuring the technology's broad uptake. The interested reader will find a review in [190].

16.2.3 REGULATION, GOVERNANCE, AND ETHICS

Many states and countries have recently allowed testing of autonomous vehicles on public roads and highways, and many that have not yet done so have begun to make provisions or announced future dates. It is clear that there is an appetite for continued progress and testing, but recent fatalities have led many to begin asking more difficult questions about where the responsibility lies when autonomous vehicles crash [191–193]. Identifying the cause of the crash may be very difficult indeed, as hardware faults, sensor or controller failures, or software glitches or algorithmic problems, wherein the decision making occurs (i.e., a potentially ethical question [194]) could be to blame in whole, in part or a combination of these.

16.2.4 OTHER AUTONOMOUS PASSENGER VEHICLES

There are numerous examples of recent self-driving passenger vehicles from around the world, ranging in terms of application from low-speed autonomous "pods" in Milton Keynes, UK, under the UKAutodrive project[2], to passenger UAVs mooted for trial in the Middle East, where Dubai is promising to trial the Ehang 184. Such technologies as flying autonomous taxis continue to raise interesting practical problems, like air traffic management for aerial passenger drones (a problem that Uber is working on with NASA) and infrastructural questions for take-off and landing.

---

[2]http://www.ukautodrive.com

16.3 OTHER AUTONOMOUS SYSTEMS

There are a multitude of additional types of autonomous systems, including those used for passenger and freight haulage, exploration, remote monitoring and detection, and industrial manufacturing and processing. These cover air-, ground-, and marine-based systems. The following subsections briefly overview some interesting types of such autonomous systems, including their major use cases.

16.3.1 AUTONOMOUS RAIL

Passenger rail systems have become more automated since the Victoria line of London's Underground system operated a Grade of Automation 2 level, meaning that trains may move between stations automatically, but a driver must be present in the cab to control the doors, detect obstacles on the track, and handle emergency situations. There are two additional levels of automation, i.e., 3 (e.g., the Docklands Light Railway in London) and 4 (e.g., the Copenhagen Metro). At level 3, a staff member is usually on board to handle emergency situations, whereas at level 4, no staff member is needed to safely operate the train.

Autonomous rail systems are also feasible for industrial applications requiring haulage over long distances. For example, Rio Tinto expects its first autonomous trains to come online in 2018 as part of its Mine of the Future program, although much of the rail infrastructure in this case is privately operated (around 1500 km). It is expected that the trains can move faster and safer under autonomous operation than with a driver at the helm. This is in addition to its autonomous trucks, which have now been in operation for around a decade. All of these systems are tied together with increasingly sophisticated machinery and processing technology, ultimately allowing all operations to be overseen and managed remotely. Effectively, autonomous vehicles, machines, and sensors allow a unified view of 16 Rio Tinto mines controlled by operators more than a thousand kilometers away, with the connected systems generating multiple terabytes of data each minute. This is an excellent example of a system of Cyber-Physical Systems in industry, where remote monitoring and control of several heterogeneous subsystems allow the overall mining and processing operation to become a more streamlined and efficient macrosystem.

16.3.2 UNMANNED AERIAL SYSTEMS

UAVs have been in use for some time in defense applications. While automatic aviation has been around for many years, e.g., autopilot systems for airplanes, over the past decade there has been a significant rise in the use of UAVs in military contexts, particularly where surveillance and remote air strikes are needed. The majority of these systems require a remote pilot to guide the aircraft, although much autonomy applies.

Unmanned aerial systems have become very common in recent years as advances in lightweight consumer drones have led to lower prices for increasingly sophisticated aircraft. The initial explosion of these UAVs has seen a major uptake in the aerial photography applications, which allows innovation far beyond imagery in terms of remote inspection and monitoring for commercial reasons.

Their autonomous operation is still not recommended for industrial or commercial use in many regions, where regulation stipulates that they may not be usually (i.e., certain permissions are feasible to obtain from the aviation authorities) beyond visual line of sight, or exceed certain distance and altitude limits.

Such aircraft pave the way for innovations across a range of industries and application areas. For example, their use in conjunction with terrestrial sensors can allow for a range of automatic data gathering systems where infrastructure networks are not available to allow direct connectivity to the Internet. This is likely to enable a host of monitoring applications in dangerous and remote locations, such as in mining, civil infrastructure, and energy production systems [195,196].

16.3.3 UNMANNED AUTONOMOUS UNDERWATER VEHICLES AND SYSTEMS

Autonomous underwater vehicles (AUVs) have been in existence since the 1950s, when the first AUV was developed at the University of Washington. Exploration and, more recently, deep sea mining benefit greatly from the development of underwater autonomous robots. Recent examples of smaller underwater robots include the SoFi robotic fish developed at MIT [197] and the European SHOAL project, which aims to use artificially intelligent robotic fish to monitor pollution more efficiently[3].

16.4 INTELLIGENT INFRASTRUCTURE

Sensing technologies are beginning to pervade most infrastructure systems for the purposes of monitoring, maintenance, and general safety. The resulting sensor data is continuously fed to central servers maintained by authorities responsible for the infrastructure, where it is examined for evidence of damage and/or decay. Examples of these include civil infrastructure, such as roads and bridges [31], water and waste distribution and management systems [198], smart energy systems [199], and traffic management systems [200].

As data is processed and made available beyond its originally intended purpose, it may be fed to additional systems where it also has significant value. For example, warning data may be made available to vehicles about civil structures, such as a bridge, if any danger is identified through its monitoring system. This can lead to additional safety through in-advance warnings, regardless of whether the vehicle is autonomous or human-controlled. Another example is where data from one sector, e.g., a traffic management system, can be used in another, in this case energy infrastructure management for load management at the grid level. The benefits of sharing data between traditionally uncoupled systems has the potential to create new business opportunities in addition to more effectively delivering legacy services. Aside from the potential benefits this can create, it also creates the possibility for abuse, which must be carefully considered.

16.5 CONVERGENCE AND SYSTEMS OF CYBER-PHYSICAL SYSTEMS

It is clear that the fusion of sensors, actuators, computation, and communication technologies can enable any method of transport to become autonomous. Autonomous vehicles and the systems they comprise together with intelligent, connected infrastructures are representative of systems of Cyber-Physical Systems. While a car or aircraft is a quintessential example of a Cyber-Physical System,

[3]https://www.roboshoal.com

where thousands of autonomous vehicles are continuously interacting with one another and potentially innumerable cloud-based application servers, managing these new systems represents a significant challenge.

This is an emerging problem certain to attract a significant amount of research and development effort. There are several driving factors to consider when thinking about such systems. It is clear that there are ways to improve efficiency, reduce environmental stress, and meet societal demands where it comes to the transportation of people and goods using all available methods, i.e., air, water, and land. On the one hand, these systems can be considered critical to innovation with regard to societal sustainability, including security, reliability, and stability, but on the other hand it is essential to consider the same parameters when it comes to the development, implementation, and management of these systems. Many of the technologies needed to safely manage a proliferation of autonomous Cyber-Physical Systems are insufficiently mature.

As discussed in Chapter 6, security is perhaps the most important factor to consider when developing and deploying connected systems. The potential for unforeseen dependencies and weaknesses to emerge when considering the nature of many application areas, particularly those that concern critical infrastructures and safety critical applications like autonomous vehicles, should not be taken lightly.

16.6 CYBER-PHYSICAL SYSTEMS CHALLENGES AND OPPORTUNITIES

This section highlights some of the key challenges in the development of future Cyber-Physical Systems without being exhaustive. Many engineering disciplines come together in the development of complex Cyber-Physical Systems like autonomous vehicles. Development cycles have thus far tended to be slow and costly. A key discipline in the design of such systems is control engineering. To date, control engineers have been very slow to tolerate any kind of nondeterministic behavior, and the most recent developments have included developing formalized methods for software engineers to interact effectively with control engineers. This has resulted in the development of design contracts, which initially specify timing requirements [201,202].

This is particularly problematic when considering desirable technologies like wireless communications and energy harvesting systems. In both cases, it is difficult to make any type of guarantees that the system will behave consistently. Wireless communications often suffer from stochastic behaviors leading lost packets, and harvesting may be transient, thus leading to energy deficiencies for some types of applications. It is also of some concern in the research community that the mathematics of physical systems, which are dynamic, have not quite yet married with the discrete nature of the mathematics of computer systems. Nevertheless, this represents a very interesting research challenge for the community, which has yet to find a systematic and principled design methodology for these types of systems; although model-based design has increased in popularity in recent years [203].

It is worth reiterating that while security is of utmost importance to ensuring the trustworthiness, and consequently societal acceptance, of autonomous systems, special care must be taken to design them with their positive and negative externalities in mind. Deploying so many parallel autonomous systems, all intertwined using Internet technologies, has the potential to result in far reaching changes to society, not all of which may be desirable or predictable. Very little such research has been conducted to date, with limited learning becoming available through studying the consequences of the social media Internet age, where for example electorates have allegedly been manipulated and nuclear facilities (behind air gaps) destroyed by combining the technologies available with social engineering. When developing the next generation of connected systems, it will be even more complicated to determine liability when unwanted situations occur, perhaps the simplest of which could be determining how and why a self-driving car happened to crash and cause harm.

LOGISTICS

17

CHAPTER OUTLINE

17.1 INTRODUCTION

Logistics is a concept which is related to movement or transportation of tangible goods or assets and in order to understand the term and its related concepts it is useful to look at a typical enterprise organization which produces physical goods. The general organization of an enterprise may differ from the provided example, however the retail-related example is mainly motivated by the dominance of the Retail Logistics in the Logistics market [204]. The importance of Logistics (as Logistics value) in the overall enterprise operation (value) varies significantly and depends mainly on the cost of the end products. The value of Logistics in an enterprise producing low-cost goods is greater than the one for an enterprise producing high-cost items. However, both types of businesses use Logistics as an essential part of the operation. There are also other types of enterprises that do not involve the production of physical goods, such as the ones in the financial sector, but these are out of the scope of this chapter.

Enterprises that produce tangible goods typically collect all the raw material and constituent parts or components needed for the end product from suppliers (also enterprises themselves), manufacture the products/goods, potentially store the finished goods in local warehouses, and finally dispatch the finished goods further to be stored and sold to customers (see Figure 17.1) [205]. The finished goods dispatch involves potentially multiple distribution centers with or without storage warehouses and eventual retail centers where the goods are presented to end consumers or customers. In the modern world of online shopping retail centers are being slowly and steadily replaced with online shop fronts with optional stock warehouses.

One important observation is that Logistics Management (LM) and the related term Supply Chain Management (SCM) do not have well-defined and widely accepted definitions, since they are generic terms to describe certain enterprise operations which may differ from enterprise to enterprise. As a result, we will first attempt to provide some definitions, which will be used throughout the chapter.

In this context, the terms Supply Chain (SC) and SCM are essential to understanding since SCM is often confused with the term LM. An SC as defined by Aitken [206] is "... a network of connected and interdependent organizations mutually and cooperatively working together to control, manage and improve the flow of materials and information from suppliers to end users". It is important to note

Internet of Things. https://doi.org/10.1016/B978-0-12-814435-0.00030-4

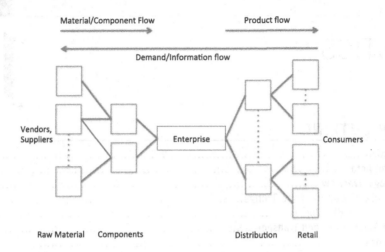

FIGURE 17.1

Typical enterprise relationships.

that not only the flow of physical materials, components or finished products is a concern in an SC but also the reverse flow of information from the consumers/customers towards the manufacturers and their suppliers. The first type of flow (physical material/components/finished goods) is typically called supply flow while the second type of flow of information is called demand flow. Ideally, manufacturing enterprises would operate in such a way so that these flows are in equilibrium, in other words, manufactured goods are produced at the appropriate time and with the right quantities so as to satisfy demand and optimize other parameters such as the overall cost of the finished products and time to market. This equilibrium is hard to achieve due to a continually changing demand which forces enterprises to take risks in predicting the demand with respect to time and quantities, and typically such risks are translated to incurred costs.

Christopher [205] defines SCM as "... the management of upstream and downstream relationships with suppliers and customers in order to deliver superior customer value at less cost to the SC as a whole". As a result, the goal of SCM is to operate the SC as close to an equilibrium as possible. Christopher [205] goes further to state that each enterprise, in general, is part of an SC which is the source of competitive advantage as now enterprises do not compete in the market *alone* but *along* with their SCs.

According to the glossary of terms [207] provided by the Council of Supply Chain Management Professionals (CSCMP)[1], Logistics is "... the process of planning, implementing, and controlling procedures for the efficient and effective transportation and storage of goods including services, and related information from the point of origin to the point of consumption for the purpose of conforming to customer requirements. This definition includes inbound, outbound, internal, and external movements." Further on, the glossary defines the term LM as part of the SCM and important activities are inbound

---

[1] http://cscmp.org

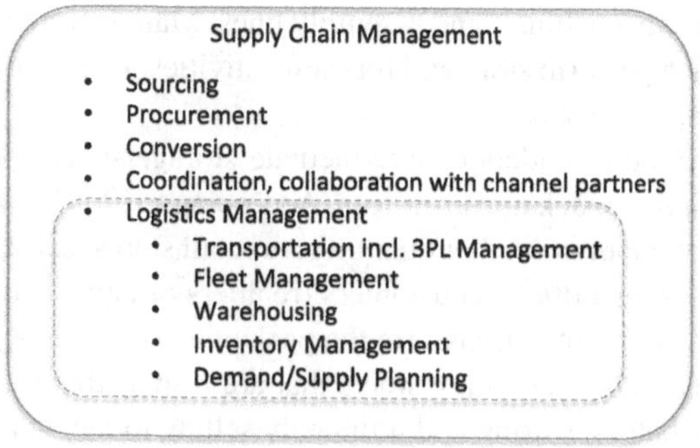

```
                  ┌─────────────────────────────────────────┐
                  │      Supply Chain Management              │
                  │   •  Sourcing                             │
                  │   •  Procurement                          │
                  │   •  Conversion                           │
                  │   •  Coordination, collaboration with channel partners │
                  │   •  Logistics Management                 │
                  │        •  Transportation incl. 3PL Management │
                  │        •  Fleet Management                │
                  │        •  Warehousing                     │
                  │        •  Inventory Management            │
                  │        •  Demand/Supply Planning          │
                  └─────────────────────────────────────────┘
```

FIGURE 17.2

Logistics Management and Supply Chain Management.

and outbound transportation management, fleet management, warehousing, materials handling, order fulfillment, logistics network design, inventory management, supply/demand planning, and management of third-party logistics (3PL) service providers. Revisiting SCM, the definition by the CSCMP glossary includes sourcing and procurement, conversion, and LM (Figure 17.2). According to the CSCMP glossary, the primary responsibility of SCM is "...linking major business functions and business processes within and across companies into a cohesive and high-performing business model". As previously stated an efficient SCM is the ultimate goal of an enterprise rather than the standalone efficiency. Some of the activities of the SCM such as procurement or conversion are mainly business related and do not touch upon the topic of the Internet of Things (IoT) while the LM activities and other SCM activities such as sourcing are more related to the physical world which IoT enables to monitor and control. As a result, this chapter focuses more on the LM rather than the whole SCM, in other words, how physical assets of an SC are flowing through the SC from suppliers to the manufacturing enterprises and towards the consumers and customers. In the process, other assets are used for facilitating this flow, such as transportation means (trucks, trains, planes, ships) and material, component, and finished suitable storage means (warehouses, distribution centers, retail storage, etc.).

17.2 ROLES AND ACTORS

The typical roles participating in an SC are the following:

- **Raw material providers**: These are the providers of raw materials needed for the manufacture of intermediate components of the finished product. Typical they own warehouses for storing the raw material, and the flow of goods is outbound.
- **Intermediate component providers**: These typically play the role of both manufacturing entities and suppliers. They typically own and operate inbound warehouses, manufacturing machines, and outbound warehouses.
- **Manufacturing enterprises**: Similar to intermediate component providers, they assemble raw materials and components into finished products. They own and operate inbound and outbound

warehouses, manufacturing machines, and assembly lines. Manufacturers may own and operate also distribution centers which perform store and forward activities, local warehousing, and repackaging of large shipments to smaller ones.

- **Distributors**: These are the providers of intermediate storage space for finished goods and store and forward and potentially repackaging services. According to [207] distributors purchase goods in large quantities and resell them in smaller quantities to retail stores, taking ownership of the finished goods and selling smaller quantities to customers (retailers or end customers). They typically own and operate warehouses which may transform the packaging of finished goods or groups.
- **Retailers**: The customer/consumer facing stores that are responsible for purchasing products from manufacturers and distributors, storing and ultimately selling to the end customers. Retailers could be physical or virtual (online stores). If retailers are large and span a large geographical area, they could own large distribution centers which potentially perform repackaging of goods. An example is a retail distribution center that (a) receives containers of finished goods full of pallets with boxes containing a certain quantity of a certain product and (b) sends containers or pallets with a distribution of product boxes to individual retail stores. The following operations are performed: (i) the containers are received and opened and the pallets are unloaded, (ii) the pallets are disaggregated and the product boxes are unloaded, (iii) new pallets, each for a target retail store, are loaded with the desired distribution of product boxes according the retail store order, and (iv) pallets are forwarded to the outbound loading docks for dispatch.
- **End customers/consumers/users**: The end customers that will purchase the finished products.
- **Transportation providers**: These are the enterprises that handle the transportation of raw materials, components, and finished goods to the manufacturers, distribution centers, retail stores, and end customers.

It is important to note that in a real SC not all roles exist, e.g., there are no distributors if the manufacturer dispatches all the finished goods directly to the retail stores or customers.

One observation is that the transportation, storage and warehousing, and inventory activities may be performed by manufacturers and retailers or outsourced to other parties, typically called Third (3rd) Party Logistics (3PL) providers. The definition of a 3PL according to the CSCMP is "...outsourcing the logistics operations of a company to a specialized company". Although this is a general definition, it captures the distinction between First Party Logistics (1PL) and Second Party Logistics (2PL), whose definitions are both not well defined. The term 1PL refers to the companies that have a "first order" interest in the products, in other words, the manufacturer or the retailer; 2PL concerns the companies that are typically used to carry these products, i.e., the different carriers by land, air, or sea. Typically in the early days of the development of the logistics market, manufacturers or retailers used to have dedicated contracts with the carrier for the transportation of products but the evolution of the market resulted in the appearance of third parties which managed on the behalf of the manufacturer or retailer the contracts with carriers, warehouse providers, etc. The main characteristic of a 3LP is that it is a different legal entity from the main interested parties (manufacturer, retailer). The different combination of transportation and warehousing services is so large that there is not a clear definition of 3PL apart from a general one. In the Logistics and SCM literature, there are also other types of roles, such as Fourth Party Logistics (4PL) and Fifth Party Logistics (5PL) providers, which are less well defined compared to 3PL and as a result are omitted from this book.

17.3 TECHNOLOGY OVERVIEW

The role of IoT is to enable cost-efficient monitoring and control of the real world "Things" or assets of interest. Logistics is also concerned with the physical movements of material, components, and finished products *as well as* the reverse flow of information. As a result, IoT could contribute to Logistics in automation of at least the reverse flow of information in an SC as well as affecting (to the extent possible) the normal flow of physical assets. This section starts with the identification of the "Things" or assets to be monitored and/or controlled followed by an outline of the main technologies applied in typical Logistics scenarios.

17.3.1 IDENTIFICATION OF THE THINGS

According to Baker et al. [204], a high percentage (more than 90%) of the total cost of Logistics, on a US or European level, is concentrated in three main activities: (a) transportation of goods (how goods are transferred from point A to point B), (b) inventory (which goods to stock, when to stock, and in which quantities), and (c) storage/warehousing (where to store goods in transit).

The potential assets depend heavily on the industry and specific industry sector, and as a result, the examples may seem arbitrary. The potential assets for a Logistics scenario are the following:

- Product-related assets:
  - **Raw materials**, for example minerals, wood, and fruit.
  - **Components**, such as electronic components, automotive components, steel parts, containers (e.g., bottles, cans), and packaging material (e.g., pallets, boxes).
  - **Finished products/goods**, such as computers, cars, bottled milk/water, and toys.
  - **Production machinery**, such as assembly lines and packaging machines.
- Storage/warehouse-related assets:
  - **Warehouse and storage facilities**, which include inbound and outbound unloading/loading areas, intermediate storage areas, unpacking and repacking areas, retail storage areas, and retail display areas.
  - **Warehouse machinery**, which includes unloading/loading machines such as forklifts, robotic trains, and repackaging machinery.
- Transportation-related assets:
  - **Means of transportation**, such as trucks, trains, airplanes, and ships.
  - **Transportation medium**, such as roads, train trucks, air, and sea.

In certain cases, human personnel may be considered as an asset which may need to be monitored in order to provide a safe working environment for humans, e.g., humans and hazardous areas in a chemical factory need to be identified and tracked in order to ensure that they are warned upon approaching an area that handles harmful gases.

17.3.2 MAIN TECHNOLOGIES

The type of assets involved in Logistics (such as transportation, warehousing, manufacturing) implies that the Logistics scenarios involve technologies and typical scenarios from other use cases, for example, fleet management, commercial building automation, and manufacturing. The core of the Logistics use cases lies in the physical flow of goods (or the lack of motion of goods, e.g., in warehouses) from suppliers to manufacturers, to retailers, and finally to customers. In other words, the core use cases are about the transformation of materials and components to finished products, the packaging/repackaging of goods, the transportation and storage of the finished products close to the customers, and finally the product sales ("track and trace"). There is a reverse flow of goods from the customer towards the retailer or manufacturer for returned items, but this case is similar to the normal flow (between the manufacturer and the customer) and is not the common case. As a result, this chapter will not cover any specific details of such a case due to the fact that the goal of returned items logistics is the same as the normal product logistics only with the roles reversed.

The SC is also about the reverse flow of information as stated before. At the beginning of the logistics era, the flow of information was also realized as a flow of paper documents from the different entities in the path. However paper documents were not standardized, information transfer was as fast (or as slow) as the product transfer, and information flow was typically unidirectional, i.e., not possible for communicating entities to interact. The advantage of today's communication infrastructure is that information flow could occur at much higher speed with electronic documents following standardized electronic formats, which are machine-readable.

One of the main technologies for enabling the reverse flow of information that captures the physical flow of material and products from suppliers to customers as well as the tracking of finished products is the GS1[2]-related technologies. GS1 is a global organization that focuses on standards for SCs and Business to Business information exchange for large industries such as Retail, Healthcare, and Transportation and Logistics. GS1 for example standardizes the product barcode or the Electronic Product Codes (EPCs).

The GS1 family of standards focuses on (a) **identifying** the products to be transported in a standardized way and standardizing the technologies of carrying such identification (e.g., barcodes and RFID tags), (b) **capturing** the identification information as the products flow in the SC either individually or in groups (e.g., pallets of goods) and associating the location, time, and individual product identification with a business context (e.g., a product moved from ordering to shipping to the customer), and finally (c) **sharing** the resulting information about the product flow from suppliers to manufacturers to customers among the involved logistics parties.

The rest of the IoT technologies, on the one hand, focus mainly on monitoring of several sensor modalities and control of applicable and relevant modalities. In other words, the rest of IoT technologies are about monitoring and potentially controlling a superset of possible assets compared to the GS1 main technologies which focus primarily on tracking and tracing product flows. However, as the set of assets or "Things" identified in LM (Section 17.3.1) is larger, the non-GS1 IoT technologies could be used to enrich the simple track and trace use cases.

The assets identified in Logistics are of two main categories:

1. The products, items, goods themselves which could be monitored with various sensor technologies in order to ensure product-specific quality (e.g., milk should not be exposed to temperatures outside a specific range) or handling. The monitoring of such assets contributes mainly to the end customer satisfaction. The use of actuation for such assets is limited but not excluded (e.g., refrigerated containers) and thus introducing sensor technologies for enhanced and enriched monitoring of goods is the main focus area of combining GS1 and non-GS1 IoT technologies.

[2]https://www.gs1.org

2. The relevant and involved assets for the transportation and storage of the products as identified in Section 17.3.1. The optimum and efficient utilization of such assets contributes to the cost-effectiveness and sustainability of the SC which in turn impacts the customer and the involved SC partners. The use of monitoring and actuation is equally important and applicable for monitoring and controlling such assets.

The use cases later in the chapter attempt to capture the use of both categories of assets since the combined monitoring of the whole SC typically enables better visibility and control of the chain than optimizing just parts of it. As a result, technologies such as Analytics and Machine Learning are extremely useful in at least monitoring and analyzing the performance of the SC end-to-end. For example, sensor data can reveal that a certain type of food is exposed to temperatures outside the normal range for some time and the result is that the expiration date might need to be updated dynamically. This could allow for some actuation to take place if the product packaging allows it (e.g., remotely configurable product details) but the impact on the efficient utilization of such products is limited. The higher impact could come from more radical actions such as the reconfiguration of the SC to push the products faster to the end customer, e.g., transporting the products through air transport instead of the keeping the original plan and transporting them through land transport. Typical SCs of businesses were formed with fixed and slow changing contracts and as a result, this kind of reconfiguration was not possible until recently with the introduction of smart contracts and blockchain technology (see Section 5.7).

17.4 EXAMPLE SCENARIO – FOOD TRANSPORT

There are numerous logistics use cases in the literature which are described in more or less detail. As a result, it was deemed better to describe a use case that includes the main logistics and IoT useful concepts in order to showcase the best of both worlds. The use case focuses on physical goods transport and more importantly food transport around the globe. The motivation is to showcase multiple modes of transport, the aggregation/disaggregation of items, boxes, pallets, containers, etc., and the transformation of raw materials to food items.

The general scenario is that a ready meal food company in a Country A uses raw ingredients locally produced as well as imported from other countries, prepares the specialty food items, packages them, refrigerates them, and dispatches them to the local or foreign supermarkets and restaurants. The use case focuses on the manufacturing, transportation, and distribution of these specialty food items.

The SC begins with the raw material (food ingredients and packaging) being transported to the food enterprise (Figure 17.3). The vegetables are transported from local producers and nearby countries by trucks while the protein ingredients, oils, and packaging are transported from further away by train transport. The raw ingredients are transported in big food containers which identify the type and quantity of ingredients (e.g., olive oil, 100 kg) while the packaging is transported in bundles of 100 items in different sizes. It is important that the raw material, as well as the prepared food, is transported in refrigerated containers, trucks or train wagons and that its exposure to environmental conditions is monitored using a related IoT sensing infrastructure. The raw ingredients are packaged in refrigerated containers equipped with humidity, temperature, and vibration sensors, which trigger the fridge or the freezer parts of the container if there is a need. When the raw ingredients are delivered to the food

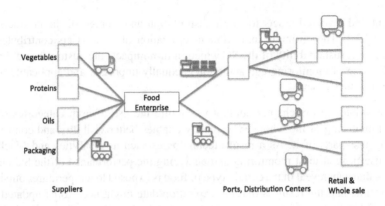

FIGURE 17.3

A food production and distribution use case.

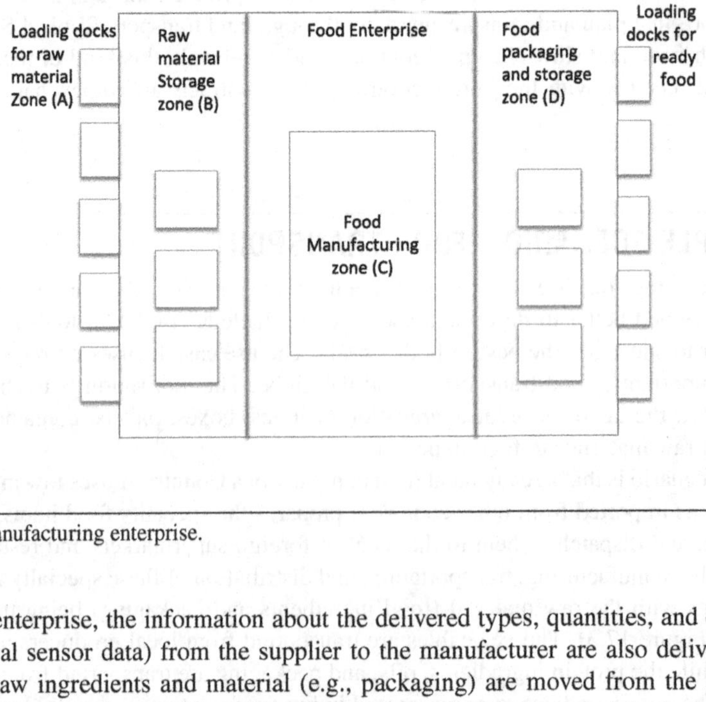

Inside the food manufacturing enterprise diagram with zones:

Loading docks for raw material Zone (A) | Raw material Storage zone (B) | Food Enterprise | Food packaging and storage zone (D) | Loading docks for ready food

Food Manufacturing zone (C)

FIGURE 17.4

Inside the food manufacturing enterprise.

manufacturing enterprise, the information about the delivered types, quantities, and condition history (mainly historical sensor data) from the supplier to the manufacturer are also delivered to the manufacturer. The raw ingredients and material (e.g., packaging) are moved from the loading docks to the storage zone (A), in Figure 17.4. The technologies used for this part of the logistic chain are the following:

- **Ordering**: The manufacturer orders from the different suppliers the desired type and amount of raw ingredients and specifies the desired environmental condition profile for the transport. In other words, the manufacturer specifies the desired temperature, humidity, and vibration value ranges within which the raw ingredients should be transported.

- **During transport**: Temperature, humidity, and vibration sensor data are collected every 10 minutes from the container of raw ingredients as well as the refrigerated container (bigger container or truck or refrigerator wagon). The data are stored locally in the ingredients' containers and remotely in the supplier IT system. The data to be stored in the supplier system are collected using wired and wireless communication technologies (Chapter 5). There is also local or remote computational logic such as edge computing (Chapter 5) or cloud computing that allows constant monitoring of the raw ingredients' conditions and sends potential alerts to the supplier if these conditions deviate from the required ones by the manufacturer.
- **Delivery**: The manufacturer retrieves the sensor data from the supplier IT system via specified APIs and checks the overall condition of the transported raw ingredients. If the conditions match the required ones upon ordering, the manufacturer instructs the transport responsible for beginning the transfer to ownership. The transport responsible transports the containers to the loading docks of the manufacturer (Zone A) and scans the containers with barcode readers or RFID readers (depending on which type of identification carriers the containers have), and the EPCIS event data are transported to the supplier EPCIS repository. The EPCIS data are also pushed to the manufacturer's EPCIS repository through Business to Business (B2B) EPCIS interfaces. The EPCIS repository of the manufacturer indicates that the expected location of the delivered goods is Zone A. The sensors of the packaging of the raw ingredients are now instructed to push the data to the manufacturer gateways and IT system as well as to continue storing the data locally.
- **Transport to Zone B**: The manufacturer moves the raw material to the storage area (Zone B) and scans the barcodes or the RFID tags of the packaging. The EPCIS repository assumes that the raw material is in Zone B. If the raw ingredients need refrigeration, they are transported and stored in refrigerated storage areas. The packaging (e.g., containers) of the raw ingredients continues to monitor the temperature, humidity, and vibration conditions and to push the data locally to the manufacturer IT system. The manufacturer also includes logic to detect deviations in the environmental conditions and dispatch alerts to personnel.

During the manufacturing process, raw ingredients are transported to the manufacturing zone C and the EPCIS repository is updated about the location of the raw ingredients. In cases when the manufacturing process stops unexpectedly or planned (e.g., the plant shuts down for the day) the raw ingredients return to the storage area (Zone B) and the relevant events are pushed to the EPCIS repository. An example event is that a container of vegetables returns to the fridge in Zone B with 50 kg less vegetables than the original capacity.

The manufacturing process results in food being produced in packages of 0.5 kg which contain a barcode or RFID which carries the information about the item. Each individual packaged item is transported to the packaging and storage Zone D. In Zone D individual items are aggregated in bigger boxes that contain an RFID which records this aggregation. An EPCIS aggregation event is also stored in the EPCIS repository. The items and boxes are now assumed to be located in Zone D. Since the food items are consumed both in the domestic and in the international market, the boxes are further aggregated into pallets (for local delivery) and multiple pallets into big containers to be shipped via train or ship to other countries. Pallets and big containers are assumed to contain temperature, humidity, and vibration sensors to monitor the conditions of the ready food items. The relevant information about the aggregation events of multiple boxes to pallets, multiple pallets to containers, and multiple containers loaded to a ship or a train are recorded to the EPCIS repository. Some of the pallets may be transported as they are via refrigerator trucks or refrigerated train wagons over short distances and delivered to a retail or wholesale enterprise. The big containers are transported to regional distribution centers (e.g., by train) where they are disaggregated, i.e., the pallets are taken out of the containers and stored locally. In a later stage, the pallets are transported to retail or wholesale enterprises where they are disaggregated further and eventually sold to retail or wholesale customers. The disaggregation, as well as the transportation of the items, boxes, pallets, and containers, generates relevant events that are stored to the relevant EP-CIS repositories of the transport companies, distribution centers, and retail/wholesale enterprises and through B2B interfaces some of these data are dispatched to the manufacturer. The retail and wholesale enterprises also get access to the sensor data of each individual item as it is disaggregated from a box, pallet, or container. Based on these data they can decide to return the items back to the manufacturer or inform the end consumer and let them make the decision.

17.5 CONCLUSIONS

This chapter outlined the logistics and SCM concepts, described the main roles and actors in a general setup, touched upon the main technologies used in a logistics scenario, and finally presented a simple use case which combined typical product tracking technologies such as RFID and barcodes and IoT monitoring and actuation technologies. Barcodes and RFIDs are carriers of identification information about products, and groups of products in several levels (boxes, pallets, containers, etc.). GS1 EPCIS contributes to the dissemination of transportation and transformation (e.g., a pallet is disaggregated to boxes of products) events within an enterprise and between businesses. However, the GS1 technologies are used for monitoring only the movement and not the condition of products. IoT technologies such as sensors on products, boxes, pallets, containers, etc. have the potential to allow more fine-grained condition monitoring for sensitive products such as food and, if available, a local action (e.g., activation of refrigeration) or the dispatch of a local and early warning to a human for further action. However, these two types of technologies (identification/tracking and IoT sensing and actuation) are not integrated low in the system but rather high in the business process level. It is expected that with the proliferation of use cases such as food transport the need for standardized solutions that combine these two types of technologies will drive a closer integration of them in unified standards.

CONCLUSIONS AND LOOKING AHEAD

18

Internet of Things (IoT) solutions addressing a wide range of different problems and objectives have taken off dramatically over the past few years due to both technology maturity and an expanding market. For instance, Wireless Sensor Network technologies that were in research laboratories a decade ago are today affordable, even at consumer-level costs, and therefore increasingly embedded in several mass low-cost solutions. IoT is also flourishing by increasingly integrating in and interacting with a multitude of systems, technologies, and people in various settings involving industry, enterprises, consumers, and society at large. Today, we are rapidly moving beyond bespoke proprietary solutions tailored for very specific problems, and we already build upon reusable and more general purpose infrastructures and tools, referring to them as IoT, IIoT, Cyber-Physical Systems, and so on.

In the same way that the PC revolutionized and democratized access to computing, these tools can now be used for real deployment by hobbyists to not only connect sensors to the Internet and the Web, but utilize sophisticated software, e.g., doing computer vision or robot control on low-cost Raspberry Pi. In addition to this maker culture, industry after industry has embarked on the Fourth Industrial Revolution [208] to capitalize on the IoT opportunities ahead; see Chapter 1.

We now witness a remarkable acceleration and paradigm shift on IoT, as illustrated in Figure 18.1. The technology evolution is driven from primarily the two perspectives of innovation and scale. Innovation now implies new technologies enabling new capabilities, e.g., algorithms for AI enabling automated operations. Scale implies cost-effective massive deployments and wide adoption, for instance, cheap sensors and low-cost everywhere available connectivity from mobile networks. As can be seen, the initial efforts focused on *connecting devices* as a technology foundation. It included the necessary hardware and software for sensors and actuator devices as well as developing and deploying network technologies and infrastructures, thus connecting devices to backend hosted applications. This enabled data collection and visualization in order to get visibility on real-world processes at fine-grained levels. It allowed remote measurement of electricity consumption in households and condition monitoring of industrial machines. There has been a rapid consolidation of technologies such as networking and device protocols, and the previously high fragmentation is decreasing with common choices across industries, including the use of IP and Web. Technology development will obviously continue to grow at an increasing pace. Embedded computing and sensing will be further miniaturized and reduced in cost, and we will see scaled and ever-increasing instrumentation of these embedded technologies in the environment around us and in the objects and goods we own and use daily.

As device-centric IoT is maturing, the next step is already underway. Here the focus is on more advanced applications that take advantage of the increasing availability of different types of IoT data. Solutions focus on various aspects of analytics, such as anomaly detection and prediction and certain capabilities of AI. Example applications include condition-based monitoring and Predictive Maintenance. Also, various types of autonomous operations of both decentralized and distributed systems are

Internet of Things. https://doi.org/10.1016/B978-0-12-814435-0.00031-6

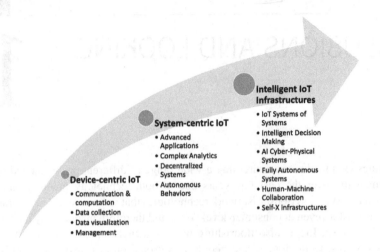

FIGURE 18.1

Evolution of IoT technology capabilities.

explored and developed. Here we see evolved industrial robots, automated process industry control, and different degrees of autonomous vehicles. What is still a characteristic is the focus on specific integrated systems of Information Technology and Operational Technology involving a finite set of tasks in a specific context such as discrete manufacturing.

However, technology-wise, we are still at the beginning rather than reaching maturity. With the increase in practical applications of Machine Intelligence, e.g., integration of control systems logic with Machine Learning and other AI technologies, new opportunities are ahead. As different IoT systems are becoming more widely deployed, these systems will form systems of systems for collaborative efforts. Data and information will increasingly be shared across systems and across domain boundaries. Semantic technologies will for this purpose be of paramount importance, as well as provenance and Quality of Information. Automation will require collaboration between different Machine Intelligence-powered systems, but also increasingly calling for seamless human–machine collaboration. This will also enable intelligent infrastructures that are adaptive to context, self-learning, and self-optimizing.

Overall, significant innovations are expected by the mere interactions among the different machines and their environments at large scale, e.g., at the smart city level or in global business networks. In addition, as IoT technology capabilities are expanding and being adopted across a number of industries, this will bring transformation of both business processes and business models; see Figure 18.2.

What has already taken place across several industries is the use of IoT to increase operational efficiency internally in organizations, such as cost reduction, increased productivity, and better utilization of expensive machinery. What today is a focus on point problems and isolated business cases is already growing into value creation and innovation, and this will increasingly involve and engage people, consumers, and enterprises. Increasingly, companies are exploring new products and new services, many times emerging as an adjacent business model by evolving the existing offering. An example of a new business model for an existing product is to sell the *use* of the product instead of the product itself. The customer then pays per use, but the product, for instance a household appliance, is not owned by

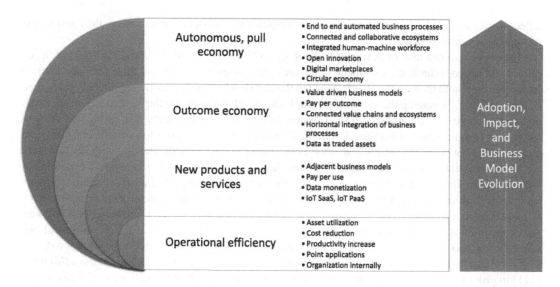

Autonomous, pull economy	• End to end automated business processes • Connected and collaborative ecosystems • Integrated human-machine workforce • Open innovation • Digital marketplaces • Circular economy	Adoption, Impact, and Business Model Evolution
Outcome economy	• Value driven business models • Pay per outcome • Connected value chains and ecosystems • Horizontal integration of business processes • Data as traded assets	
New products and services	• Adjacent business models • Pay per use • Data monetization • IoT SaaS, IoT PaaS	
Operational efficiency	• Asset utilization • Cost reduction • Productivity increase • Point applications • Organization internally	

FIGURE 18.2

IoT-enabled transformation.

the user but by a third party that could be the manufacturer of the product. This model is enabled by IoT that can monitor use and also ensure that the appliance is operational at all times. IoT solutions themselves also undergo this transformation, i.e., instead of selling hardware and software for an IoT application, it can be delivered in various XaaS models, like an IoT SaaS or IoT PaaS.

The next step is towards an outcome economy (see [3]), where the business model has shifted from *pay per use* to *pay per outcome*, i.e., it is rather the value offered than the use of a product that drives the business. Here we also see how actors get connected across value chains including how data and information are flowing horizontally and business processes get integrated across organizations. We already see some information marketplaces emerging, and that is also one important step towards IoT and outcome-based economies, where the delivered measurable results and the value they entail are the driving force. A further step is a fully autonomous economy that is self-operational. The current industrial model is centered around anticipating a market demand and to plan and execute accordingly, what sometimes is referred to as a "push economy". In a fully digital economy with all actors and markets being online as well as business processes connected and automated, one can talk about a "pull economy" [209], where products and services can be fully and flexibly tailored and assembled on the fly based on individual needs.

Looking beyond technology and business toward society at large, needs and innovation building on IoT will not decrease but instead grow in importance and magnitude as global issues become increasingly urgent. Handling scarcity of natural resources, reducing the impact on the climate and environment, and improving living situations due to increasing urbanization through smart city developments are big issues, but it is now within reach to apply technology to properly manage these issues with the engagement of people and businesses. One prospect to address these bigger issues is the circular economy; see [210]. The circular economy is a regenerative system that minimizes the consumption of finite resources and waste.

The above outlined dual evolution of technology and business will drive horizontal reuse of the deployed embedded technologies, connecting the multitude of small devices and things into the Internet. In combination with the trends towards data and the cloud, AI technologies are significant components of the future IoT. Managing the increasing complexity of information and the need to automate actions and control of real-world assets will require new technology developments that go beyond what we today refer to as analytics or Machine Learning. Hence we inexorably move towards the creation of, interconnection between, and interaction among large ecosystems, that will interface with the real world via IoT and mediate between the stakeholders and the IoT in a value-driven way that will benefit all involved parties. The progress, however, will also bring new challenges, especially pertaining to privacy, security, and ethics, since now the machines will move from passive data generators to proactive actors within business and society, becoming on par with humans. Such global high-impact effects will need the cooperation among the different stakeholders, as IoT should be treated as a socio-technical phenomenon and therefore its implications evade technology, society, law, and ethics. Businesses will need to train their employees to new IoT technologies, e.g., via Massive Open Online Courses (MOOCs) [211] in order to capitalize on the benefits offered. One can claim that we still are at the dawn of an era, especially when we attempt to grasp the socio-technical implications.

For us, the current evolution and steps in the deployment of IoT is just the beginning of a truly connected, intelligent, and sustainable world that will emerge in the coming decade.

ETSI M2M

CHAPTER OUTLINE

Appendix A contains a summary of the ETSI Machine to Machine architecture and interfaces. Since the architecture and interface specifications are merged to oneM2M specifications and evolved since the conclusion of ETSI work in 2012, the material of this chapter is only of historical importance.

A.1 INTRODUCTION

ETSI in 2009 formed a Technical Committee (TC) on M2M topics aimed at producing a set of standards for communication among machines from an end-to-end viewpoint. The Technical Committee consisted of representatives from telecom network operators, equipment vendors, administrations, research bodies, and specialist companies. The ETSI M2M specifications are based on specifications from ETSI as well as other standardization bodies such as the Internet Engineering Task Force (IETF), 3rd Generation Partnership Project (3GPP), Open Mobile Alliance (OMA), and Broadband Forum (BBF). ETSI M2M produced the first release of the M2M standards in early 2012, while in the middle of 2012 seven of the leading Information and Communications Technology (ICT) standards organizations (ARIB, TTC, ATIS, TIA, CCSA, ETSI, TTA) formed a global organization called oneM2M Partnership Project (oneM2M) in order to develop M2M specifications, promote the M2M business, and ensure the global functionality of M2M systems. The ETSI M2M work was concluded after the formation of oneM2M and therefore the material in this appendix is included just for completeness.

A.1.1 ETSI M2M HIGH-LEVEL ARCHITECTURE

Figure A.1 shows the high-level ETSI M2M architecture. This high-level architecture is a combination of both a functional and a topological view showing some Functional Groups clearly associated with pieces of physical infrastructure (e.g., M2M Devices, Gateways) while other Functional Groups lack specific topological placement. There are two main domains, a network domain and a device and gateway domain. The boundary between these conceptually separated domains is the topological border between the physical devices and gateways and the physical communication infrastructure (Access network).

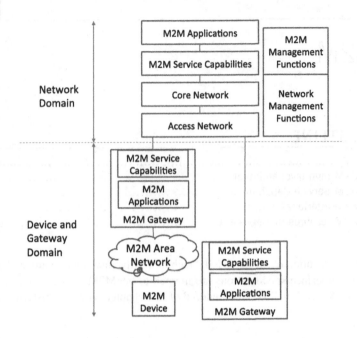

FIGURE A.1

ETSI M2M High-Level Architecture (redrawn from ETSI [32]). Copyright European Telecommunications Standards Institute 2013. Further use, modification, copy and/or distribution is strictly prohibited.

The Device and Gateway Domain contains the following functional/topological entities:

- **M2M Device**: This is the device of interest for an M2M scenario, for example, a device with a temperature sensor. An M2M Device contains M2M Applications and M2M Service Capabilities. An M2M device connects to the Network Domain either directly or through an M2M Gateway:
    - Direct connection: The M2M Device is capable of performing registration, authentication, authorization, management, and provisioning to the Network Domain. Direct connection also means that the M2M device contains the appropriate Physical Layer to be able to communicate with the Access Network.
    - Through one or more M2M Gateway: This is the case when the M2M device does not have the appropriate Physical Layer, compatible with the Access Network technology, and therefore it needs a network domain proxy. Moreover, a number of M2M devices may form their own local M2M Area Network that typically employs a different networking technology from the Access Network. The M2M Gateway acts as a proxy for the Network Domain and performs the procedures of authentication, authorization, management, and provisioning. An M2M Device could connect through multiple M2M Gateways.
- **M2M Area Network**: This is typically a Local Area Network (LAN) or a Personal Area Network (PAN) and provides connectivity between M2M Devices and M2M Gateways. Typical networking technologies are IEEE 802.15.1 (Bluetooth), IEEE 802.15.4 (ZigBee, IETF 6LoWPAN/RoLL/CoRE), MBUS, KNX (wired or wireless), PLC, etc.

- **M2M Gateway**: The device that provides connectivity for M2M Devices in an M2M Area Network towards the Network Domain. The M2M Gateway contains M2M Applications and M2M Service Capabilities. The M2M Gateway may also provide services to other legacy devices that are not visible to the Network Domain.

The Network Domain contains the following functional/topological entities:

- **Access Network**: This is the network that allows the devices in the Device and Gateway Domain to communicate with the Core Network. Example Access Network Technologies are fixed (xDSL, HFC) and wireless (Satellite, GERAN, UTRAN, E-UTRAN WLAN, WiMAX).
- **Core Network**: Examples of Core Networks are 3GPP Core Network and ETSI TISPAN Core Network. It provides the following functions:
  - IP connectivity.
  - Service and Network control.
  - Interconnection with other networks.
  - Roaming.
- **M2M Service Capabilities**: These are functions exposed to different M2M Applications through a set of open interfaces. These functions use underlying Core Network functions, and their objective is to abstract the network functions for the sake of simpler applications. More details about the specific service capabilities are provided later in the appendix.
- **M2M Applications**: These are the specific M2M applications (e.g., smart metering) that utilize the M2M Service Capabilities through the open interfaces.
- **Network Management Functions**: These are all the necessary functions to manage the Access and Core Network (e.g., Provisioning, Fault Management, etc.).
- **M2M Management Functions**: These are the necessary functions required to manage the M2M Service Capabilities on the Network Domain while the management of an M2M Device or Gateway is performed by specific M2M Service Capabilities. There are two M2M Management functions:
  - M2M Service Bootstrap Function (MSBF): The MSBF facilitates the bootstrapping of permanent M2M service layer security credentials in the M2M Device or Gateway and the M2M Service Capabilities in the Network Domain. In the Network Service Capabilities Layer, the Bootstrap procedures perform, among other procedures, provisioning of an M2M Root Key (secret key) to the M2M Device or Gateway and the M2M Authentication Server (MAS).
  - MAS: This is the safe Execution Environment where permanent security credentials such as the M2M Root Key are stored. Any security credentials established on the M2M Device or Gateway are stored in a secure environment such as a trusted platform module.

An important observation regarding the ETSI M2M functional architecture is that it focuses on the high-level specification of functionalities within the M2M Service Capabilities Functional Groups and the open interfaces between the most relevant entities, while avoiding specifying in detail the internals of M2M Service Capabilities. However, the interfaces are specified in different levels of detail, from abstract to a specific mapping of an interface to a specific protocol (e.g., HTTP [212], IETF CoAP [213]). The most relevant entities in the ETSI M2M architecture are the M2M Nodes and M2M Applications. An M2M Node can be a Device M2M, Gateway M2M, or Network M2M Node (Figure A.2). An M2M Node is a logical representation of the functions on an M2M Device, Gateway, and Network that should at least include a Service Capability Layer (SCL) Functional Group.

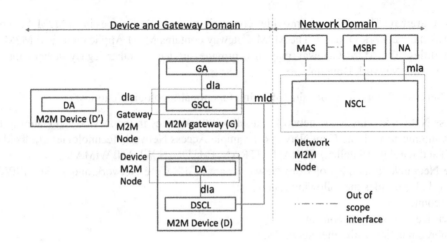

FIGURE A.2

M2M Service Capabilities, M2M Nodes, and Open Interfaces [32].

An M2M Application is the main application logic that uses the Service Capabilities to achieve the M2M system requirements. The application logic can be deployed on a Device (Device Application, DA), Gateway (Gateway Application, GA) or Network (Network Application, NA). The SCL is a collection of functions that are exposed through the open interfaces or reference points mIa, dIa, and mId [214]. Because the main topological entities that SCL can be deployed on are the Device, Gateway, and Network Domains, there are three types of SCL: DSCL (Device Service Capabilities Layer), GSCL (Gateway Service Capabilities Layer), and NSCL (Network Service Capabilities Layer). SCL functions utilize underlying networking capabilities through technology-specific interfaces. For example, an NSCL using a 3GPP type of access network uses 3GPP communication service interfaces. The ETSI M2M Service Capabilities are recommendations of Functional Groups for building SCLs, but their implementation is not mandatory, while the implementation of the interfaces mIa, dIa, and mId is mandatory for a compliant system. It is worth repeating that from the point of view of the ETSI M2M architecture, an M2M device can be either capable of supporting the mId interface (towards the NSCL) or the dIa interface (towards the GSCL). The specification actually distinguishes these two types of devices, i.e., device D and device D' (D prime), respectively.

A.1.2 ETSI M2M SERVICE CAPABILITIES

All the possible Service Capabilities (where "x" is N(etwork), G(ateway), and D(evice)) are shown in Figure A.3.

1. Application Enablement (xAE). The xAE Service Capability is an application facing functionality and typically provides the implementation of the respective interface: NAE implements the mIa interface and the GAE and DAE implement the dIa interface. The xAE includes registration of applications (xA) to the respective xSCL; for example, a Network Application towards the NSCL. In certain configurations xAE enables xAs to exchange messages to each other; for example, multiple Device Applications associated with the same M2M Gateway can exchange messages through

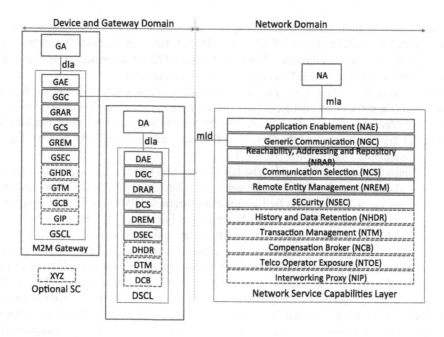

FIGURE A.3

M2M Capabilities for different M2M Nodes.

the GAE. In certain configurations security operations such as authentication and authorization of applications are also performed by xAE.

2. Generic Communication (xGC). The NGC is the single point of contact for communication towards the GSCL and DSCL. It provides transport session establishment and negotiation of security mechanisms, potentially secure transmission of messages, and reporting of errors such as transmission errors. The GSC/DSC is the single point of contact for communication with the NSCL, and they both perform similar operations to the NGC (e.g., secure message transmissions to NSCL). The GSC performs a few more functions, such as relaying of messages to/from NSCL from/to other SCs in the GSCL, and handles name resolution for the requests within the M2M Area Network.

3. Reachability, Addressing, and Repository (xRAR). This is one of the main service capabilities of the ETSI M2M architecture. The NRAR hosts mappings of M2M Device and Gateway names to reachability information (routable address information such as IP address and reachability status of the device such as up or down) and scheduling information relating to reachability, such as whether an M2M Device is reachable between 10 and 11 o'clock. It provides group management (creation/update/deletion) for groups of M2M Devices and Gateways, stores application (DA, GA, NA) data, and manages subscriptions to these data, stores registration information for NA, GSCL, and DSCL, and manages events (subscription notifications). The GRAR provides similar functionality to the NRAR, such as maintaining mappings of the names of M2M Devices or groups to reachability information (routable addresses, reachability status, and reachability scheduling), storing DA, GA, NSCL registration information, storing DA, GA, NA, GSCL, NSCL data and

managing subscriptions about them, managing groups of M2M Devices, and managing events. Similar to NRAR and GRAR, the DRAR stores DA, GA, NA, DSCL, and NSCL data and manages subscriptions about these data, stores DA registration and NSCL information, and provides group management for groups of M2M Devices and event management.

4. Communication Selection (xCS): This capability allows each xSCL to select the best possible communication network when there is more than one choice or when the current choice becomes unavailable due to communication errors. The NCS provides such a selection mechanism based on policies for reaching an M2M Device or Gateway, while the GCS/DCS provides a similar selection mechanism for reaching the NSCL.

5. Remote Entity Management (xREM). The NREM provides management capabilities such as Configuration Management (CM) for M2M Devices and Gateways (e.g., installs management objects in device and gateways), collects performance management (PM) and Fault Management (FM) data and provides them to NAs or M2M Management Functions, and performs Device Management to M2M Devices and Gateways such as firmware and software (application, SCL software) updates, device configuration, and M2M Area Network configuration. The GREM acts as a management client for performing management operations to devices using the DREM and a remote proxy for NREM to perform management operations to M2M Devices in the M2M Area Network. Examples of proxy operations are mediation of NREM-initiated software updates and handling management data flows from NREM to sleeping M2M Devices. The DREM provides the CM, PM, and FM counterpart on the device (e.g., start collecting radio link performance data) and provides the device-side software and firmware update support.

6. SECurity (xSEC). These capabilities provide security mechanisms such as M2M Service Bootstrap, key management, mutual authentication, key agreement (NSEC performs mutual authentication and key agreement while the GSEC and DESC initiate the procedures) and potential platform integrity mechanisms.

7. History and Data Retention (xHDR). The xHDR capabilities are optional capabilities, in other words, they are deployed when required by operator policies. These capabilities provide data retention support to other xSCL capabilities (which data to retain) as well as messages exchanged over the respective reference points.

8. Transaction Management (xTM). This set of capabilities is optional and provides support for atomic transactions of multiple operations. An atomic transaction involves three steps: (a) propagation of a request to a number of recipients, (b) collection of responses, and (c) commitment or roll back whether all the transactions successfully completed or not.

9. Compensation Broker (xCB). This capability is optional and provides support for brokering M2M-related requests and compensation between a Customer and a Service Provider. In this context a Customer and a Service Provider is an M2M Application.

10. Telco Operator Exposure (NTOE). This is also an optional capability and provides exposure of the Core Network service offered by a Telecom Network Operator.

11. Interworking Proxy (xIP). This capability is an optional capability and provides mechanisms for connecting non-ETSI M2M Devices and Gateways to ETSI SCLs. NIP provides mechanisms for non-ETSI M2M Devices and Gateways to connect to NSCL while GIP provides the functionality for noncompliant M2M Devices to connect to GSCL via the reference point dIa, and the DIP provides the necessary mechanisms to connect noncompliant devices to DSCL via the dIa reference point.

A.1.3 ETSI M2M INTERFACES

The main interfaces mIa, dIa, and mId [214] can be briefly described as follows:

- mIa: This is the interface between a Network Application and the Network Service Capabilities Layer (NSCL). The procedures supported by this interface are (among others) registration of a Network Application to the NSCL, request to read/write information to NSCL, GSCL, or DSCL, request for Device Management actions (e.g., software updates), and subscription to and notification of specific events.
- dIa: This is the interface between a Device Application and (D/G)SCL or a Gateway Application and the GSCL. The procedures supported by this interface are (among others) registration of a Device/Gateway Application to the GSCL, registration of a Device Application to the DSCL, request to read/write information to NSCL, GSCL, or DSCL, and subscription to and notification of specific events.
- mId: This is the interface between the NSCL and the GSCL or the DSCL. The procedures supported by this interface are (among others) registration of a Device/Gateway SCL to the NSCL, request to read/write information to NSCL, GSCL, or DSCL, and subscription to and notification of specific events.

A.1.4 ETSI M2M RESOURCE MANAGEMENT

The ETSI M2M architecture assumes that applications (DA, GA, NA) exchange information with SCLs by performing Create/Read/Update/Delete (CRUD) operations on a number of Resources following the RESTful (Representational State Transfer) architecture paradigm [212]. One of the principles of this paradigm is that representations of uniquely addressed resources are transferred from the entity that hosts these resources to the requesting entity. In the ETSI M2M architecture, all the state information maintained in the SCLs is modeled as a resource structure that architectural entities operate on. A very simplified view of the resource structure is a collection of containers of information structured in hierarchical manner following a corresponding hierarchy of a unique naming structure. In addition to the CRUD operations, ETSI M2M defines two more operations: NOTIFY and EXECUTE. The NOTIFY operation is triggered upon a change in the representation of a resource and results in a notification sent to the entity that originally subscribed to monitor changes to the resource in question. This operation is not an orthogonal operation to the CRUD set, but can be implemented by an UPDATE operation from the resource host towards the requesting entity. The EXECUTE operation is not orthogonal either, but can be implemented by an UPDATE operation with no parameters from the requesting entity to a specific resource. When a requesting entity issues an EXECUTE operation towards a specific resource, the specific resource executes a specific task.

The following example in Figure A.4 demonstrates how an ETSI M2M entity communicates with another entity using the CRUD and NOTIFY operations. Assume that a DA is programmed to send a sensor measurement to a Network Application (NA). The DA using the DSCL updates the representation of a specific resource (Ra) residing on the NSCL (steps 1 and 2 in Figure A.4). The NA has configured the NSCL to be notified when the specific resource is updated, in which case the NA reads the updated representation (steps 4 and 5, Figure A.4).

The root of the hierarchical resource tree is the <sclBase> resource and contains all the other resources hosted by the SCL. The root has a unique identifier. In case the RESTful architecture is implemented in a real system by using web resources, the <sclBase> has an absolute Universal

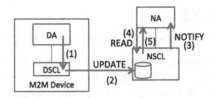

FIGURE A.4

Communication between DA and NA using the SCLs.

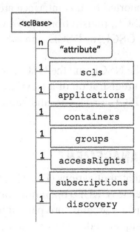

FIGURE A.5

The top-level structure of the sclBase resource (redrawn from ETSI [32]). Copyright European Telecommunications Standards Institute 2013. Further use, modification, copy and/or distribution is strictly prohibited.

Resource Identifier (URI), for example, "`http://m2m.operator1.com/some/path/to/base`". The top-level structure of the <sclBase> resource is shown in Figure A.5.

The different fonts used in this figure denote different information semantics. A term between the symbols "<" and ">" denotes an arbitrary resource name; for example, <sclBase> in Figure A.5. A term within quotes (" ") denotes a placeholder for one or more fixed names. In this specific case, "attribute" represents a member of a fixed list of attributes for the resource <sclBase>. A term annotated in `Courier New` font such as `scls` denotes a literal resource name used by the specification as is. The <sclBase> is structured as a tree with different branches, each of which is annotated with a designation of its cardinality. For example, the <sclBase> contains n attributes, one `scls` resource, one `applications` resource, etc. For more information about the resource structure of ETSI M2M specification, please refer to [32].

The ETSI M2M specification also describes an HTTP and a CoAP [212,213] binding for the RESTful resources stored in the SCLs, as well as for the implementation of the mId interface.

Abbreviations

1PL	1st Party Logistics
2PL	2nd Party Logistics
3GPP	Third (3rd) Generation Partnership Project
3PL	3rd Party Logistics
4PL	4th Party Logistics
5PL	5th Party Logistics
6LoRH	6LoWPAN Routing Header
6LoWPAN	IPv6 over Lower Power Wireless Personal Area Networks
6top	6TiSCH Operational sublayer
AAA	Authentication, Authorization, Accounting
ACE	Authentication and Authorization for Constrained Environments
ACM	Association of Computer Machinery
ADC	Analog-to-Digital Converter
AECC	Automotive Edge Computing Consortium
AES	Advanced Encryption Standard
AI	Artificial Intelligence
AIOTI	Alliance for the IoT Innovation
ALG	Application Layer Gateway
ANSI	American National Standards Institute
API	Application Programming Interface
ARIB	Association of Radio Industries and Businesses
ARIMA	Autoregressive Integrated Moving Average
ARM	Architectural Reference Model
AS	Authorization Server
AT	ATtention
ATIS	Alliance for Telecommunications Industry Solutions
ATM	Automated Teller Machines
B2B	Business to Business
B2B2C	Business to Business to Consumer
B2C	Business to Consumer
BACnet	Building Automation and Control network
BAN	Body Area Network
BAS	Building Automation System
BBF	Broadband Forum
BEMI	Bi-directional Energy Management Interface
BI	Business Intelligence
BLE	Bluetooth Low Energy
BoF	Birds of a Feather
BOM	Bill of Materials
BPMN	Business Process Model and Notation
BPSK	Binary Phase Shift Keying
BSS	Business Support System
BT	Bluetooth
BTLE	Bluetooth Low Energy
BTMesh	Bluetooth Mesh
CAN	Control Area Network
CAPEX	Capital Expenditures

CAT	Catalog service
CBC-MAC	Cipher Block Chaining Message Authentication Code
CBC	Cipher Block Chaining
CBOR	Concise Binary Object Representation
CBV	Core Business Vocabulary
CCM	Counter with CBC-MAC (Cipher Block Chaining – Message Authentication Code)
CCSA	China Communications Standards Association
CD	Committee Draft
CDN	Content Delivery Network
CENELEC	European Committee for Electrotechnical Standardization
CEP	Complex Event Processing
CFB	Cipher Feedback mode
CIM	Context Information Model
CIPART	Cloud Intelligent Protection At Run-Time
CM	Conceptual Model
CM	Configuration Management
CoAP	Constrained Application Protocol
CoMI	CoAP Management Interface
CONET	Cooperating Objects
CoRE	Constrained RESTful Environments
COSE	CBOR Object Signing and Encryption
CPE	Customer-Premises Equipment
CPS	Cyber-Physical Systems
CPSPWG	Cyber-Physical Systems Public Working Group
CPU	Central Processing Unit
CRISP-DM	Cross Industry Standard Process for Data Mining
CRM	Customer Relationship Management
CRUD	Create/Retrieve/Update/Delete
CSCMP	Council of Supply Chain Management Professionals
CSS	Chirp Spread Spectrum
CTR	Counter mode
CUD	Create/Update/Delete
CWMP	CPE WAN Management Protocol
DA	Device Application
DAE	see xAE for x=(D)evice
DAG	Directed Acyclic Graph
DAO	Destination Advertisement Object
DARPA	Defense Advanced Research Projects Agency
DATE	Design Automation and Test in Europe
DB	Database
DC	Data Center
DCS	Distributed Control System
DDM	Device and Data Management
DECT	Digital Enhanced Cordless Telecommunications
DG	Distributed Generation
DH	Diffie–Hellman
DHCP	Dynamic Host Configuration Protocol
DIO	DAG Information Object
DIP	see xIP for x=(D)evice
DIS	DAG Information Solicitation
DM	Data Management
DM	Data Mining
DM	Device Management

DNCP	Distributed Node Consensus Protocol
DNS	Domain Name System
DODAG	Destination Oriented Directed Acyclic Graph
DoS	Denial of Service
DPWS	Device Profile for Web Services
DRAR	see xRAR for x=(D)evice
DREM	see xREM for x=(D)evice
DSA	Digital Signature Algorithm
DSC	see xSC for x=(D)evice
DSCL	Device Service Capabilities Layer
DSL	Digital Subscriber Line
DSSS	Direct Sequence Spread Spectrum
DTLS	UDP Transport Layer Security
E-UTRAN	Evolved Universal Mobile Telecommunications System (UMTS) Terrestrial Radio Access Network
EC-GSM-IoT	Extended Coverage GSM IoT
EC-GSM	Extended Coverage GSM
EC	European Commission
ECB	Electronic Codebook
ECC	Elliptic Curve Cryptography
ECDH	Elliptic Curve Diffie–Hellman
ECDSA	Elliptic Curve Digital Signature Algorithm
ECIES	Elliptic Curve Integrated Encryption Scheme
EDGE	Enhanced Data rates for Global Evolution
EDHOC	Ephemeral Diffie–Hellman Over COSE
EE	Execution Environment
EIF	European Internet Foundation
EPC	Electronic Product Code
EPCIS	Electronic Product Code Information Services
EPSRC	Engineering and Physical Sciences Research Council
ERP	Enterprise Resource Planning
ETS	Error-Trend-Seasonal
ETSI-M2M	ETSI Machine-to-Machine
ETSI	European Telecommunications Standards Institute
EU	European Union
EV	Electric Vehicle
EXI	Efficient XML Interchange
FaaS	Function as a Service
FC	Functional Component
FCAPS	Fault, Configuration, Accounting, Performance, Security
FG	Functional Group
FM	Fault Management
FoF	Factory of the Future
FP	Framework Program
FQDN	Fully Qualified Domain Name
FSMA	Food Safety and Modernization Act
GA	Gateway Application
GAE	see xAE for x=(G)ateway
GCS	see xCS for x=(G)ateway
GDPR	General Data Protection Regulation
GDSN	Global Data Synchronization Network
GERAN	GSM EDGE Radio Access Network
GIP	see xIP for x=(G)ateway
GIS	Geographic Information System

GPIO	General Purpose I/O
GPRS	General Packet Radio Service
GPS	Global Positioning System
GRAR	see xRAR for x=(G)ateway
GREM	see xREM for x=(G)ateway
GSC	see xSC for x=(G)ateway
GSCL	Gateway Service Capabilities Layer
GSEC	see xSEC for x=(G)ateway
GSM	Global System for Mobile Communications (originally Groupe Spécial Mobile)
GSMA	GSM Association
GSR	Galvanic Skin Response
GTIN	Global Trade Item Number
GVC	Global Value Chain
GW	Gateway
HAN	Home or Building Area Networks
HC	Header Compression
HDFS	Hadoop File System
HFC	Hybrid Fiber Coaxial
HIP	Host Identity Protocol
HLA	High Level Architecture
HMI	Human–Machine Interaction
HNCP	Home Networking Control Protocol
HTTP	Hypertext Transfer Protocol
HTTPS	HTTP over TLS
HVAC	Heating, Ventilation, and Air Conditioning
HW	Hardware
I-GVC	Information-Driven Global Value Chain
I^2C	Inter-Integrated Circuit
I2C	Inter-Integrated Circuit
I4.0	Industrie 4.0/Industry 4.0
IaaS	Infrastructure as a Service
ICMP	Internet Control Message Protocol
ICT	Information and Communications Technology
ID	Identity
IEC	International Electrotechnical Commission
IEEE	Institute of Electrical and Electronic Engineers
IEM	Integration and Energy Management (System)
IERC	Internet of Things European Research Cluster
IETF	Internet Engineering Task Force
IG	Interest Group
IGMP	Internet Group Management Protocol
IIAF	Industrial Internet of Things Analytics Framework
IIC	Industrial Internet Consortium
IICF	Industrial Internet Connectivity Framework
IIoT	Industrial Internet of Things
IIRA	Industrial Internet Reference Architecture
IISF	Industrial Internet Security Framework
IKEv2	Internet Key Exchange version 2
IoS	Internet of Services
IoT-A	Internet of Things Architecture
IoT-GSI	IoT Global Standards Initiative
IoT-i	IoT Initiative
IoT	Internet of Things

IP	Internet Protocol
IPSec	Internet Protocol Security
IPSO	IP for Smart Objects
ISA	International Society of Automation
ISG	Industry Specification Group
ISIM	IP Multimedia Subscriber Identity Module
ISM	Industrial, Scientific, Medical (band)
ISO	International Organization for Standardization
ISP	Internet Service Provider
IT	Information Technology
ITU-T	International Telecommunication Union Telecommunication sector
ITU	International Telecommunications Union
JAR	Java ARchive
JCA-IoT	Joint Coordination Activity on IoT
JCA-NID	Joint Coordination Activity on Network Aspects of Identification Systems
JOSE	JSON Object Signing and Encryption
JSON-LD	JSON Linked Data
JSON	JavaScript Object Notation
JTC	Joint Technical Committee
KDD	Knowledge Discovery in Databases
KMF	Knowledge Management Framework
KPI	Key Performance Indicator
LAN	Local Area Network
LBS	Location Based Services
LCM	Lifecycle Management
LDU	Local Discovery Unit
LE	Low Energy
LED	Light Emitting Diode
LM	Logistics Management
LoRa	Long Range
LP-WAN	Low Power Wide Area Networking
LPM	Low Power Mode
LPWA	Low Power Wide Area
LR-WPAN	Low-Rate Wireless Personal Area Network
LTE-M	Long-Term Evolution, Category M1
LTE	Long-Term Evolution
LTN	Low Throughput Network
LwM2M	Lightweight Machine-to-Machine
M-BUS	Meter Bus
M2B	Machine to Business
M2M	Machine-to-Machine
MAC	Medium Access Control
MAC	Message Authentication Code
MAN	Metropolitan Area Network
MAS	M2M Authentication Server
MCIM	Machine Communications Identity Module
MD	Message Digest
MEC	Multiaccess Edge Computing
MEMS	Microelectro-Mechanical System
MES	Manufacturing Execution System
MI	Machine Intelligence
MIB	Management Information Base
MIC	Message Integrity Code

MII	Manufacturing Integration and Intelligence
MIPS	Million Instructions Per Second
ML	Machine Learning
MLD	Multilevel Discovery
MNO	Mobile Network Operator
MOO	Multiobjective Optimization
MPL	Multicast Protocol for Low-Power and Lossy Networks
MPP	Massively Parallel Processing
MQTT	Message Queuing Telemetry Transport
MS	Master–Slave
MS	Microsoft
MSBF	M2M Service Bootstrap Function
MTC	Machine Type Communication
NA	Network Application
NAE	see xAE for x=(N)etwork
NAN	Neighborhood Area Network
NASA	National Aeronautics and Space Administration
NB-IoT	Narrowband IoT
NCS	see xCS for x=(N)etwork
ND	Neighbor Discovery
NETCONF	Network Configuration Protocol
NFC	Near Field Communication
NFV	Network Function Virtualization
NGC	see xGC for x=(N)etwork
NIC	National Intelligence Council
NIP	see xIP for x=(N)etwork
NIST	National Institute of Standards and Technology
NRAR	see xRAR for x=(N)etwork
NREM	see xREM for x=(N)etwork
NSCL	Network Service Capabilities Layer
NSEC	see xSEC for x=(N)etwork
NTOE	Network Telecom Operator Exposure
OASIS	Organization for the Advancement of Structured Information Standards
OAuth	Web Authorization Protocol
OCB	Outside the Context of a Basic Service Set
OCF	Open Connectivity Foundation
OFB	Output Feedback (mode)
OGC	Open Geospatial Consortium
OIC	Open Interconnect Forum
OLE	Object Linking and Embedding
OMA-DM	OMA Device Management
OMA	Open Mobile Alliance
OMNA	Open Mobile Naming Authority
ONS	Object Naming System
OPC-UA	OLE for Process Control Unified Architecture
OPC	OLE for Process Control
OPC	Open Platform Communications
OPEX	Operating Expenditure
OS	Operating System
OSCORE	Object Security for Constrained RESTful Environments
OSGi	Open Service Gateway initiative
OSI	Open System Interconnection
OSS	Operations Support System

OT	Operational Technology
OWL	Web Ontology Language
P2P	Peer-to-Peer
PaaS	Platform as a Service
PAN	Personal Area Network
PC	Personal Computer
PCBC	Propagating Cipher-Block Chaining (mode)
PDDL	Planning Domain Definition Language
PdM	Predictive Maintenance
PHY	Physical Layer
PID	Proportional Integral Derivative
PKI	Public Key Infrastructure
PLC	Power Line Communication
PLC	Programmable Logic Controller
PLM	Product Lifecycle Management
PM	Performance Monitoring
POS	Point of Sales
PRB	Physical Resource Block
PS	Participatory Sensing
PSI	Public Sector Information (EU legislation)
PV	Photovoltaic Panel
PWM	Pulse-Width Modulation
QoI	Quality of Information
QoS	Quality of Service
QR	Quick Response
RADIUS	Remote Authentication Dial-In User Service
RAM	Random Access Memory
RAMI 4.0	Reference Architectural Model Industrie 4.0
RD	Resource Directory
RDF	Resource Description Framework
REST	Representational State Transfer
RF	Radio Frequency
RFC	Request For Comments
RFIC	Radio Frequency Integrated Circuit
RFID	Radio Frequency Identification
RoLL	(IETF) Routing over Low Power and Lossy Networks
ROOF	Real-time Onsite Operations Facilitation
RPL	IPv6 Routing Protocol for Low Power and Lossy Networks
RPM	Revolutions Per Minute
RPMA	Random Phase Multiple Access
RS	Recommend Standard
RS	Resource Server
RSA	Rivest–Shamir–Adleman algorithm
RT	Real-Time
RTOS	Real-Time Operating System
SA	Service and System Aspects
SaaS	Software as a Service
SAN	Sensor/Actuator Network
SAN	Storage Area Network
SAW	Surface Acoustic Wave
SC	Service Capability
SC	Supply Chain
SCADA	Supervisory Control and Data Acquisition

SCL	Service Capability Layer
SCM	Supply Chain Management
SDK	Software Development Kit
SDN	Software Defined Networking
SDO	Standards Developing Organization
SEMMA	Sample, Explore, Modify, Model, and Assess
SenML	Sensor Measurement Lists
SensorML	Sensor Model Language
SEP	Smart Energy Profile
SG	Study Group
SG20	Study Group 20
SHA	Secure Hash Algorithm
SIA	SOCRADES Integration Architecture
SIG	Special Interest Group
SIM	Subscriber Identity Module
SKA	Symmetric Key Algorithm
SLA	Service Level Agreement
SLO	Service Level Objective
SMHI	Sveriges Meterologiska och Hydrologiska Institut (Swedish Meteorological and Hydrological Institute)
SMS	Short Messaging Service
SNMP	Simple Network Management Protocol
SNO	SigFox Network Operator
SOA	Service Oriented Architecture
SOAP	Simple Object Access Protocol
SoC	System-on-a-Chip
SOS	Sensor Observation Service
SoS	System of Systems
SPI	Serial Peripheral Interface
SPS	Sensor Planning Service
SQL	Structured Query Language
SRA	Strategic Research Agenda
SUN	Smart Utility Network
SW	Software
SWE	Sensor Web Enablement
SWEET	Semantic Web for Earth and Environmental Terminology
TC	Technical Committee
TCP	Transmission Control Protocol
TD	Thing Description
TDMA	Time Division Multiple Access
TFL	Transport For London
TG	Technical Group
TIA	Telecommunications Industry Association
TISPAN	Telecoms and Internet converged Services and Protocols for Advanced Networks
TLS	Transport Layer Security
TLV	Type-Length-Value
TML	Transducer Model Language
TP	Token Passing
TR	Technical Report
TRE	Trusted Environment
TS	Technical Specification
TSCH	Timeslotted Channel Hopping
TTA	Telecommunications Technology Association
TTC	Telecommunication Technology Committee

UART	Universal Asynchronous Receiver Transmitter
UAV	Unmanned Aerial Vehicle
UDP	User Datagram Protocol
UI	User Interface
UICC	Universal Integrated Circuit Card
ULE	Ultra Low Energy
UML	Unified Modeling Language
UMTS	Universal Mobile Telecommunications System
UNB	Ultra Narrow Band
UP	Urban Prototyping
UPnP	Universal Plug and Play
URI	Universal Resource Identifier
URL	Uniform Resource Locator
US	United States (of America)
USD	US Dollar
USDL	Universal Service Description Language
USFDA	US Food and Drug Administration
USIM	Universal Subscriber Identity Module
UTRAN	Universal Mobile Telecommunications System (UMTS) Terrestrial Radio Access Network
UWB	Ultra-Wide Band
V2I	Vehicle to Infrastructure
V2V	Vehicle-to-Vehicle
VAV	Variable Air Volume Box
VM	Virtual Machine
VNF	Virtual Network Function
W3C	World Wide Web Consortium
WADL	Web Application Description Language
WAN	Wide Area Network
WEF	World Economic Forum
WG	Working Group
Wi-Fi	Wireless Fidelity
Wi-SUN	Wireless Smart Ubiquitous Networks
WiMAX	Worldwide Interoperability for Microwave Access
WLAN	Wireless Local Area Network
WoT	Web of Things
WPAN	Wireless Personal Area Network
WS	Web Service(s)
WSAN	Wireless Sensor/Actuator Network
WSDL	Web Services Description Languages
WSN	Wireless Sensor Network
WWAN	Wireless Wide Area Network
WWW	World Wide Web
xA	x Application, x=(D)evice, (G)ateway, (N)etwork
XaaS	Everything (X) as a Service
xAE	x Application Enablement, x=(D)evice, (G)ateway, (N)etwork
xCB	x Compensation Broker, x=(D)evice, (G)ateway, (N)etwork
xCS	x Communication Selection, x=(D)evice, (G)ateway, (N)etwork
xGC	x Generic Communication, x=(D)evice, (G)ateway, (N)etwork
xHDR	x History and Data Retention, x=(D)evice, (G)ateway, (N)etwork
xIP	x Interworking Proxy, x=(D)evice, (G)ateway, (N)etwork
XML	Extensible Markup Language
xRAR	x Reachability, Addressing, and Repository, x=(D)evice, (G)ateway, (N)etwork
xREM	x Remote Entity Management, x=(D)evice, (G)ateway, (N)etwork

xSC	x Service Capabilities
XSD	XML Schema Definition
xSEC	x Security, x=(D)evice, (G)ateway, (N)etwork
xTM	x Transaction Management, x=(D)evice, (G)ateway, (N)etwork
YANG	Data Modeling Language for NETCONF

Bibliography

[1] Höller J, Tsiatsis V, Mulligan C, Karnouskos S, Avesand S, Boyle D. From machine-to-machine to the internet of things: introduction to a new age of intelligence. Elsevier. ISBN 978-0-12-407684-6, 2014. Available from: http://www.amazon.com/From-Machine---Machine-Internet-Things/dp/012407684X/.

[2] The Ericsson mobility report. Available from: https://www.ericsson.com/en/mobility-report.

[3] Industrial internet of things: unleashing the potential of connected products and services. Tech. rep.; 2015. Available from: http://www3.weforum.org/docs/WEFUSA_IndustrialInternet_Report2015.pdf.

[4] Manyika J, Chui M, Bisson P, Woetzel J, Dobbs R, Bughin J, et al. The internet of things: mapping the value beyond the hype. Tech. rep.; 2015. Available from: https://goo.gl/V2fnJm.

[5] Presser M, Barnaghi P, Eurich M, Villalonga C. The SENSEI project: integrating the physical world with the digital world of the network of the future. IEEE Communications Magazine 2009;47(4):1–4. https://doi.org/10.1109/mcom.2009.4907403.

[6] Plantagon. Available from: http://www.plantagon.com/.

[7] US Food & Drug Administration. FDA Food Safety Modernization Act (FSMA). Available from: https://www.fda.gov/Food/GuidanceRegulation/FSMA/default.htm.

[8] National Intelligence Council. Global trends. Available from: https://www.dni.gov/index.php/global-trends-home.

[9] European Internet Forum. The digital world in 2030. Available from: http://www.eifonline.org/digitalworld2030.

[10] Singh S. New mega trends. UK: Palgrave Macmillan; 2012.

[11] Manyika J, Chui M, Bisson P, Dobbs R, Bughin J, Marrs A. Disruptive technologies: advances that will transform life, business, and the global economy. Tech. rep.; 2013. Available from: https://www.mckinsey.com/business-functions/digital-mckinsey/our-insights/disruptive-technologies.

[12] Mulligan C. The communications industries in the era of convergence. Abingdon, Oxon New York: Routledge. ISBN 9780415584845, 2012.

[13] Nester Research. Internet of Things (IoT) market: global demand, growth analysis & opportunity outlook 2023; 2018.

[14] Mulligan CEA. The communications industries in the era of convergence. Routledge. ISBN 978-1138686960, 2011.

[15] Gereffi G. A commodity chains framework for analyzing global industries; 1999. http://www.ids.ac.uk/ids/global/conf/pdfs/gereffi.pdf.

[16] Global Value Chains. Available from: http://www.globalvaluechains.org, 2011.

[17] Moore JF. The death of competition: leadership and strategy in the age of business ecosystems. Harper Business. ISBN 0-88730-850-3, 1996.

[18] Rozanski N, Woods E. Software systems architecture: working with stakeholders using viewpoints and perspectives. 2nd edition. Addison-Wesley Professional. ISBN 032171833X, 2011.

[19] Carrez F, Bauer M, Boussard M, Bui N, Jardak C, Loof JD, et al. SENSEI deliverable D1.5 – final architectural reference model for the IoT v3.0, Internet of Things Architecture IoT-A. EC research project. Available from: http://www.meet-iot.eu/deliverables-IOTA/D1_5.pdf, 2013.

[20] SENSEI project. Integrating the physical with the digital world of the network of the future. FP7-ICT. European Commission, Project ID: 215923. Available from: https://cordis.europa.eu/project/rcn/85429_en.html, 2010.

[21] Pastor A, Ho E, Magerkurth C, Martín G, Sáinz I, et al. IoT-A deliverable D6.2 – updated requirements list, Internet of Things Architecture IoT-A. EC Research Project. Available from: http://www.meet-iot.eu/deliverables-IOTA/D6_2.pdf, 2011.

[22] Magerkurth C, Segura AS, Vicari N, Boussard M, Meyer S. IoT-A deliverable D6.3 – final requirements list, Internet of Things Architecture IoT-A. EC Research Project. Available from: http://www.meet-iot.eu/deliverables-IOTA/D6_3.pdf, 2013.

[23] European Telecommunications Standards Institute Machine to Machine Technical Committee (ETSI M2M TC). ETSI TS 102 689 Machine to Machine communications (M2M); M2M service requirements. Available from: http://www.etsi.org/deliver/etsi_ts/102600_102699/102689/01.01.01_60/ts_102689v010101p.pdf, 2010.

[24] European Telecommunications Standards Institute. ETSI TR 103 375 v1.1.1 (2016–19) SmartM2M: IoT standards landscape and future evolutions. Available from: http://www.etsi.org/deliver/etsi_tr/103300_103399/103375/01.01.01_60/tr_103375v010101p.pdf, 2016.

[25] European Commission. Smart Grid Mandate 490 – standardization mandate to European Standardisation Organisations (ESOs) to support European smart grid deployment. Tech. rep.; 2011. Available from: http://ec.europa.eu/growth/tools-databases/mandates/index.cfm?fuseaction=search.detail&id=475#.

[26] Paradiso JA, Starner T. Energy scavenging for mobile and wireless electronics. IEEE Pervasive Computing 2005;4(1):18–27. https://doi.org/10.1109/MPRV.2005.9.

[27] Mitcheson PD, Yeatman EM, Rao GK, Holmes AS, Green TC. Energy harvesting from human and machine motion for wireless electronic devices. Proceedings of the IEEE 2008;96(9):1457–86. https://doi.org/10.1109/JPROC.2008.927494.

[28] Magno M, Boyle D, Brunelli D, O'Flynn B, Popovici E, Benini L. Extended wireless monitoring through intelligent hybrid energy supply. IEEE Transactions on Industrial Electronics 2014;61(4):1871–81. https://doi.org/10.1109/TIE.2013.2267694.

[29] Boyle DE, Newe T. On the implementation and evaluation of an elliptic curve based cryptosystem for Java enabled wireless sensor networks. Sensors and Actuators A: Physical 2009;156(2):394–405. https://doi.org/10.1016/j.sna.2009.10.012. Available from: http://www.sciencedirect.com/science/article/pii/S0924424709004361.

[30] Shi W, Cao J, Zhang Q, Li Y, Xu L. Edge computing: vision and challenges. IEEE Internet of Things Journal 2016;3(5):637–46. https://doi.org/10.1109/JIOT.2016.2579198.

[31] Boyle D, Magno M, O'Flynn B, Brunelli D, Popovici E, Benini L. Towards persistent structural health monitoring through sustainable wireless sensor networks. In: 2011 seventh international conference on intelligent sensors, sensor networks and information processing; 2011. p. 323–8.

[32] European Telecommunications Standards Institute Machine to Machine Technical Committee (ETSI M2M TC). ETSI TS 102 690 Machine to Machine communications (M2M) functional architecture. Available from: http://www.etsi.org/deliver/etsi_ts/102600_102699/102690/01.02.01_60/ts_102690v010201p.pdf, 2013.

[33] Jain P, Hedman P, Zisimopoulos H. Machine type communications in 3GPP systems. IEEE Communications Magazine 2012;50(11).

[34] Liberg O, Sundberg M, Wang E, Bergman J, Sachs J. Cellular Internet of Things: technologies, standards, and performance. Academic Press; 2017.

[35] Raza U, Kulkarni P, Sooriyabandara M. Low power wide area networks: an overview. IEEE Communications Surveys Tutorials 2017;19(2):855–73. https://doi.org/10.1109/COMST.2017.2652320.

[36] Vlasios T, Alexander G, Tim B, Frederic M, Jesus B, Martin B, et al. The sensei real world internet architecture. Stand Alone. Towards the future internet; 2010. p. 247–56.

[37] European Commission. European legislation on re-use of public sector information. Available from: https://ec.europa.eu/digital-single-market/en/european-legislation-reuse-public-sector-information, 2017.

[38] Russell S, Norvig P. Artificial intelligence: a modern approach. 3rd edition. Upper Saddle River, NJ, USA: Prentice Hall Press; 2009. ISBN 0136042597, 9780136042594.

[39] Cheng B, Zhang J, Hancke GP, Karnouskos S, Colombo AW. Industrial cyberphysical systems: realizing cloud-based big data infrastructures. IEEE Industrial Electronics Magazine 2018;12(1):25–35. https://doi.org/10.1109/mie.2017.2788850.

[40] Allemang D, Hendler J. Semantic web for the working ontologist: effective modeling in RDFS and OWL. Waltham, MA: Morgan Kaufmann/Elsevier. ISBN 978-0123859655, 2011.

[41] Holler J, Tsiatsis V, Mulligan C. Toward a machine intelligence layer for diverse industrial IoT use cases. IEEE Intelligent Systems 2017;32(4):64–71. https://doi.org/10.1109/mis.2017.3121543.

[42] Fowler M. Microservices: a definition of this new architectural term. Available from: https://martinfowler.com/articles/microservices.html, 2014.

[43] IERC. IoT semantic interoperability: research challenges, best practices, recommendations and next steps. Tech. rep.; 2015. Available from: http://www.internet-of-things-research.eu/pdf/IERC_Position_Paper_IoT_Semantic_Interoperability_Final.pdf.

[44] Cho JH, Wang Y, Chen IR, Chan KS, Swami A. A survey on modeling and optimizing multi-objective systems. IEEE Communications Surveys & Tutorials 2017;19(3):1867–901. https://doi.org/10.1109/comst.2017.2698366.

[45] Anderson N, Diab WW, French T, Harper KE, Lin SW, Nair D, et al. Industrial internet of things analytics framework. Tech. rep. IIC:PUB:T3:V1.00:PB:20171023; 2017. Available from: http://www.iiconsortium.org/industrial-analytics.htm.

[46] Kreutz D, Ramos FMV, Verissimo PE, Rothenberg CE, Azodolmolky S, Uhlig S. Software-defined networking: a comprehensive survey. Proceedings of the IEEE 2015;103(1):14–76. https://doi.org/10.1109/jproc.2014.2371999.

[47] Mijumbi R, Serrat J, Gorricho JL, Bouten N, Turck FD, Boutaba R. Network function virtualization: state-of-the-art and research challenges. IEEE Communications Surveys & Tutorials 2016;18(1):236–62. https://doi.org/10.1109/comst.2015.2477041.

[48] Mell P, Grance T. The NIST definition of cloud computing. Special Publication. Available from: http://csrc.nist.gov/publications/nistpubs/800-145/SP800-145.pdf, 2011.

[49] Roberts M. Serverless architectures. Available from: https://martinfowler.com/articles/serverless.html, 2016.

[50] Lynn T, Rosati P, Lejeune A, Emeakaroha V. A preliminary review of enterprise serverless cloud computing (function-as-a-service) platforms. In: 2017 IEEE international conference on cloud computing technology and science (CloudCom). IEEE; 2017. p. 162–9.

[51] Iorga M, Feldman L, Barton R, Martin MJ, Goren N, Mahmoudi C. Fog computing conceptual model. Tech. rep.; 2018.

[52] Reznik A, Arora R, Cannon M, Cominardi L, Featherstone W, Frazao R, et al. Developing software for multi-access edge computing. White Paper. Available from: http://www.etsi.org/images/files/ETSIWhitePapers/etsi_wp20_MEC_SoftwareDevelopment_FINAL.pdf, 2017.

[53] Consortium O. OpenFog reference architecture for fog computing. Tech. rep.; 2017. OpenFog Consortium publication No. OPFRA001.020817. Available from: https://www.openfogconsortium.org/ra/.

[54] Persson P, Angelsmark O. Kappa: serverless IoT deployment. In: Proceedings of the 2nd international Workshop on Serverless Computing – WoSC '17, WoSC '17. New York, NY, USA: ACM. ISBN 978-1-4503-5434-9, 2017. p. 16–21.

[55] Karnouskos S. Efficient sensor data inclusion in enterprise services. Datenbank-Spektrum 2009;9(28):5–10.

[56] Karnouskos S, Vilaseñor V, Handte M, Marrón PJ. Ubiquitous integration of cooperating objects. International Journal of Next-Generation Computing (IJNGC) 2011;2(3).

[57] Karnouskos S. Stuxnet worm impact on industrial cyber-physical system security. In: 37th annual conference of the IEEE industrial electronics society (IECON 2011); 2011. p. 4490–4.

[58] Marrón PJ, Karnouskos S, Minder D, Ollero A, editors. The emerging domain of cooperating objects. Springer; 2011. 271 pp.

[59] Karnouskos S, Marrón PJ, Fortino G, Mottola L, Martínez-de Dios JR. Applications and markets for cooperating objects. SpringerBriefs in electrical and computer engineering. Springer. ISBN 978-3-642-45400-4, 2014.

[60] Sfar AR, Natalizio E, Challal Y, Chtourou Z. A roadmap for security challenges in internet of things. Digital Communications and Networks 2017. https://doi.org/10.1016/j.dcan.2017.04.003.

[61] Karnouskos S, Kerschbaum F. Privacy and integrity considerations in hyperconnected autonomous vehicles. Proceedings of the IEEE 2018;106(1):160–70. https://doi.org/10.1109/jproc.2017.2725339.

[62] Business Process Model and Notation (BPMN). Available from: http://www.bpmn.org/.

[63] Tranquillini S, Spiess P, Daniel F, Karnouskos S, Casati F, Oertel N, et al. Process-based design and integration of wireless sensor network applications. In: 10th international conference on Business Process Management (BPM). Springer Berlin Heidelberg; 2012. p. 134–49.

[64] Daniel F, Eriksson J, Finne N, Fuchs H, Gaglione A, Karnouskos S, et al. makesense: real-world business processes through wireless sensor networks. In: 4th international workshop on networks of cooperating objects for smart cities 2013 (CONET/UBICITEC 2013). CEUR-WS.org; 2013. p. 58–72. Available from: http://ceur-ws.org/Vol-1002/paper6.pdf.

[65] Spiess P, Karnouskos S. Maximizing the business value of networked embedded systems through process-level integration into enterprise software. In: Second International Conference on Pervasive Computing and Applications (ICPCA 2007); 2007. p. 536–41.

[66] Vasseur JP, Dunkels A. Interconnecting smart objects with IP: the next internet. San Francisco, CA, USA: Morgan Kaufmann Publishers Inc.; 2010. ISBN 0123751659, 9780123751652.

[67] Karnouskos S, Somlev V. Performance assessment of integration in the cloud of things via web services. In: IEEE International Conference on Industrial Technology (ICIT 2013); 2013. p. 1988–93.

[68] Mottola L, Picco GP, Opperman FJ, Eriksson J, Finne N, Fuchs H, et al. makeSense: simplifying the integration of wireless sensor networks into business processes. IEEE Transactions on Software Engineering 2018:1. https://doi.org/10.1109/tse.2017.2787585.

[69] Spiess P, Karnouskos S, de Souza LMS, Savio D, Guinard D, Trifa V, et al. Reliable execution of business processes on dynamic networks of service-enabled devices. In: 7th IEEE international conference on industrial informatics INDIN 2009; 2009. p. 533–8.

[70] Schneier B. Applied cryptography: protocols, algorithms, and source code in C. John Wiley & Sons; 2007.

[71] Law YW, Doumen J, Hartel P. Survey and benchmark of block ciphers for wireless sensor networks. ACM Transactions on Sensor Networks (TOSN) 2006;2(1):65–93.

[72] Cheung RC, Luk W, Cheung PY. Reconfigurable elliptic curve cryptosystems on a chip. In: Proceedings of the conference on design, automation and test in Europe-volume 1. IEEE Computer Society; 2005. p. 24–9.

[73] Wander AS, Gura N, Eberle H, Gupta V, Shantz SC. Energy analysis of public-key cryptography for wireless sensor networks. In: Pervasive computing and communications, 2005. PerCom 2005. Third IEEE international conference on. IEEE; 2005. p. 324–8.

[74] Amish P, Vaghela V. Detection and prevention of wormhole attack in wireless sensor network using AOMDV protocol. Procedia Computer Science 2016;79:700–7.

[75] Adat V, Gupta B. Security in internet of things: issues, challenges, taxonomy, and architecture. Telecommunication Systems 2018;67(3):423–41.

[76] Karlof C, Wagner D. Secure routing in wireless sensor networks: attacks and countermeasures. Ad Hoc Networks 2003;1(2–3):293–315.

[77] Antonakakis M, April T, Bailey M, Bernhard M, Bursztein E, Cochran J, et al. Understanding the mirai botnet. In: 26th USENIX security symposium (USENIX security 17). Vancouver, BC: USENIX Association. ISBN 978-1-931971-40-9, 2017. p. 1093–110. Available from: https://www.usenix.org/conference/usenixsecurity17/technical-sessions/presentation/antonakakis.

[78] Schneier B. Lessons from the dyn ddos attack. Schneier on Security Blog 2016;8.

[79] Schaad J. CBOR Object Signing and Encryption (COSE). RFC 8152 (proposed standard); 2017. Available from: https://www.rfc-editor.org/rfc/rfc8152.txt.

[80] Winter T, editor, Thubert P, editor, Brandt A, Hui J, Kelsey R, Levis P, et al. RPL: IPv6 routing protocol for low-power and lossy networks. RFC 6550 (proposed standard); 2012. Available from: https://www.rfc-editor.org/rfc/rfc6550.txt.

[81] Alliance L. Lorawan specification. LoRa Alliance 2015.

[82] Lasota PA, Fong T, Shah JA, et al. A survey of methods for safe human–robot interaction. Foundations and Trends® in Robotics 2017;5(4):261–349.

[83] Iso I. Iso 10218-1: 2011: robots and robotic devices—safety requirements for industrial robots—part 1: robots. Geneva, Switzerland: International Organization for Standardization; 2011.

[84] Committee SORAVS, et al. Taxonomy and definitions for terms related to on-road motor vehicle automated driving systems. SAE International; 2014.

[85] Boyle DE, Yates DC, Yeatman EM. Urban sensor data streams: London 2013. IEEE Internet Computing 2013;17(6):12–20.

[86] Miller B, Rowe D. A survey scada of and critical infrastructure incidents. In: Proceedings of the 1st annual conference on research in information technology. ACM; 2012. p. 51–6.

[87] Nicol DM. Hacking the lights out. Scientific American 2011;305(1):70–5.

[88] Sicari S, Rizzardi A, Grieco LA, Coen-Porisini A. Security, privacy and trust in internet of things: the road ahead. Computer Networks 2015;76:146–64.

[89] Sadeghi AR, Wachsmann C, Waidner M. Security and privacy challenges in industrial internet of things. In: Proceedings of the 52nd annual design automation conference. ACM; 2015. p. 54.

[90] Roman R, Zhou J, Lopez J. On the features and challenges of security and privacy in distributed internet of things. Computer Networks 2013;57(10):2266–79.

[91] Smart NP, Rijmen V, Gierlichs B, Paterson K, Stam M, Warinschi B, et al. Algorithms, key size and parameters report. European Union Agency for Network and Information Security; 2014. p. 1–95. Available from: https://www.enisa.europa.eu/publications/algorithms-key-sizes-and-parameters-report/at_download/fullReport.

[92] Covington MJ, Fogla P, Zhan Z, Ahamad M. A context-aware security architecture for emerging applications. In: Computer security applications conference, 2002. Proceedings. 18th annual. IEEE; 2002. p. 249–58.

[93] Chen D, Chang G, Sun D, Li J, Jia J, Wang X. TRM-IoT: a trust management model based on fuzzy reputation for internet of things. Computer Science and Information Systems 2011;8(4):1207–28.

[94] Vermesan O, Friess P. Internet of things: converging technologies for smart environments and integrated ecosystems. Rover Publishers. ISBN 978-87-92982-73-5, 2013. Available from: http://www.internet-of-things-research.eu/pdf/Converging_Technologies_for_Smart_Environments_and_Integrated_Ecosystems_IERC_Book_Open_Access_2013.pdf.

[95] Internet Protocol for Smart Objects (IPSO) Alliance, IPSO Smart Object Committee. IPSO SmartObject guideline, smart objects starter pack 1.0; 2014. Available from: https://www.ipso-alliance.org/so-starter-pack/.

[96] Internet Protocol for Smart Objects (IPSO) Alliance, IPSO Smart Object Committee. IPSO SmartObject guideline, smart objects expansion pack; 2015. Available from: https://www.ipso-alliance.org/so-expansion-pack/.

[97] Open Mobile Alliance (OMA). Lightweight machine to machine technical specification V1.0.2. Available from: http://www.openmobilealliance.org/release/LightweightM2M/V1_0_2-20180209-A/OMA-TS-LightweightM2M-V1_0_2-20180209-A.pdf, 2018.

[98] Shelby Z, Vial M, Koster M, Groves C, Zhu J, Silverajan B. Reusable interface definitions for constrained RESTful environments draft-ietf-core-interfaces-10. Tech. rep.; 2017. Available from: https://datatracker.ietf.org/doc/html/draft-ietf-core-interfaces-10.

[99] Veillette M, der Stok PV, Pelov A, Bierman A. CoAP management interface draft-ietf-core-comi-01. Tech. rep.; 2017. Available from: https://datatracker.ietf.org/doc/html/draft-ietf-core-comi-01.

[100] Jennings C, Shelby Z, Arkko J, Keränen A, Bormann C. Media types for sensor measurement lists (SenML) draft-ietf-core-senml-11. Tech. rep.; 2017. Available from: https://datatracker.ietf.org/doc/html/draft-ietf-core-senml-11.

[101] Shelby Z, Koster M, Bormann C, der Stok PV, Amsüss C. CoRE resource directory draft-ietf-core-resource-directory-12. Tech. rep.; 2017. Available from: https://tools.ietf.org/html/draft-ietf-core-resource-directory-12.

[102] Vial M. CCoRE mirror server draft-vial-core-mirror-server-01. Tech. rep.; 2013. Available from: https://tools.ietf.org/html/draft-vial-core-mirror-server-01.

[103] Adolphs P, Bedenbender H, Dirzus D, Ehlich M, Epple U, Hankel M, et al. Reference architecture model industrie 4.0 (RAMI4.0). Tech. rep.; 2015. Available from: https://goo.gl/3DcvkQ.

[104] Cooperation among two key leaders in the industrial internet. Available from: https://goo.gl/d9m643, 2016.

[105] Pai M. Interoperability between IIC architecture & industry 4.0 reference architecture for industrial assets. Tech. rep.; 2016. Available from: https://www.infosys.com/engineering-services/white-papers/Documents/industrial-internet-consortium-architecture.pdf.

[106] ISO/IEC/IEEE systems and software engineering – architecture description; 2011.

[107] ISO/IEC/IEEE international standard – systems and software engineering – system life cycle processes; 2015.

[108] Industrial Internet Consortium. The industrial internet of things volume G1: reference architecture V1.8. Tech. rep.; 2017. Available from: http://www.iiconsortium.org/IIRA.htm.

[109] Kruchten P. The 4+1 view model of architecture. IEEE Software 1995;12(6):42–50. https://doi.org/10.1109/52.469759.

[110] Rowley J. The wisdom hierarchy: representations of the DIKW hierarchy. Journal of Information Science 2007;33(2):163–80. https://doi.org/10.1177/0165551506070706.

[111] NASA Jet Propulsion Laboratory. Semantic Web for Earth and Environmental Terminology (SWEET) ontologies. Available from: http://sweet.jpl.nasa.gov/ontology.

[112] De S, Barnaghi P, Bauer M, Meissner S. Service modelling for the internet of things. In: 2011 Federated Conference on Computer Science and Information Systems (FedCSIS); 2011. p. 949–55.

[113] Martín G, Meissner S, Dobre D, Thoma M. IoT-A deliverable D2.1 – resource description specification, Internet of Things Architecture IoT-A. EC Research Project. Available from: http://www.meet-iot.eu/deliverables-IOTA/D2_1.pdf, 2012.

[114] Gruschka N, Gessner D, Serbanati A, Segura AS, Olivereau A, Saied YB, et al. SENSEI deliverable D4.2 – concepts and solutions for privacy and security in the resolution infrastructure, Internet of Things Architecture IoT-A. EC Research Project. Available from: http://www.meet-iot.eu/deliverables-IOTA/D4_2.pdf, 2012.

[115] Shirey R. Internet security glossary, version 2. Tech. rep.; 2007.

[116] Moskowitz R, Nikander P. Host Identity Protocol (HIP) architecture. Tech. rep.; 2006.

[117] Industrial Internet Consortium. The industrial internet of things volume G5: connectivity framework V1.0. Tech. rep.; 2017. Available from: http://www.iiconsortium.org/IICF.htm.

[118] Industrial Internet Consortium. Industrial internet of things volume G4: security framework V1.0. Tech. rep.; 2016. Available from: http://www.iiconsortium.org/IISF.htm.

[119] Industrial Internet Consortium. The industrial internet of things volume G8: vocabulary V2.0. Tech. rep.; 2017. Available from: http://www.iiconsortium.org/vocab/index.htm.

[120] Ogata K. Modern control engineering. Pearson. 5th edition. ISBN 0136156738, 2010.

[121] Yiğitler H, Jäntti R, Kaltiokallio O, Patwari N. Detector based radio tomographic imaging. Available from: https://arxiv.org/abs/1604.03083, 2016.

[122] Boyle DE, Kiziroglou ME, Mitcheson PD, Yeatman EM. Energy provision and storage for pervasive computing. IEEE Pervasive Computing 2016;15(4):28–35. https://doi.org/10.1109/MPRV.2016.65.

[123] Jammes F, Karnouskos S, Bony B, Nappey P, Colombo AW, Delsing J, et al. Promising technologies for soa-based industrial automation systems. In: Industrial cloud-based cyber-physical systems: the IMC-AESOP approach. Springer; 2014. p. 89–109.

[124] Márquez AC, Díaz VGP, Fernández JFG, editors. Advanced maintenance modelling for asset management. Springer International Publishing; 2018.

[125] Heng FL, Zhang K, Goyal A, Chaudhary H, Hirsch S, Kim Y, et al. Integrated analytics system for electric industry asset management. IBM Journal of Research and Development 2016;60(1):2:1–2:12. https://doi.org/10.1147/jrd.2015.2475955.

[126] Muller A, Marquez AC, Iung B. On the concept of e-maintenance: review and current research. Reliability Engineering & System Safety 2008;93(8):1165–87. https://doi.org/10.1016/j.ress.2007.08.006.

[127] Cannata A, Karnouskos S, Taisch M. Dynamic e-maintenance in the era of SOA-ready device dominated industrial environments. In: IMS – manufacturing technology platform (M4SM). London: Springer; 2009. p. 411–9.

[128] Macchi M, Martínez LB, Márquez AC, Fumagalli L, Granados MH. Value assessment of e-maintenance platforms. In: Advanced maintenance modelling for asset management. Springer International Publishing; 2017. p. 371–85.

[129] Haller S, Karnouskos S. CoBIs: collaborative business items. In: Rabe M, Mihók P, editors. New technologies for the intelligent design and operation of manufacturing networks, chap. CoBIs: collaborative business items. Fraunhofer IRB Verlag. ISBN 978-3-8167-7520-1, 2007. p. 201–2.

[130] Decker C, Spiess P, de Souza LMs, Beigl M, Nochta Z. Coupling enterprise systems with wireless sensor nodes: analysis, implementation, experiences and guidelines. In: Workshop on Pervasive Technology Applied (PTA) at the international conference on pervasive computing. ISBN 978-3-00-018411-6, 2006. p. 393–400. Available from: http://www.ibr.cs.tu-bs.de/dus/publications/pta2006.pdf.

[131] Karnouskos S, Spiess P. Towards enterprise applications using wireless sensor networks. In: Cardoso J, Cordeiro J, Filipe J, editors. 9th International Conference on Enterprise Information Systems (ICEIS). ISBN 978-972-8865-90-0, 2007. p. 230–6.

[132] Karnouskos S, Guinard D, Savio D, Spiess P, Baecker O, Trifa V, et al. Towards the real-time enterprise: service-based integration of heterogeneous SOA-ready industrial devices with enterprise applications. IFAC Proceedings Volumes 2009;42(4):2131–6. https://doi.org/10.3182/20090603-3-ru-2001.0551.

[133] Colombo AW, Karnouskos S. Towards the factory of the future: a service-oriented cross-layer infrastructure. In: ICT shaping the world: a scientific view. European Telecommunications Standards Institute (ETSI), John Wiley and Sons. ISBN 9780470741306, 2009. p. 65–81.

[134] Boyd A, Noller D, Peters P, Salkeld D, Thomasma T, Gifford C, et al. SOA in manufacturing – guidebook. Tech. rep.; 2008. Available from: ftp://public.dhe.ibm.com/software/plm/pdif/MESA_SOAinManufacturingGuidebook.pdf.

[135] Karnouskos S, Bangemann T, Diedrich C. Integration of legacy devices in the future SOA-based factory. IFAC Proceedings Volumes 2009;42(4):2113–8. https://doi.org/10.3182/20090603-3-ru-2001.0487.

[136] Leitão P, Colombo AW, Karnouskos S. Industrial automation based on cyber-physical systems technologies: prototype implementations and challenges. Computers in Industry 2016;81:11–25. https://doi.org/10.1016/j.compind.2015.08.004.

[137] Hohpe G, Woolf B. Enterprise integration patterns: designing, building, and deploying messaging solutions. Boston, MA, USA: Addison-Wesley Longman Publishing Co., Inc.. ISBN 0321200683, 2003.

[138] Cannata A, Karnouskos S, Taisch M. Evaluating the potential of a service oriented infrastructure for the factory of the future. In: 8th international conference on industrial informatics (INDIN); 2010. p. 592–7.

[139] Trappey AJC, Trappey CV, Govindarajan UH, Sun JJ, Chuang AC. A review of technology standards and patent portfolios for enabling cyber-physical systems in advanced manufacturing. IEEE Access 2016;4:7356–82. https://doi.org/10.1109/access.2016.2619360.

[140] Taisch M, Colombo AW, Karnouskos S, Cannata A. SOCRADES roadmap: the future of SOA-based factory automation. Tech. rep.; 2009. Available from: http://www.socrades.net/Documents/objects/file1274836528.pdf.

[141] Colombo AW, Karnouskos S, Mendes JM. Factory of the future: a service-oriented system of modular, dynamic reconfigurable and collaborative systems. In: Benyoucef L, Grabot B, editors. Artificial intelligence techniques for networked manufacturing enterprises management. Springer. ISBN 978-1-84996-118-9, 2010. p. 459–81.

[142] Karnouskos S, Savio D, Spiess P, Guinard D, Trifa V, Baecker O. Real world service interaction with enterprise systems in dynamic manufacturing environments. In: Benyoucef L, Grabot B, editors. Artificial intelligence techniques for networked manufacturing enterprises management. Springer. ISBN 978-1-84996-118-9, 2010. p. 423–57.

[143] SAP Manufacturing Integration and Intelligence (SAP MII). Available from: https://www.sap.com/germany/products/manufacturing-intelligence-integration.html.

[144] Colombo AW, Bangemann T, Karnouskos S, Delsing J, Stluka P, Harrison R, et al., editors. Industrial cloud-based cyber-physical systems: the IMC-AESOP approach. Springer. ISBN 978-3-319-05623-4, 2014. Available from: http://www.springer.com/engineering/production+engineering/book/978-3-319-05623-4.

[145] Delsing J. IoT automation: arrowhead framework. CRC Press. ISBN 9781315350868, 2017.

[146] Karnouskos S, Colombo AW, Bangemann T. Trends and challenges for cloud-based industrial cyber-physical systems. In: Industrial cloud-based cyber-physical systems: the IMC-AESOP approach. Springer; 2014. p. 231–40.

[147] Karnouskos S, Colombo AW. Architecting the next generation of service-based SCADA/DCS system of systems. In: 37th annual conference of the IEEE industrial electronics society (IECON 2011); 2011. p. 359–64.

[148] Bangemann T, Karnouskos S, Camp R, Carlsson O, Riedl M, McLeod S, et al. State of the art in industrial automation. In: Industrial cloud-based cyber-physical systems: the IMC-AESOP approach. Springer; 2014. p. 23–47.

[149] Colombo AW, Karnouskos S, Kaynak O, Shi Y, Yin S. Industrial cyberphysical systems: a backbone of the fourth industrial revolution. IEEE Industrial Electronics Magazine 2017;11(1):6–16. https://doi.org/10.1109/mie.2017.2648857.

[150] Karnouskos S, Colombo AW, Bangemann T, Manninen K, Camp R, Tilly M, et al. A SOA-based architecture for empowering future collaborative cloud-based industrial automation. In: 38th annual conference of the IEEE industrial electronics society (IECON 2012); 2012. p. 5766–72.

[151] Karnouskos S, Colombo AW, Bangemann T, Manninen K, Camp R, Tilly M, et al. The IMC-AESOP architecture for cloud-based industrial CPS. In: Industrial cloud-based cyber-physical systems: the IMC-AESOP approach. Springer; 2014. p. 49–88.

[152] BDI. Internet of energy: ICT for energy markets of the future. Tech. rep.; 2010. BDI publication No. 439. Available from: https://www.iese.fraunhofer.de/content/dam/iese/en/mediacenter/documents/BDI_initiative_IoE_us-IdE-Broschuere_tcm27-45653.pdf.

[153] Appelrath HJ, Kagermann H, Mayer C, editors. Future energy grid: migration to the internet of energy. Germany: acatech – National Academy of Science and Engineering. ISBN 978-91-87253-06-5, 2012. Available from: http://www.acatech.de/fileadmin/user_upload/Baumstruktur_nach_Website/Acatech/root/de/Publikationen/Projektberichte/EIT-ICT-Labs_final_acatech-Study_AS_121106_Einzelseiten_final.pdf.

[154] Greer C, Wollman DA, Prochaska DE, Boynton PA, Mazer JA, Nguyen CT, et al. NIST framework and roadmap for smart grid interoperability standards, release 3.0. Tech. rep.; 2014. Available from: http://www.nist.gov/smartgrid/upload/NIST-SP-1108r3.pdf.

[155] Vingerhoets P, Chebbo M, Hatziargyriou N. The digital energy system 4.0. Tech. rep.; 2016. Available from: http://www.etip-snet.eu/wp-content/uploads/2017/04/ETP-SG-Digital-Energy-System-4.0-2016.pdf.

[156] Gangale F, Vasiljevska J, Covrig CF, Mengolini A, Fulli G. Smart grid projects outlook 2017: facts, figures and trends in Europe. Tech. rep.; 2017. Available from: https://ses.jrc.ec.europa.eu/sites/ses.jrc.ec.europa.eu/files/publications/sgp_outlook_2017-online.pdf.

[157] Karnouskos S. Future smart grid prosumer services. In: IEEE international conference on Innovative Smart Grid Technologies (ISGT 2011); 2011. p. 1–2.

[158] Karnouskos S, Terzidis O. Towards an information infrastructure for the future internet of energy. In: Kommunikation in Verteilten Systemen (KiVS 2007) conference. VDE Verlag. ISBN 978-3-8007-2980-7, 2007. p. 55–60.

[159] SmartGrids SRA 2035: strategic research agenda update of the SmartGrids SRA 2007 for the needs by the year 2035. Tech. rep.; 2012. Available from: http://www.etip-snet.eu/wp-content/uploads/2017/04/sra2035.pdf.

[160] Karnouskos S. The cooperative internet of things enabled smart grid. In: 14th IEEE International Symposium on Consumer Electronics (ISCE2010); 2010. p. 1–6.

[161] Karnouskos S. Demand side management via prosumer interactions in a smart city energy marketplace. In: IEEE international conference on Innovative Smart Grid Technologies (ISGT 2011); 2011. p. 1–7.

[162] Goncalves Da Silva P, Ilić D, Karnouskos S. The impact of smart grid prosumer grouping on forecasting accuracy and its benefits for local electricity market trading. Smart Grid, IEEE Transactions on 2014;5(1):402–10. https://doi.org/10.1109/tsg.2013.2278868.

[163] Zhang K, Mao Y, Leng S, Maharjan S, Zhang Y, Vinel A, et al. Incentive-driven energy trading in the smart grid. IEEE Access 2016;4:1243–57. https://doi.org/10.1109/access.2016.2543841.

[164] Borenstein S. Effective and equitable adoption of opt-in residential dynamic electricity pricing. Review of Industrial Organization 2012;42(2):127–60. https://doi.org/10.1007/s11151-012-9367-3.

[165] Karnouskos S, Goncalves Da Silva P, Ilić D. Assessment of high-performance smart metering for the web service enabled smart grid. In: Proceeding of the second joint WOSP/SIPEW international conference on performance engineering – ICPE '11. ACM Press; 2011. p. 133–44.

[166] Ilić D, Karnouskos S, Wilhelm M. A comparative analysis of smart metering data aggregation performance. In: IEEE 11th International Conference on Industrial Informatics (INDIN); 2013. p. 434–9.

[167] European Commission. Smart grids and meters. Available from: https://ec.europa.eu/energy/en/topics/markets-and-consumers/smart-grids-and-meters.

[168] Karnouskos S, Weidlich A, Kok K, Warmer C, Ringelstein J, Selzam P, et al. Field trials towards integrating smart houses with the smart grid. In: 1st international ICST conference on E-energy. Springer; 2010. p. 114–23.

[169] Dimeas A, Drenkard S, Hatziargyriou N, Karnouskos S, Kok K, Ringelstein J, et al. Smart houses in the smart grid: developing an interactive network. IEEE Electrification Magazine 2014;2(1):81–93. https://doi.org/10.1109/mele.2013.2297032.

[170] Marqués A, Serrano M, Karnouskos S, Marrón PJ, Sauter R, Bekiaris E, et al. NOBEL – a Neighborhood Oriented Brokerage ELectricity and monitoring system. In: 1st international ICST conference on E-energy. Springer; 2010. p. 187–96.

[171] Karnouskos S, Ilić D, Goncalves Da Silva P. Assessment of an enterprise energy service platform in a smart grid city pilot. In: IEEE 11th international conference on industrial informatics (INDIN); 2013. p. 24–9.

[172] Ilić D, Goncalves Da Silva P, Karnouskos S, Griesemer M. An energy market for trading electricity in smart grid neighbourhoods. In: 6th IEEE international conference on digital ecosystem technologies – complex environment engineering (IEEE DEST-CEE); 2012. p. 1–6.

[173] Ilić D, Karnouskos S, Goncalves Da Silva P. Sensing in power distribution networks via large numbers of smart meters. In: The third IEEE PES Innovative Smart Grid Technologies (ISGT) Europe; 2012. p. 1–6.

[174] Höglund J, Ilić D, Karnouskos S, Sauter R, Goncalves Da Silva P. Using a 6LoWPAN smart meter mesh network for event-driven monitoring of power quality. In: Third IEEE international conference on smart grid communications (SmartGridComm); 2012. p. 448–53.

[175] Karnouskos S, Goncalves Da Silva P, Ilić D. Developing a web application for monitoring and management of smart grid neighborhoods. In: IEEE 11th international conference on industrial informatics (INDIN); 2013. p. 408–13.

[176] Karnouskos S. Query-driven smart grid city management. ERCIM News 2014;98:33–4. Available from: http://ercim-news.ercim.eu/en98/special/query-driven-smart-grid-city-management.

[177] Garcia JJ, Cardenas JJ, Enrich R, Ilić D, Karnouskos S, Sauter R. Smart city energy management via monitoring of key performance indicators. In: Challenges of implementing active distribution system management, CIRED workshop 2014; 2014. p. 1–15. Paper 0263. Available from: http://www.cired.net/publications/workshop2014/papers/CIRED2014WS_0263_final.pdf.

[178] Goldman J, Shilton K, Burke J, Estrin D, Hansen M, Ramanathan N, et al. Participatory sensing: a citizen-powered approach to illuminating the patterns that shape our world. White Paper. Available from: https://www.wilsoncenter.org/sites/default/files/participatory_sensing.pdf, 2009.

[179] Eisenman SB, Miluzzo E, Lane ND, Peterson RA, Ahn GS, Campbell AT. Bikenet: a mobile sensing system for cyclist experience mapping. ACM Transactions on Sensor Networks 2009;6(1):1–39. https://doi.org/10.1145/1653760.1653766.

[180] Sheth A. Citizen sensing, social signals, and enriching human experience. IEEE Internet Computing 2009;13(4):87–92. https://doi.org/10.1109/mic.2009.77.

[181] Sheth A, Jadhav A, Kapanipathi P, Lu C, Purohit H, Smith GA, et al. Twitris: a system for collective social intelligence. In: Encyclopedia of social network analysis and mining. New York: Springer. ISBN 978-1-4614-6170-8, 2014. p. 2240–53.

[182] Wang D, Abdelzaher T, Kaplan L. Social sensing: building reliable systems on unreliable data. 1st edition. San Francisco, CA, USA: Morgan Kaufmann Publishers Inc.; 2015. ISBN 0128008679, 9780128008676.

[183] Wang D, Szymanski BK, Abdelzaher T, Ji H, Kaplan L. The age of social sensing. Available from: https://arxiv.org/abs/1801.09116, 2018.

[184] Lee EA. Cyber physical systems: design challenges. In: Object oriented real-time distributed computing (ISORC), 2008 11th IEEE international symposium on. IEEE; 2008. p. 363–9.

[185] Takagi K, Morikawa K, Ogawa T, Saburi M. Road environment recognition using on-vehicle lidar. In: 2006 IEEE intelligent vehicles symposium; 2006. p. 120–5.

[186] Wei J, Snider JM, Kim J, Dolan JM, Rajkumar R, Litkouhi B. Towards a viable autonomous driving research platform. In: Intelligent Vehicles symposium (IV), 2013 IEEE. IEEE; 2013. p. 763–70.

[187] Ozguner U, Stiller C, Redmill K. Systems for safety and autonomous behavior in cars: the darpa grand challenge experience. Proceedings of the IEEE 2007;95(2):397–412. https://doi.org/10.1109/JPROC.2006.888394.

[188] Hank P, Müller S, Vermesan O, Van Den Keybus J. Automotive ethernet: in-vehicle networking and smart mobility. In: Proceedings of the conference on design, automation and test in Europe. EDA Consortium; 2013. p. 1735–9.

[189] Fürst S, Mössinger J, Bunzel S, Weber T, Kirschke-Biller F, Heitkämper P, et al. Autosar–a worldwide standard is on the road. In: 14th international VDI congress electronic systems for vehicles, vol. 62. 2009. p. 5.

[190] Longo M, Zaninelli D, Viola F, Romano P, Miceli R, Caruso M, et al. Recharge stations: a review. In: Ecological vehicles and renewable energies (EVER), 2016 eleventh international conference on. IEEE; 2016. p. 1–8.

[191] Marchant GE, Lindor RA. The coming collision between autonomous vehicles and the liability system. Santa Clara Law Review 2012;52:1321.

[192] Hevelke A, Nida-Rümelin J. Responsibility for crashes of autonomous vehicles: an ethical analysis. Science and Engineering Ethics 2015;21(3):619–30.

[193] Lin P. Why ethics matters for autonomous cars. In: Autonomous driving. Springer; 2016. p. 69–85.

[194] Gogoll J, Müller JF. Autonomous cars: in favor of a mandatory ethics setting. Science and Engineering Ethics 2017;23(3):681–700.

[195] Qin Y, Boyle D, Yeatman E. A novel protocol for data links between wireless sensors and UAV based sink nodes. In: 2018 IEEE 4th World Forum on Internet of Things (WF-IoT); 2018. p. 371–6.

[196] Mitcheson PD, Boyle D, Kkelis G, Yates D, Saenz JA, Aldhaher S, et al. Energy-autonomous sensing systems using drones. In: 2017 IEEE SENSORS; 2017. p. 1–3.

[197] Katzschmann RK, DelPreto J, MacCurdy R, Rus D. Exploration of underwater life with an acoustically controlled soft robotic fish. Science Robotics 2018;3(16):eaar3449. https://doi.org/10.1126/scirobotics.aar3449.

[198] Kartakis S, Fu A, Mazo M, McCann JA. Communication schemes for centralized and decentralized event-triggered control systems. IEEE Transactions on Control Systems Technology 2017:1–14. https://doi.org/10.1109/TCST.2017.2753166.

[199] Ilic MD, Xie L, Khan UA, Moura JM. Modeling of future cyber–physical energy systems for distributed sensing and control. IEEE Transactions on Systems, Man, and Cybernetics-Part A: Systems and Humans 2010;40(4):825–38.

[200] Gubbi J, Buyya R, Marusic S, Palaniswami M. Internet of Things (IoT): a vision, architectural elements, and future directions. Future Generations Computer Systems 2013;29(7):1645–60.

[201] Sangiovanni-Vincentelli A, Damm W, Passerone R. Taming Dr. Frankenstein: contract-based design for cyber-physical systems. European Journal of Control 2012;18(3):217–38.

[202] Derler P, Lee EA, Tripakis S, Törngren M. Cyber-physical system design contracts. In: Proceedings of the ACM/IEEE 4th international conference on cyber-physical systems. ACM; 2013. p. 109–18.

[203] Jensen JC, Chang DH, Lee EA. A model-based design methodology for cyber-physical systems. In: Wireless Communications and Mobile Computing Conference (IWCMC), 2011 7th International. IEEE; 2011. p. 1666–71.

[204] Baker P, Croucher P, Rushton A. The handbook of logistics and distribution management. Kogan Page. ISBN 9780749476786, 2017. Available from: https://www.koganpage.com/product/the-handbook-of-logistics-and-distribution-management-9780749476779.

[205] Christopher M. Logistics & supply chain management. FT Publishing International. ISBN 9781292083827, 2016. Available from: http://www.informit.com/store/logistics-supply-chain-management-9781292083797.

[206] Aitken J. Supply chain integration within the context of a supplier association: case studies of four supplier associations. Ph.D. thesis. Cranfield University; 1998. Available from: http://dspace.lib.cranfield.ac.uk/handle/1826/9990.

[207] Council of Supply Chain Professionals. Supply chain terms and glossary. Available from: http://cscmp.org/CSCMP/Educate/SCM_Definitions_and_Glossary_of_Terms/CSCMP/Educate/SCM_Definitions_and_Glossary_of_Terms.aspx?hkey=60879588-f65f-4ab5-8c4b-6878815ef921, 2013.

[208] Schwab K. The fourth industrial revolution: what it means, how to respond. World Economic Forum (WEF). Available from: https://www.weforum.org/agenda/2016/01/the-fourth-industrial-revolution-what-it-means-and-how-to-respond/, 2016.

[209] Bollier D. When push comes to pull: the new economy and culture of networking technology. Tech. rep.; 2005. Available from: http://www.bollier.org/files/aspen_reports/2005InfoTechText.pdf.

[210] Intelligent assets: unlocking the circular economy potential. Tech. rep.; 2015. Available from: http://www3.weforum.org/docs/WEF_Intelligent_Assets_Unlocking_the_Cricular_Economy.pdf.

[211] Karnouskos S. Massive open online courses (MOOCs) as an enabler for competent employees and innovation in industry. Computers in Industry 2017;91:1–10. https://doi.org/10.1016/j.compind.2017.05.001.

[212] Fielding RT. Architectural styles and the design of network-based software architectures. Ph.D. thesis. Irvine: University of California; 2000. Available from: https://www.ics.uci.edu/~fielding/pubs/dissertation/fielding_dissertation.pdf.

[213] Shelby Z, Hartke K, Bormann C. The Constrained Application Protocol (CoAP). Tech. rep.; 2014.

[214] European Telecommunications Standards Institute Machine to Machine Technical Committee (ETSI M2M TC). ETSI TS 102 921 Machine-to-Machine communications (M2M) mIa, dIa and mId interfaces. Available from: http://www.etsi.org/deliver/etsi_ts/102900_102999/102921/01.02.01_60/ts_102921v010201p.pdf, 2013.

[215] International Telecommunication Union Telecom. Overview of the Internet of things, ITU-T Recommendation Y.2060. Available from: https://www.itu.int/rec/T-REC-Y.2060-201206-I, 2012.

Index

Printed in the United States
By Bookmasters